A GUIDE TO
CHALCOGEN-NITROGEN CHEMISTRY

Tristram Chivers
University of Calgary, Canada

A GUIDE TO CHALCOGEN-NITROGEN CHEMISTRY

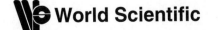

World Scientific

NEW JERSEY · LONDON · SINGAPORE · BEIJING · SHANGHAI · HONG KONG · TAIPEI · CHENNAI

Published by

World Scientific Publishing Co. Pte. Ltd.

5 Toh Tuck Link, Singapore 596224

USA office: 27 Warren Street, Suite 401-402, Hackensack, NJ 07601

UK office: 57 Shelton Street, Covent Garden, London WC2H 9HE

British Library Cataloguing-in-Publication Data
A catalogue record for this book is available from the British Library.

A GUIDE TO CHALCOGEN-NITROGEN CHEMISTRY

ISBN 981-256-095-5

Printed in Singapore by World Scientific Printers (S) Pte Ltd

Preface

The quintessential chalcogen-nitrogen compound tetrasulfur tetranitride, S_4N_4, was first detected by Gregory in 1835 just ten years after the discovery of benzene. Its unusual structure, like that of benzene, was not elucidated for over 100 years. The application of diffraction techniques revealed the unusual cage arrangement with two weak cross-ring sulfur–sulfur interactions. The details of the electronic structure of this fascinating molecule are still a matter of debate today.

Pioneering work in Germany, especially by the groups of Becke-Goehring, Weiss and Glemser, in the middle of the previous century uncovered a rich chemistry for inorganic sulfur–nitrogen systems. Their early efforts were notable because of the unavailability of many modern physical techniques for structural characterization that are commonplace today. The book by Goehring entitled *"Ergebnisse und Probleme der Chemie der Schewfelstickstoffverbindungen"* deserves special mention for the stimulus that it provided to subsequent workers in the field.

The polymer, $(SN)_x$, was first obtained in 1910 and its metallic character was noted. However, it was the discovery in 1973 by Labes that a polymer comprised only of non-metallic elements behaves as a superconductor at 0.26 K that sparked widespread interest in sulfur-nitrogen (S–N) chemistry. A year earlier Banister proposed that planar S-N heterocycles belong to a class of *"electron-rich aromatics"* that conform to the well known Hückel $(4n + 2)\pi$-electron rule of organic chemistry. This suggestion, which was based on simple electron-counting concepts, provided an additional impetus for both experimental and theoretical investigations of S–N systems. The classic book in this field *"The Inorganic Heterocyclic Chemistry of Sulfur, Nitrogen and*

Phosphorus" by Heal covered developments up to the end of the 1970s. This opus contributed authoritative insights into the fascinating chemistry of S–N compounds. It used a descriptive approach that drew attention to the many facets of the synthesis, structures and reactions that were poorly understood at that time.

In the first chapter of his book Heal states:

"Indeed, the reaction chemistry of these substances (i.e., SN compounds) *deserves to rank with that of boranes for novelty and interest".*

In the past twenty-five years the field of S–N chemistry has reached maturity as a result of contributions from many countries, notably Germany, the U.K., Canada, Japan and the United States. The combination of structural studies, primarily through X-ray crystallography, spectroscopic information and molecular orbital calculations has provided reasonable rationalizations of the structure-reactivity relationships of these fascinating compounds. The unusual structures and properties of S–N compounds have attracted the attention of numerous theoretical chemists, who continue to address the "aromatic' character of binary S–N systems.   Interfaces with other areas of chemistry *e.g.*, materials chemistry, organic synthesis, biochemistry and coordination chemistry have been established and are under active development. For example, materials with unique magnetic and conducting properties that depend on intermolecular chalcogen–nitrogen interactions between radical species have been designed. Some carbon-nitrogen-sulfur heterocycles exhibit magnetic behaviour that is of potential significance in the construction of organic data recording devices. In another area of materials chemistry, polymers involving both S–N and P–N linkages in the backbone have been used as components of matrices for oxygen sensors in the aerospace industry. In a biological setting, S-nitrosothiols (RSNO) have emerged as important species in the storage and transport of nitric oxide. As NO donors these sulfur–nitrogen compounds have potential medical applications in the treatment of blood circulation problems. In a different, but fascinating, context, thionitrite anions $[S_xNO]^-$ (x = 1,2) are implicated in the gunpowder reaction through an explosive decomposition.

The chemistry of selenium– and tellurium–nitrogen compounds has progressed more slowly but, in the last ten years, there have been numerous developments in these areas also, as a result of the creative contributions of both inorganic and organic synthetic chemists. Significant differences are apparent in the structures, reactivities and properties of these heavier chalcogen derivatives, especially in the case of tellurium. In addition, the lability of Se–N and Te–N bonds has led to applications of reagents containing these reactive functionalities in organic syntheses and, as a source of elemental chalcogen, in the production of metal chalcogenide semi-conductors.

In addition to providing a modern account of developments in chalcogen–nitrogen chemistry, including a comparison of sulfur systems with those of the heavier chalcogens, these interfaces will provide a major focus of this monograph. As implied by the inclusion of *"A Guide to ...."* in the title, it is not intended that the coverage of the primary literature will be comprehensive. Rather it provides an overview of the field with an emphasis on general concepts. Each chapter is designed to be self-contained, but there are extensive cross-references between chapters. By the use of selected examples, it is hoped that a reader, who is unfamiliar with or new to the field, will be able to gain an appreciation of the subtleties of chalcogen–nitrogen chemistry. A complete list of review articles is given at the end of Chapter 1. Key references to the primary literature are identified at the end of each chapter for the reader who wishes to pursue an individual topic in detail. The literature is covered up to mid-2004. Apart from two notable exceptions, the coverage of sulfur-nitrogen chemistry in standard inorganic (and organic) chemistry textbooks is sparse and usually limited to brief comments about the neutral binary compounds S_2N_2, S_4N_4 and $(SN)_x$. Those exceptions are the second edition of *"Chemistry of the Elements"* by N.N. Greenwood and A. Earnshaw (Butterworth-Heinemann, 1997) and the 34th edition of *"Inorganic Chemistry"* by Hollemann–Wiberg (Academic Press, 2001), which devote 26 and 14 pages, respectively, to this topic. It is hoped that the information in this book will be helpful to those who wish to go beyond the standard textbook treatment of various aspects of this important subject.

Acknowledgements

The first draft of this book was written during the tenure of a Killam Resident Fellowship at the University of Calgary in the fall term, 2003. The author is grateful for the financial support that provided release from other duties in order to focus on this project.

Subsequent drafts were composed after the receipt of input from various international experts on individual chapters in their areas of expertise. The author acknowledges, with gratitude, helpful (and encouraging) comments from the following individual scientists: Professor R. T. Boeré (University of Lethbridge, Canada), Professor N. Burford (Dalhousie University, Canada), Dr P. Kelly (Loughborough University, England), Professor R. S. Laitinen (University of Oulu, Finland), Professor Dr. R. Mews (Universität Bremen, Germany), Professor R. T. Oakley (University of Waterloo, Canada), Professor J. Passmore (University of New Brunswick, Canada), Professor K. E. Preuss (University of Guelph, Canada), Dr J. R. Rawson (University of Cambridge, England), Professor H. B. Singh (Indian Institute of Technology, Bombay, India), Professor I. Vargas-Baca (McMaster University, Canada). Their perceptive suggestions have enhanced the quality and accuracy of the final version of this monograph substantially. Nevertheless, there are undoubtedly shortcomings in the form of errors or omissions for which the author is entirely responsible.

Special thanks are accorded to Dr Dana Eisler (University of Calgary), who not only prepared all the structural drawings, figures and schemes, but also diligently proof-read the penultimate version of the manuscript. Professor Richard Oakley provided the idea and created the graphics for the design on the cover page. This representation of poly(sulfur nitride) is most appropriate in view of the impetus that the discovery of the properties of this unique polymer had on the field of chalcogen-nitrogen chemistry. The unfailing advice of the editor, Serene Ong, is sincerely appreciated. The author also wishes to recognize Nyuk-Wan Wong for assistance with formatting the text and Jamie Ritch for help with searching the literature.

Finally, I thank Sue for her continuing support and patience.

Tris Chivers
Calgary, August 2004

Contents

Chapter 1

INTRODUCTION

1.1 General Considerations

Oxygen, sulfur, selenium and tellurium are members of Group 16 in the Periodic Table. These elements, which are known as chalcogens, are isovalent. Consequently, some similarities between nitrogen-oxygen and other chalcogen-nitrogen species might be expected. On the other hand the fundamental properties of the chalcogens differ significantly and these differences can be expected to lead to disparities between the structures and properties of N–O molecules and ions and those of their heavier chalcogen-nitrogen counterparts. The major contributors to these differences are discussed below.

(a) The Pauling electronegativities for N, O, S, Se and Te are 3.04, 3.44, 2.58, 2.55 and 2.10, respectively. As a result an N–O bond will be polarized with a partial negative charge on oxygen, whereas an E–N (E = S, Se, Te) bond will be polarized in the opposite direction.

(b) The preference for σ-bonding over π-bonding is a common feature of the chemistry of the heavier p-block elements. In this context sulfur–nitrogen π-bonds are weaker than oxygen–nitrogen π-bonds and there is an increasing reluctance to form multiple bonds to nitrogen for the heavier chalcogens. In their elemental forms oxygen and nitrogen exist as the diatomic molecules O=O and N≡N, respectively, whereas sulfur forms a cyclic S_8 molecule with only

1

single S–S bonds. More generally, sulfur has a much stronger tendency to catenate than oxygen.

(c) Both oxygen and sulfur can adopt a variety of oxidation states in the formation of molecules or ions. These range from –2 to +2 for oxygen and from –2 to +6 for sulfur. The ability to achieve oxidation states >2, especially +4 and +6, gives rise to sulfur–nitrogen species that have no N–O analogues. The higher oxidation states are less stable for Se and Te.

(d) The size of Group 16 elements (chalcogens) increases down the series O<S<Se<Te. The covalent radii are 0.73, 1.03, 1.17 and 1.35 Å, respectively.

1.2 Binary Species

Binary molecules will be considered first as examples of the differences between N–O and S–N species. The most common oxides of nitrogen are the gases nitric oxide, NO, nitrous oxide, N_2O, and nitrogen dioxide, NO_2. The latter dimerizes to N_2O_4 in the liquid or solid states. The sulfur analogues of these binary systems are all unstable (Section 5.2). Thiazyl monomer has only a transient existence in the gas phase, although it can be stabilized through bonding to a transition metal. The dimer S_2N_2, a colourless solid, exists as a square planar molecule while the tetramer S_4N_4, which forms orange crystals, adopts an unusual cage structure. Dinitrogen sulfide is unstable above 160 K and decomposes to give N_2 and elemental sulfur. There are, however, some similarities for binary cationic species. For example, salts of the sulfur analogues of the well-known nitrosyl and nitronium cations, $[NS]^+$ and $[NS_2]^+$, are easily prepared. These species are important reagents in S–N chemistry (Sections 5.3.1 and 5.3.2).

There are notable differences in both structures and stabilities for binary N–O and S–N anions (Section 5.4). The most common oxo-anions of nitrogen are the nitrite $[NO_2]^-$ and the nitrate anion $[NO_3]^-$; the latter has a branched chain structure **1.1**. The sulfur analogue of nitrite is unstable, but salts of the [NS$_3$]⁻ anion can be isolated. By contrast to nitrate, the latter is believed to have an unbranched structure [SNSS]⁻ with a sulfur-sulfur bond **1.2** (Section 5.4.2). Peroxynitrite [ONOO]⁻, formed by the reaction of NO with the superoxide radical anion O$_2$·⁻, can be isolated as a tetramethylammonium salt, but it is less stable than nitrate. The tendency of sulfur to catenate is further manifested by the structure of the deep blue [NS$_4$]⁻ anion (**1.3**), which has two sulfur–sulfur bonds. The structures of [NS$_x$]⁻ (x = 3, 4) also reflect the higher electronegativity of N compared to S, because the unbranched arrangement allows for the dispersal of negative charge on N as well as S. In the branched structure of [NO$_3$]⁻ the negative charge is accommodated entirely on the more electronegative O atoms.

1.1   **1.2**   **1.3**

An interesting consequence of the structural differences between N–O and N–S anions is that the latter almost invariably behave as chelating ligands towards a single metal site. By contrast, the nitrate ion is able to function as a bridging ligand in a variety of coordination modes, as well as a chelating bidentate ligand, as a result of its branched structure.

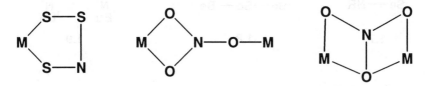

Fig. 1.1   Typical coordination modes for [NS$_3$]⁻ and [NO$_3$]⁻

1.3   Cyclic Chalcogen Imides

Cyclic chalcogen imides in which sulfur is in the formal +2 oxidation state (or lower) can, in the case of sulfur, act as a source of binary S–N anions via deprotonation (Section 6.2). These heterocycles are structurally related to *cyclo*-S_8 by the replacement of a sulfur atom by one or more NH groups, as illustrated by the examples S_7NH (**1.4**) and $S_4(NH)_4$ (**1.5**) shown below. In the case of selenium, a wider range of ring sizes, including five, six, eight and fifteen-membered rings are known (**1.6-1.9**) (Section 6.3). With the exception of the six-membered ring, for which a tellurium analogue is also known, these cyclic systems contain Se–Se linkages reflecting, as in the case of the cyclic sulfur imides, the tendency of the heavier chalcogens to catenate.

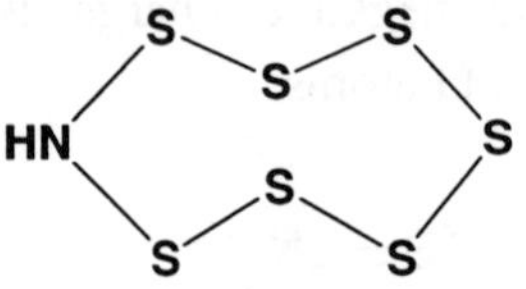

1.4 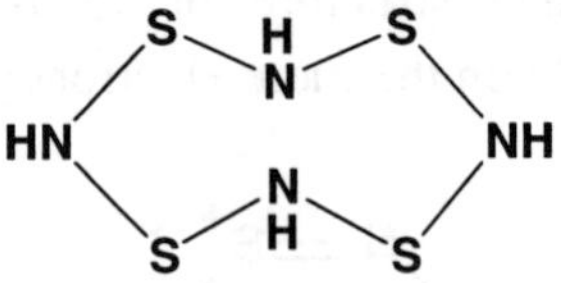   **1.5** 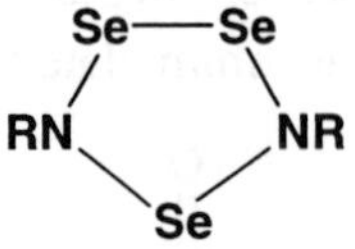   **1.6**

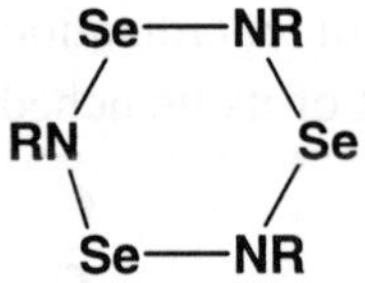

1.7 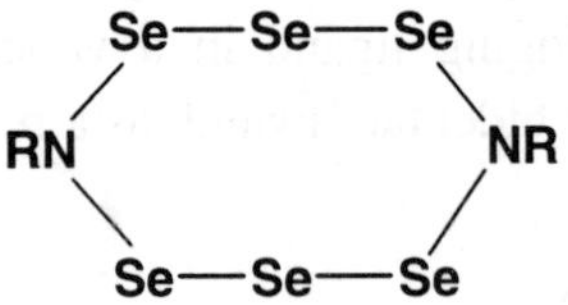   **1.8** 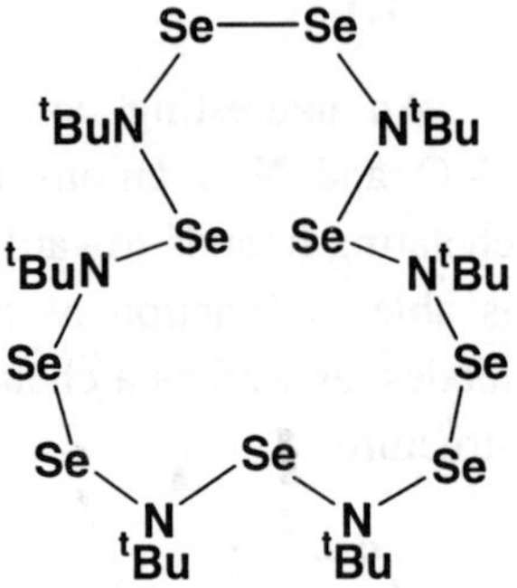   **1.9**

1.4   Organic Derivatives

Sulfur analogues of well known organic compounds with NO (nitroso) or NO_2 (nitro) functionalities also exhibit different stabilities and/or structures compared to those of RNO or RNO_2. For example, there are no stable *C*-thionitroso compounds RN=S (R = alkyl, aryl), although *N,N*′-dimethylthionitrosoamine Me_2NNS can be isolated (Section 10.2).

The stability of this derivative is attributed to the strong mesomeric effect of the NMe$_2$ group, which changes the nature of the N=S bond. The selenium analogues RNSe have only been detected as transient species.

In the case of RNS$_2$ compounds, a branched structure **1.10**, analogous to those of nitro compounds, is unstable. However, the N-thiosulfinyl (RN=S=S) isomer **1.11** can be stabilized, for example by use of a sterically hindered aniline (Section 10.3). The isovalent thionylamines RNSO (**1.12**) are known as thermally stable species for a wide variety of R groups (Section 9.6). On the other hand *S*-nitrosothiols (or thionitrites), RSNO (**1.13**), in which an alkyl or aryl substituent is attached to the sulfur atom of a thionitro group have a thermally labile S-N bond. This class of chalcogen-nitrogen compound is especially interesting because of their important role in the biological storage and transportation of nitric oxide (Section 9.7).

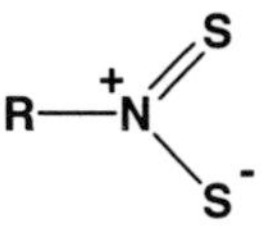

| **1.10** | **1.11** | **1.12** | **1.13** |

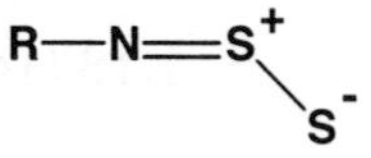

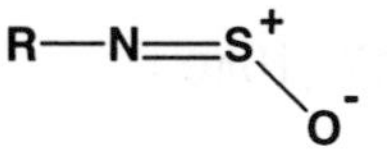

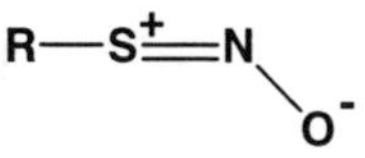

1.5 Unsaturated Sulfur-Nitrogen Ring Systems

Perhaps the most notable difference between S–N and N–O compounds is the existence of a wide range of cyclic compounds for the former. As indicated by the examples illustrated below, these range from four- to ten-membered ring systems and include cations and anions as well as neutral systems (**1.14-1.18**) (Sections 5.2-5.4). Interestingly, the most stable systems conform to the well known Hückel (4n + 2)π-electron rule. By using a simple electron-counting procedure (each S atom contributes two electrons and each N atom provides one electron to the π-system in these planar rings) it can be seen that stable entities include species with n = 1, 2 and 3.

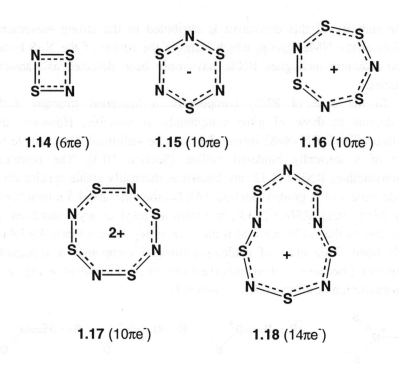

1.14 (6πe⁻)   **1.15 (10πe⁻)**   **1.16 (10πe⁻)**

1.17 (10πe⁻)   **1.18 (14πe⁻)**

1.6   Intramolecular π*–π* Interactions

A unique feature of chalcogen–nitrogen systems is the widespread occurrence of weak intra- or inter-molecular chalcogen•••chalcogen bonds via π*–π* interactions. Typically these sulfur•••sulfur bonds are >2.40 Å, *cf.* d(S–S) = 2.05 Å in *cyclo*-S_8. The classic example of the former type of interaction is tetrasulfur tetranitride (**1.19**). The planar S_4N_4 molecule, which would be a twelve π-electron biradical system, undergoes a Jahn-Teller distortion to form the closed-shell (diamagnetic) cage structure with two transannular S•••S bonds (Section 5.2.6). The selenium analogue Se_4N_4 has a similar structure. Ring systems with a single cross-ring S•••S bond (**1.20**) may result when two antipodal sulfur atoms in S_4N_4 are replaced by a group that contributes only one π-electron to the π-system, *e.g.*, PR_2 (R = alkyl, aryl) (Section 13.2.1) or CR (Section 12.4.1). In the case of the latter substituent, the resulting structure is dependent on the electronic properties of the R substituent on carbon. The electron-donating Me_2N group induces a folded structure (**1.20**) whereas the phenyl derivative is a planar, delocalised ten π-electron ring (**1.21**).

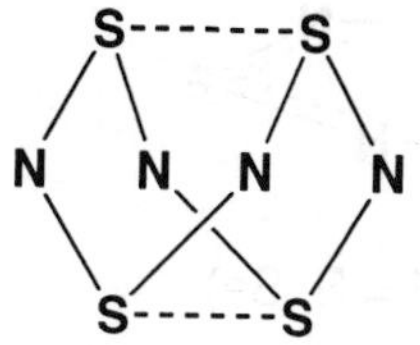

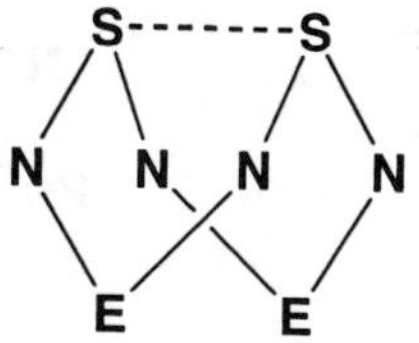

E = CNMe$_2$, PR$_2$

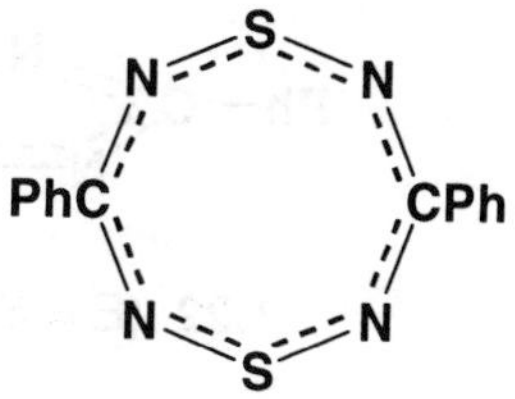

1.19　　　　　　　**1.20**　　　　　　　**1.21** (10πe$^-$)

1.7　Intermolecular π^*–π^* Interactions

Intermolecular chalcogen–chalcogen interactions occur for cyclic radical species, *e.g.*, the seven π-electron systems $[RCN_2E_2]^\cdot$ (dichalcogenadiazolyl) (Section 11.3.1) and $[(PhC)_2N_3E]^\cdot$ (chalcogenatriazinyl) (E = S, Se) (Section 12.2.1), which normally dimerize in the solid state to give structures of the type **1.22** and **1.23**, respectively. In the latter case this type of bonding is similar to that observed in the dithionite dianion $[S_2O_4]^{2-}$, in which two sulfur dioxide radical anions are linked by a sulfur•••sulfur bond of 2.39 Å. In the five-membered dithiadiazolyl systems two intermolecular chalcogen•••chalcogen interactions occur in most cases. Molecular orbital descriptions of various examples of intermolecular π^*–π^* interactions are given in Section 4.7. The solid-state magnetic and conducting properties of this class of S–N heterocycles can be adjusted by the judicious choice of the substituent attached to carbon. For example, monomeric structures with exceptional magnetic behaviour have been observed for polyfluorinated aryl derivatives of the type $[4-XC_6F_4CN_2S_2]^\cdot$ (Section 11.3.1).

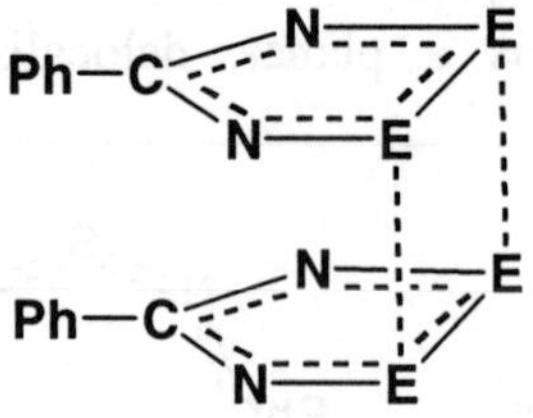

1.22 (E = S, Se) 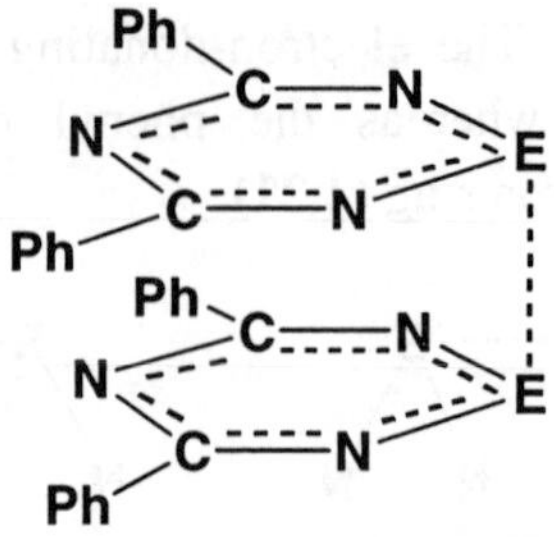   **1.23** (E = S, Se)

1.8   Chalcogen-Nitrogen Chains and Polymers

Polymer and chain formation is another property of chalcogen–nitrogen compounds that distinguishes them from their oxygen analogues. In addition to the unique, superconducting poly(sulfur nitride) $(SN)_x$ (**1.24**) (Section 14.2), a variety of poly(thiazyl) chains such as RS_5N_4R (**1.25**) (Section 14.3) have been characterized. Interest in these chains stems from their possible use as models for the behaviour of $(SN)_x$ and as components in molecular materials, *e.g.*, as molecular wires.

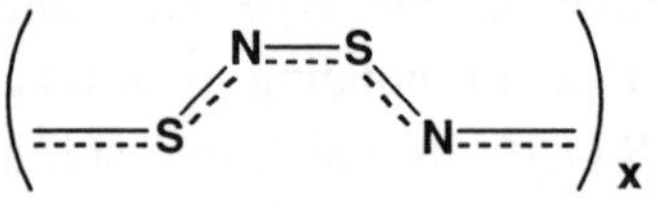

1.24 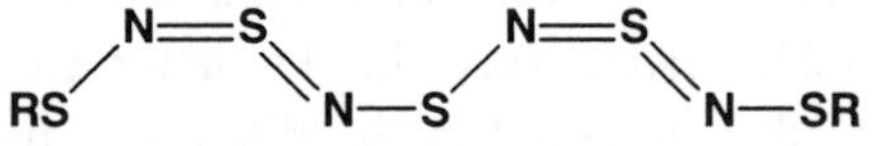   **1.25**

1.9   Chalcogen–Nitrogen Compounds with the Chalcogen in Higher Oxidation States

The accessibility of the +4 and +6 oxidation states for sulfur and, to a lesser extent, selenium gives rise to both acyclic and cyclic molecules that have no parallels in N–O chemistry. Thus there is an extensive chemistry of chalcogen diimides RN=E=NR (E = S, Se, Te) (Section 10.4). In the case of Te these unsaturated molecules form dimeric structures reflecting the increasing reluctance for the heavier chalcogens to form multiple bonds to nitrogen. The acyclic molecule N≡SF₃, containing a sulfur–nitrogen triple bond, is also a stable molecule with a well-studied chemistry (Section 8.3). Stable heterocyclic systems with sulfur in a high oxidation state include the cyclothiazyl halides $(NSX)_n$ (**1.26**, X = Cl, n = 3; x = F, n = 3,4) (Sections 8.7 and 8.8) and the trimeric sulfanuric halides $[NS(O)X]_3$ (**1.27**, X = Cl, F) (Section 8.9).

1.26   **1.27**

For the purpose of this book, chalcogen–nitrogen compounds will be defined as compounds in which the number of heteroatoms exceeds the number of carbon atoms in the compound (excluding the carbon atoms that are in substituents attached to nitrogen or the chalcogen). Consequently, cyclic systems that would fall more appropriately under the rubric of organic heterocycles are not included. This comparison of chalcogen-nitrogen compounds with the more familiar nitrogen–oxygen species is followed by three general chapters outlining (a) the methods for the formation of chalcogen–nitrogen bonds, (b) the techniques used for the characterization of chalcogen–nitrogen compounds, and (c) the insights into their structures, properties and reactivities provided by molecular orbital calculations. The remainder of the chapters deal with different classes of chalcogen–nitrogen compounds starting with binary systems (Chapter 5), cyclic chalcogen imides (Chapter 6) and metal complexes (Chapter 7). Chapters 8 and 9 are concerned with two important classes of reagents, chalcogen-nitrogen halides and chalcogen–nitrogen–oxygen compounds. This is followed by a discussion of acyclic organochalcogen–nitrogen compounds (Chapter 10). The propensity of carbon–halcogen–nitrogen systems to form long-lived radicals together with related solid-state materials that exhibit interesting conducting or magnetic properties are the focus of Chapter 11. Six-membered and larger carbon-nitrogen-chalcogen systems are discussed in the subsequent chapter (Chapter 12). Chalcogen–nitrogen heterocycles containing other *p*-block elements, primarily phosphorus, are the subject of Chapter 13. Chapter 14 begins with a discussion of the unique polymer $(SN)_x$ and then moves on to an examination of S–N chains and other types of S–N polymers. A final chapter (Chapter 15) is devoted to a consideration of the influence of intramolecular interactions on the structures and reactivities of chalcogen-nitrogen compounds in organic and biochemical systems.

This introductory chapter will conclude with a bibliographic listing of the previous books that have been devoted primarily (or entirely) to chalcogen-nitrogen chemistry together with a compilation of the reviews and book chapters that are either general in scope or, as indicated by their titles, focus on a particular aspect of the subject. The reviews listed are limited to those that have appeared since (or at about the same time as) the publication of Heal's book. They are organized in chronological order under sub-headings showing the particular aspect of chalcogen-nitrogen chemistry that is discussed.

Books

M. Goehring, *Ergebnisse und Probleme der Chemie der Schwefelstickstoffverbindungen*, Akademie-Verlag, Berlin (1957).

H.G. Heal, *The Inorganic Heterocyclic Chemistry of Sulfur, Nitrogen and Phosphorus*, Academic Press, London (1980).

Gmelin Handbook of Inorganic Chemistry, 8ᵗʰ ed., Sulfur-Nitrogen Compounds, Part 2: Compounds with Sulfur of Oxidation Number IV, Springer-Verlag, Berlin (1985).

Reviews and Book Chapters

(a) General Reviews

H. W. Roesky, Cyclic Sulfur–Nitrogen Compounds, *Adv. Inorg. Chem. Radiochem.*, **22**, 239 (1979).

A. J. Banister, Electron-Rich Thiazenes, *Phosphorus and Sulfur*, **6**, 421 (1979).

H. W. Roesky, Structure and Bonding in Cyclic Sulfur–Nitrogen Compounds. *Angew. Chem. Int. Ed. Engl.* **18**, 91 (1979).

R. Gleiter, Structure and Bonding in Cyclic Sulfur–Nitrogen Compounds: Molecular Orbital Considerations, *Angew. Chem. Int. Ed. Engl.*, **20**, 444 (1981).

H. W. Roesky, Results and Perspectives in Sulfur and Nitrogen Chemistry, *Comments Inorg. Chem.*, **1**, 183 (1981).

H. W. Roesky, The Sulfur–Nitrogen Bond, in A. Senning (ed.) *Sulfur in Organic Chemistry*, Marcel Dekker, Inc., Vol 4, pp 15-45 (1982).

A. J. Banister and N. R. M. Smith, Some Experiments in Sulfur–Nitrogen Chemistry, *J. Chem. Ed.*, **59**, 1058 (1982).

T. Chivers, Electron-Rich Sulfur–Nitrogen Heterocycles, *Acc. Chem. Res.*, **17**, 166 (1984)

T. Chivers, Synthetic Methods and Structure-Reactivity Relationships in Electron-rich Sulfur-Nitrogen Rings and Cages, *Chem. Rev.*, **85**, 341 (1985).

J. L. Morris and C. W. Rees, Organic Poly(Sulfur–Nitrogen) Chemistry, *Chem. Soc. Rev.*, **15**, 1 (1986).

T. Chivers, Sulfur–Nitrogen Heterocycles, in I. Haiduc and D. B. Sowerby (ed.) *The Chemistry of Inorganic Homo- and Heterocycles*, Vol. 2, Academic Press, London, pp. 793-870 (1987).

R. T. Oakley, Cyclic and Heterocyclic Thiazenes, *Prog. Inorg. Chem.*, **36**, 1 (1988).

K. F. Preston and L. H. Sutcliffe, ESR Spectroscopy of Free Radicals Containing Sulfur Linked to Nitrogen, *Mag. Reson. Chem.*, **28**, 189 (1990).

A. W. Cordes, R. C. Haddon and R. T. Oakley, Heterocyclic Thiazyl and Selenazyl Radicals: Synthesis and Applications in Solid-State Architecture, in R. Steudel (ed.) *The Chemistry of Inorganic Heterocycles*, Elsevier, pp. 295-322 (1992).

A. J. Banister and J. M. Rawson, Some Synthetic and Structural Aspects of Dithiadiazoles and Related Compounds, in R. Steudel (ed.) *The Chemistry of Inorganic Heterocycles*, Elsevier, pp. 323-348 (1992).

C. W. Rees, Polysulfur–Nitrogen Heterocyclic Chemistry, *J. Heterocycl. Chem.*, 29, 639 (1992).

R. T. Oakley, Chemical Binding Within and Between Inorganic Rings: The Design and Synthesis of Molecular Conductors, *Can. J. Chem.*, 71, 1775 (1993).

T. Chivers, Sulfur–Nitrogen Compounds, in R. B. King (ed.) *Encyclopedia of Inorganic Chemistry*, Vol.7, John Wiley, pp. 3988-4009 (1994).

J. M. Rawson, A. J. Banister and I. Lavender, The Chemistry of Dithiadiazolylium and Dithiadiazolyl Rings, *Adv. Heterocycl. Chem.*, 62, 137 (1995).

A. Haas, Some Recent Developments in Chalcogen Heterocyclic Chemistry, *Adv. Heterocycl. Chem.*, 71, 115 (1998).

J. M. Rawson and G. D. McManus, Benzo-fused Dithiazolyl Radicals: From Chemical Curiosities to Materials Chemistry, *Coord. Chem. Rev.*, 189, 135 (1999).

T. Torroba, Poly(Sulfur–Nitrogen) Heterocycles via Sulfur Chlorides and Nitrogen Reagents, *J. Prakt. Chem.*, 341, 99 (1999).

R. T. Boeré and T. L. Roemmele, Electrochemistry of Redox-active Group 15/16 Heterocycles, *Coord. Chem. Rev.*, 210, 369 (2000).

A. Haas, The General Course of Reactions and Structural Correlations in Sulfur–Nitrogen Chemistry Discussed in Terms of Carbon and Sulfur(IV) Equivalence, *J. Organomet. Chem.*, 623, 3 (2001).

J. M. Rawson and F. Palacio, Magnetic Properties of Thiazyl Radicals, *Structure and Bonding*, 100, 93 (2001).

(b) Small Sulfur–Nitrogen Species

K. K. Pandey, D. K. M. Raju, H. L. Nigam and U. C. Agarwala, Chemistry of the Thionitrosyl Group, *Proc. Ind. Nat. Sci. Acad.*, 48A, 16 (1982).

T. Chivers, The Chemistry of Compounds Containing the SNSO or SNSS Chromophores, *Sulfur Reports*, **7**, 89 (1986).

C. Wentrup and P. Kambouris, N-Sulfides: Dinitrogen Sulfide, Thiofulminic Acid and Nitrile Sulfides, *Chem. Rev.*, **91**, 363 (1991).

S. Parsons and J. Passmore, Rings, Radicals and Synthetic Metals: The Chemistry of [SNS]$^+$, *Acc. Chem. Res.*, **27**, 101 (1994).

R. Mews, P. G. Watson and E. Lork, Three-coordinate Sulfur(VI)–Nitrogen Species: An Attempt to Breathe Some New Life into an Old Topic, *Coord. Chem. Rev.*, **158**, 233 (1997).

D. L. H. Williams, The Chemistry of *S*-Nitrosothiols, *Acc. Chem. Res.*, **32**, 869 (1999).

(c) Sulfur–Nitrogen Anions

R. P. Sharma and R. D. Verma, Recent Developments in the Chemistry of Sulfur–Nitrogen Compounds: Sulfur-Nitrogen Cyclic Anions, *J. Sci. Ind. Res.*, **40**, 34 (1981).

T. Chivers and R. T. Oakley, Sulfur–Nitrogen Anions and Related Compounds, *Topics Curr. Chem.*, **102**, 119 (1982).

(d) Metal Complexes of Sulfur–Nitrogen Ligands

K. Vrieze and G. van Koten, Coordination and Activation of the N=S Bond in Sulfur Diimines RN=S=NR and Sulfinylanilines RN=S=O by Metal Atoms, *J. Royal Netherlands Soc.*, **99**, 145 (1980).

H. W. Roesky and K. K. Pandey, Transition-Metal Thionitrosyl and Related Complexes, *Adv. Inorg. Chem. Radiochem.*, **26**, 337 (1983).

K. Dehnicke and U. Müller, Coordination Chemistry of S_2N_2, *Transition Met. Chem.*, **10**, 361 (1985).

K. Dehnicke and U. Müller, Chlorothionitrene Complexes of Transition Metals, *Comments Inorg. Chem.*, **4**, 213 (1985).

P. F. Kelly and J. D. Woollins, The Preparation and Structure of Complexes Containing Simple Inorganic Sulfur–Nitrogen Ligands, *Polyhedron*, **5**, 607 (1986).

T. Chivers and F. Edelmann, Transition-Metal Complexes of Inorganic Sulfur–Nitrogen Ligands, *Polyhedron*, **5**, 1661 (1986).

M. Herberhold, Small Reactive Sulfur-Nitrogen Compounds and Their Transition-Metal Complexes, *Comments Inorg. Chem.*, **7**, 53 (1988).

P. F. Kelly, A. M. Z. Slawin, D. J. Williams and J. D. Woollins, Caged Explosives: Metal-stabilized Chalcogen Nitrides. *Chem. Soc. Rev.*, 246 (1992).

J. D. Woollins, The Preparation and Structure of Metalla-sulfur/selenium Nitrogen Complexes and Cages, in R. Steudel (ed.) *The Chemistry of Inorganic Heterocycles*, Elsevier, pp. 349-372 (1992).

K. K. Pandey, Coordination Chemistry of Thionitrosyl, Thiazate, Disulfidothionitrate, Sulfur Monoxide and Disulfur Monoxide, *Prog. Inorg. Chem.*, **40**, 445 (1992).

A. F. Hill, Organotransition Metallic Chemistry of SO_2 Analogs, *Adv. Organomet. Chem.*, **36**, 159 (1994).

A. J. Banister, I. May, J. M. Rawson and J. N. B. Smith, Dithiadiazolyls as Heterocyclic Chelators: A Radical Approach to Coordination Chemistry, *J. Organomet. Chem.*, **550**, 241 (1998).

R. Fleischer and D. Stalke, A New Route to Sulfur Polyimido Anions $[S(NR)_n]^{m-}$: Reactivity and Coordination Behaviour, *Coord. Chem. Rev.*, **176**, 431 (1998).

D. Stalke, Polyimido Sulfur Anions $[S(NR)_n]^{m-}$: Free Radicals and Coordination Behaviour, *Proc. Ind. Acad. Sci. Chem. Sci.*, **112**, 155 (2000).

(e) Polymers and Chains

M. M. Labes, P. Love and L. F. Nichols, Poly(sulfur nitride): A Metallic, Superconducting Polymer, *Chem. Rev.*, **79**, 1 (1979).

I. Manners, Sulfur–Nitrogen–Phosphorus Polymers, *Coord. Chem. Rev.*, **137**, 109 (1994).

J. M. Rawson and J. J. Longridge, Sulfur–Nitrogen Chains: Rational and Irrational Behaviour, *Chem. Soc. Rev.*, 53 (1997).

A. J. Banister and I. B. Gorrell, Poly(sulfur nitride): The First Polymeric Metal, *Adv. Mater.*, **10**, 1415 (1998).

(f) Sulfur–Nitrogen Halides

R. Mews, Nitrogen–Sulfur–Fluorine Ions, *Adv. Inorg. Chem. Radiochem.*, **19**, 185 (1976).

O. Glemser and R. Mews, Chemistry of NSF and NSF_3: A Quarter Century of N-S-F Chemistry, *Angew. Chem. Int. Ed. Engl.*, **19**, 883 (1980).

(g) Phosphorus–Nitrogen–Chalcogen Systems

J. C. van de Grampel, Investigations on the Six-membered Inorganic Ring Systems, $[(NPCl_2)(NSOCl)_2)]$ and $[(NPCl_2)_2(NSOCl)]$, *Rev. Inorg. Chem.*, **3**, 1 (1981).

T. Chivers, D. D. Doxsee, M. Edwards and R. W. Hilts, Diphosphadithia- and Diphosphadiselena-tetrazocines and Their Derivatives, in R. Steudel (ed.) *The Chemistry of Inorganic Ring Systems*, Elsevier, pp. 271-294 (1992).

J. C. van de Grampel, Selected Chemistry of Cyclophosphazenes and Cyclothiaphosphazenes, *Coord. Chem. Rev.*, **112**, 247 (1992).

T. Chivers and R. W. Hilts, Coordination Chemistry of Cyclic and Acyclic PNS and PNSe Ligands, *Coord. Chem. Rev.*, **137**, 201 (1994).

A. R. McWilliams, H. Dorn and I. Manners, New Inorganic Polymers Containing Phosphorus, in J-P. Majoral (ed.) *New Aspects in Phosphorus Chemistry III: Topics Curr. Chem.*, **229**, 141-167 (2002).

(h) Selenium– and Tellurium–Nitrogen Compounds

M. Björgvinsson and H. W. Roesky, The Structures of Compounds Containing Se–N and Te–N Bonds, *Polyhedron*, **10**, 2353 (1991).

T. M. Klapötke, Binary Selenium–Nitrogen Species and Related Compounds, in R. Steudel (ed.) *The Chemistry of Inorganic Ring Systems*, Elsevier, pp. 409-427 (1992).

T. Chivers and D.D. Doxsee, Heterocyclic Selenium– and Tellurium–Nitrogen Compounds, *Comments Inorg. Chem.* **15**, 109 (1993).

T. Chivers, Selenium–Nitrogen Chemistry, *Main Group Chemistry News*, **1**, 6 (1993).

A. Haas, J. Kasprowski, and M. Pryka, Tellurium– and Selenium–Nitrogen Compounds: Preparation, Characterization and Properties, *Coord. Chem. Rev.*, **130**, 301 (1994).

T. Chivers, X. Gao, N. Sandblom and G. Schatte, Recent Developments in Tellurium–Nitrogen Chemistry, *Phosphorus, Sulfur and Silicon*, **136-138**, 11 (1998).

I. D. Sadekov and V. I. Minkin, Tellurium–Nitrogen Containing Heterocycles, *Adv. Heterocycl. Chem.*, **79**, 1 (2001).

A. Haas, Acyclic and Heterocyclic Te–S–N Compounds: A Review of Recent Publications, *J. Organomet. Chem.*, **646**, 80 (2002).

Chapter 2

FORMATION OF CHALCOGEN–NITROGEN BONDS

This chapter will provide an overview of the methods available for the formation of chalcogen–nitrogen bonds with an emphasis on the synthesis of the most important reagents. The use of these reagents for further development of chalcogen–nitrogen chemistry will be described in subsequent chapters for specific types of chalcogen–nitrogen compounds. Synthetic methods that are limited to a single example are not included, unless they are part of the comparison of the synthesis of S–N reagents with that of the less widely studied Se and Te systems. Some informative general reviews of synthetic methods are given in references 1–6.

2.1 From Ammonia and Ammonium Salts

The cyclocondensation reaction of ammonia (or ammonium salts) with sulfur halides is the most important route to S–N heterocycles, since it provides an easy preparation of several key starting materials (Scheme 2.1).[2] The polarity of the solvent has a remarkable influence on the outcome of this reaction. When conducted in carbon tetrachloride or methylene chloride, this reaction is the standard synthesis of S_4N_4 which, in turn, is used for the preparation of numerous S–N ring systems (Section 5.2.6). The reaction of ammonia with S_2Br_2 has been adapted for the synthesis of small quantities of ^{33}S-labelled S_4N_4.[7] The selenium analogue Se_4N_4 is prepared by the reaction of $SeBr_4$ with ammonia at high pressure.[8]

17

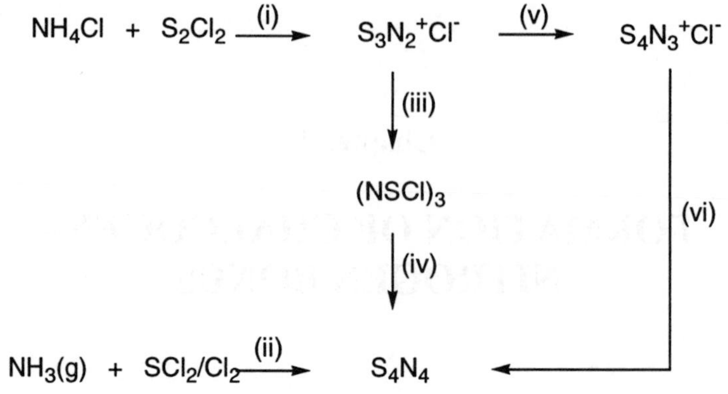

$$NH_4Cl + S_2Cl_2 \xrightarrow{(i)} S_3N_2^+Cl^- \xrightarrow{(v)} S_4N_3^+Cl^-$$

$$\downarrow (iii)$$

$$(NSCl)_3 \qquad\qquad (vi)$$

$$\downarrow (iv)$$

$$NH_3(g) + SCl_2/Cl_2 \xrightarrow{(ii)} S_4N_4 \longleftarrow$$

Scheme 2.1 Preparation of S–N heterocycles from ammonia or ammonium salts and sulfur halides: (i) reflux (ii) CCl_4 or CH_2Cl_2 (iii) Cl_2 (iv) Fe, Hg or Ph_3Sb (v) S_2Cl_2 (vi) KI

The sulfur–nitrogen halide [S_3N_2Cl]Cl, another important reagent in S–N chemistry (Section 8.6), is formed as an intermediate in the synthesis of S_4N_4. It is best prepared, however, by the treatment of S_2Cl_2 with sulfur and ammonium chloride at 150-160°C.[9a] This method is especially important for the synthesis of ^{15}N-labelled S–N compounds for NMR studies (Section 3.2), since [S_3N_2Cl]Cl can be chlorinated with SO_2Cl_2 to give $(NSCl)_3$, which is reduced to S_4N_4 by $SbPh_3$.[9b]

When the reaction of S_2Cl_2 with ammonia is carried out in a polar solvent, *e.g.*, DMF, the hydrolysis of the reaction mixture with aqueous HCl produces a mixture of the cyclic sulfur imides S_7NH, 1,3-, 1,4- and 1,5-$S_6(NH)_2$ and 1,3,5- and 1,3,6-$S_5(NH)_3$, which can be separated by chromatography on silica gel using CS_2 as eluant (Section 6.2.1).[10]

2.2 From Amines and Amido–Lithium or Sodium Reagents

The reactions of amines or amido-lithium compounds with chalcogen halides provide a convenient source of a variety of important chalcogen-nitrogen reagents. The reactions of $LiN(SiMe_3)_2$ with SCl_2, Se_2Cl_2 or $TeCl_4$ in hexane or diethyl ether produce the chalcogen(II) diamides $E[N(SiMe_3)_2]_2$ (E = S, Se, Te)[11, 12] in good yields (Eq. 2.1-2.3).

$$2LiN(SiMe_3)_2 + SCl_2 \rightarrow S[N(SiMe_3)_2]_2 + 2LiCl \qquad (2.1)$$

$$2LiN(SiMe_3)_2 + Se_2Cl_2 \rightarrow Se[N(SiMe_3)_2]_2 + 2LiCl + 1/8Se_8 \qquad (2.2)$$

$$4LiN(SiMe_3)_2 + TeCl_4 \rightarrow Te[N(SiMe_3)_2]_2 + 4LiCl \qquad (2.3)$$

The reactivity of the Si–N bonds in these reagents can be utilized in further synthesis of chalcogen-nitrogen compounds. For example, the reaction of the sulfur derivative with an equimolar mixture of SCl_2 and SO_2Cl_2 provides a convenient source of S_4N_4.[11] In a similar vein, the reaction of $LiN(SiMe_3)_2$ with a mixture of Se_2Cl_2 and $SeCl_4$ (1:4 molar ratio) provides a safer alternative to the explosive compound Se_4N_4 than the $SeBr_4$-NH_3 reaction.[13]

The treatment of $LiN(SiMe_3)_2$ with aryl tellurenyl iodides gives stable *N,N'*-bis(trimethylsilyl)tellurenamides that react with acetylenes to give acetylenyl tellurides (Eq. 2.4).[14]

$$ArTeI + LiN(SiMe_3)_2 \rightarrow ArTeN(SiMe_3)_2 + LiI \qquad (2.4)$$

N,N'- Bis(trimethylsilyl)sulfur(IV) diimide $Me_3SiN=S=NSiMe_3$ is an especially versatile source of the N=S=N functionality in the formation of both acyclic and cyclic S–N compounds. It is conveniently prepared by the reaction of $NaN(SiMe_3)_2$ and thionyl chloride (Eq. 2.5).[15]

$$2NaN(SiMe_3)_2 + SOCl_2 \rightarrow Me_3SiNSNSiMe_3 + (Me_3Si)_2O + 2NaCl \qquad (2.5)$$

The selenium analogue $Me_3SiN=Se=NSiMe_3$ can be prepared by the analogous reaction of $LiN(SiMe_3)_2$ and $SeOCl_2$, but it is thermally unstable.[16] The tellurium analogue is unknown, but the *tert*-butyl derivative, which is a dimer $^tBuNTe(\mu$-N^{t}Bu$)_2TeN^tBu$, can be prepared in excellent yield by the reaction of lithium *tert*-butylamide with $TeCl_4$ in THF (Eq. 2.6).[17] It has proved to be an excellent reagent for the development of tellurium(IV)–nitrogen chemistry (Section 10.4). The formation of the Te(IV) diimide in this reaction is noteworthy because the reaction of $TeCl_4$ with the lithium derivatives of secondary amines, *e.g.*, $LiNMe_2$[18] or $LiN(SiMe_3)_2$[12], results in reduction to Te(II).

$$8LiNH^tBu + 2TeCl_4 \rightarrow {}^tBuNTe(\mu\text{-}N^tBu)_2TeN^tBu + 4LiCl + 4^tBuNH_2 \quad (2.6)$$

Compounds containing the structural units $-N=SCl_2$, $N=S=O$ or $N=S=S$ may be prepared from primary amines and sulfur halides. For example, the reaction of $^tBuNH_2$ with SCl_2 in boiling hexane produces $^tBuNSCl_2$.[19] The classical route for the preparation of sulfinyl derivatives RNSO is shown in Eq. 2.7.[20] The dimeric selenium analogue $OSe(\mu\text{-}N^tBu)_2SeO$ is obtained in a similar manner.[21]

$$RNH_2 + SOCl_2 \rightarrow RNSO + 2HCl \quad\quad (2.7)$$

N-Thiosulfinylanilines are conveniently prepared by condensing the corresponding aniline with S_2Cl_2 (Eq. 2.8).[22]

$$ArNH_2 + S_2Cl_2 \rightarrow \quad ArNSS + 2HCl \quad\quad (2.8)$$

$$(Ar = 4\text{-}Me_2NC_6H_4)$$

The selenium(II) dihalide $SeCl_2$, prepared by oxidizing elemental selenium with SO_2Cl_2, can be stabilized in THF solution.[23] The reaction of $SeCl_2$ with primary amines, *e.g.*, $^tBuNH_2$, provides a rich source of both acyclic and cyclic selenium halides depending on the stoichiometry of the reagents. With a 2:3 molar ratio the acyclic imidoselenium dichlorides $ClSe[N(^tBu)Se]_nCl$ (n = 1, 2) are the major products, whereas the predominant products in the 1:3 reaction are the cyclic selenium imides $Se_3(N^tBu)_2$ and $Se_3(N^tBu)_3$ (Section 6.3).[24]

A general route to benzo-2,1,3-selenadiazoles involves the condensation reaction between 1,2-diaminobenzene and selenium tetrachloride. This method has also been used recently for the synthesis of 4,5,6,7-tetrafluoro-2,1,3-benzotelluradiazole (Eq. 2.9).[25]

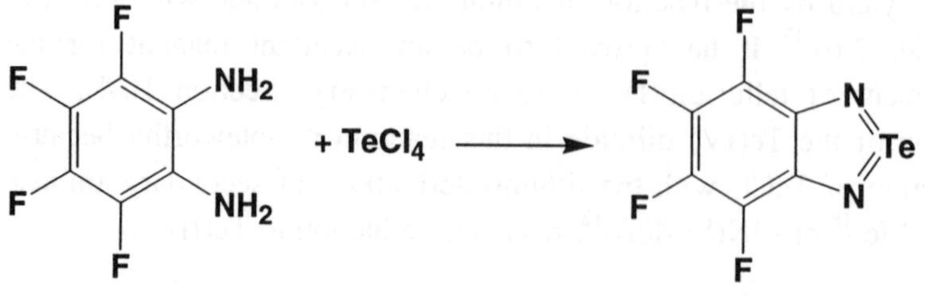

In a related cyclocondensation six-membered rings containing the NSeN functionality may be prepared by the reaction of imidoylamidines with $SeCl_4$ (Eq. 2.10).[26]

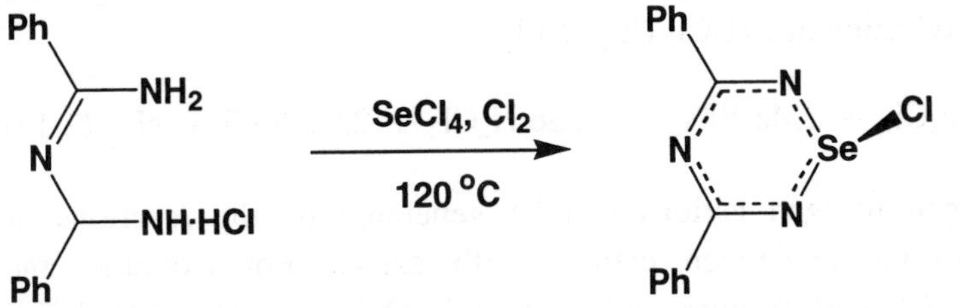

(2.10)

The cyclocondensation route is also effective for the synthesis of planar $C_2N_4S_2$ rings from benzamidines and sulfur dichloride in the presence of a strong base (Eq. 2.11).[27]

$$PhC(NH)(NH_2) + 3SCl_2 + 6DBU \rightarrow PhC(NSN)_2CPh + 6DBU \cdot HCl$$

$$(DBU = diazabicycloundecane) \qquad\qquad + 1/8S_8 \qquad (2.11)$$

Tellurium(VI)–nitrogen bonds can be generated by the reaction of hexamethyldisilazane with tellurium hexafluoride (Eq. 2.12).[28] The product $(Me_3SiNH)TeF_5$ is a useful precursor for a variety of $NTeF_5$ compounds. By contrast, SF_6 is inert towards Si–N reagents.

$$(Me_3Si)_2NH + TeF_6 \rightarrow (Me_3SiNH)TeF_5 + Me_3SiF \qquad (2.12)$$

2.3 From Azides

Reactions of ionic or covalent azides with chalcogen halides or, in the case of sulfur, with the elemental chalcogen provide an alternative route to certain chalcogen–nitrogen compounds. For example, the reaction of sodium azide with *cyclo*-S_8 in hexamethylphosphoric triamide is a more convenient synthesis of S_7NH than the S_2Cl_2 reaction (Section 6.2.1).[29] Moreover, the azide route can be used for the preparation of 50% ^{15}N-enriched S_7NH.

The polymer $(SN)_x$ may also be made in a two-step process using azide reagents. Thus the reaction of S_2Cl_2 with sodium azide in acetonitrile followed by treatment of the $[S_3N_2]Cl$ produced with Me_3SiN_3 yields $(SN)_x$ as a black solid.[30] By contrast, the explosive and insoluble black compound $Se_3N_2Cl_2$, which probably contains the $[Se_3N_2Cl]^+$ cation, is prepared by the treatment of Se_2Cl_2 with trimethylsilyl azide in CH_2Cl_2 (Eq. 2.13).[31]

$$3Se_2Cl_2 + 2Me_3SiN_3 \rightarrow 2Se_3N_2Cl_2 + 2Me_3SiCl + N_2 \quad (2.13)$$

Chalcogen-nitrogen cations can be generated by the reactions of homopolyatomic chalcogen cations with azides. For example, the reaction of S_8^{2+} with sodium azide in liquid SO_2 is a source of $[NS_2^+]$.[32] Remarkably, the reaction of Te_4^{2+} with potassium azide in SO_2 gives the tris(azido)tellurium(IV) cation $[Te(N_3)_3][SbF_6]$ as a stable salt.[33]

Organosulfur(II) azides evolve N_2 spontaneously at room temperature. For example, the reaction of CF_3SCl with Me_3SiN_3 produces the, presumably cyclic, tetramer $(CF_3SN)_4$.[34] Covalent arylselenium azides may be generated from arylselenium(II) halides and Me_3SiN_3 and used as *in situ* reagents, but they decompose to the corresponding diselenides ArSeSeAr at room temperature. A notable exception is the derivative $2\text{-}Me_2NCH_2C_6H_4SeN_3$, prepared by using AgN_3 instead of Me_3SiN_3, which is stabilized by intramolecular $N{\rightarrow}Se$ coordination (Section 15.6).[35] In contrast to the thermal instability of covalent organoselenium azides, a wide range of stable dialkyltellurium(IV) diazides and alkyl or aryl-tellurium(IV) triazides have been prepared by the metathesis of the corresponding fluorides with Me_3SiN_3 in CH_2Cl_2 (Eq. 2.14).[36a] The binary tellurium(IV) azide $Te(N_3)_4$ is likewise obtained by metathesis of TeF_4 (in $CFCl_3$)[36b] or TeF_6 (in CH_3CN)[36c] with trimethylsilyl azide (Section 5.2.9).

$$R_2TeF_2 + 2Me_3SiN_3 \rightarrow R_2Te(N_3)_2 + 2Me_3SiF \quad (2.14)$$

$$(R = Me, Et, {}^nPr, {}^iPr, Cy)$$

The cyclocondensation of trimethylsilyl azide with a bis(sulfenyl chloride) is an efficient synthesis of dithiazolium cations (Section 11.3.5) (Eq. 2.15).[37]

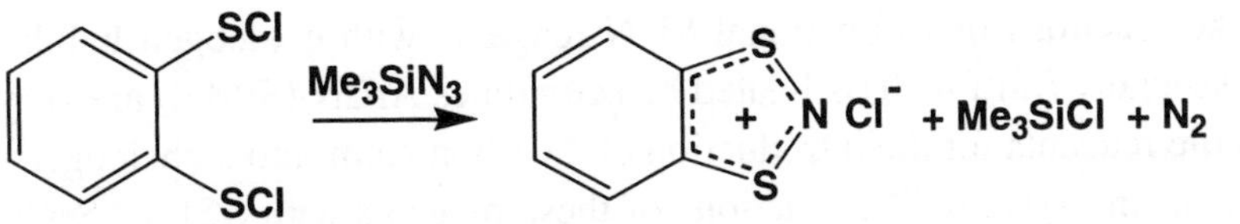

$$\text{(2.15)}$$

The use of azide reagents is also important for the synthesis of cyclic sulfur(VI)–nitrogen systems. The reaction of $SOCl_2$ with sodium azide in acetonitrile at $-35°C$ provides a convenient preparation of the trimeric sulfanuric chloride $[NS(O)Cl]_3$ (Eq. 2.16).[38a] Thionyl azide, $SO(N_3)_2$ is generated by the heterogeneous reaction of thionyl chloride vapour with silver azide (Eq. 2.17).[38b] This thermally unstable gas was characterized *in situ* by photoelectron spectroscopy. The phenyl derivative of the six-membered ring $[NS(O)Ph]_3$ can be prepared from lithium azide and $PhS(O)Cl$.[39]

$$3NaN_3 + 3SOCl_2 \rightarrow [NS(O)Cl]_3 + 3NaCl \qquad (2.16)$$

$$2AgN_3 + SOCl_2 \rightarrow SO(N_3)_2 + 2AgCl \qquad (2.17)$$

2.4 From Silicon–Nitrogen or Tin–Nitrogen Reagents

Condensation reactions involving the combination of chalcogen halides and reagents with one or more Si–N functionalities represent a very versatile approach to a wide range of cyclic and acyclic chalcogen–nitrogen compounds. The driving force for these reactions is normally the formation of volatile Me_3SiCl , which is easily separated from the other reaction products.

The outcome of these reactions is often markedly dependent on the chalcogen. For example, the monofunctional iminophosphorane reagent $Ph_3P=NSiMe_3$ reacts with SCl_2, in a 2:1 molar ratio, to give the sulfonium salt $[S(NPPh_3)_3]Cl$, presumably via a disproportionation process.[40] By contrast, both mono and bis-N=PPh$_3$ substituted derivatives are obtained by the reaction of this reagent with ECl_4 (E = Se, Te) (Eq. 2.18).[41, 42]

$$nPh_3PNSiMe_3 + ECl_4 \rightarrow (Ph_3PN)_nECl_{4-n} + nMe_3SiCl \qquad (2.18)$$
$$(n= 1, 2)$$

The reactions of trifunctional Si–N reagents with chalcogen halides are especially fruitful. Trisilylated benzamidines $ArCN_2(SiMe)_3$ are very versatile reagents for the introduction of a carbon atom into a chalcogen–nitrogen ring system. The reactions of these reagents with SCl_2 or $SeCl_2$ (generated *in situ*) give rise to the five-membered cyclic cations $[ArCN_2E_2]^+$ (E = S, Se) (Section 11.3.1)[44] (*cf.* the formation of eight-membered $C_2N_4S_2$ rings from the reactions of benzamidines with SCl_2) (Section 12.4.1). In the reaction with $TeCl_4$ only one mole of Me_3SiCl is eliminated to give a four-membered CN_2Te ring (Scheme 2.2).[45]

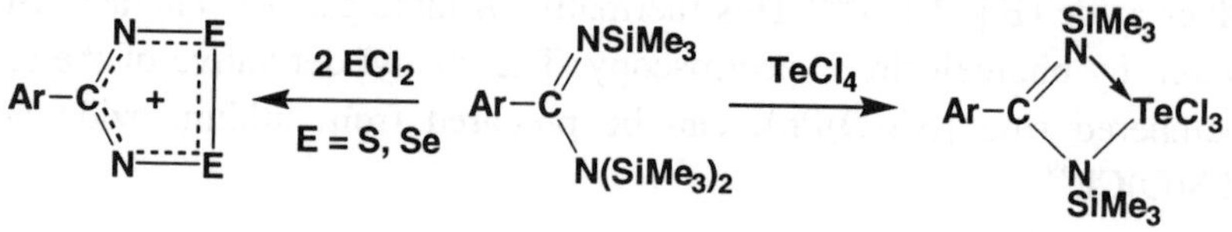

Scheme 2.2   Reactions of trisilylated benzamidines with chalcogen halides

The reactions of trisilylated benzamidines with benzenechalcogenyl halides are of particular interest. The initially formed trisubstituted derivatives are thermally unstable and decompose by a radical process to generate the dark blue diazenes *trans*-PhEN(Ph)CN=NC(Ph)NEPh (E = S, Se) (Scheme 2.3) (Section 15.2).[46]

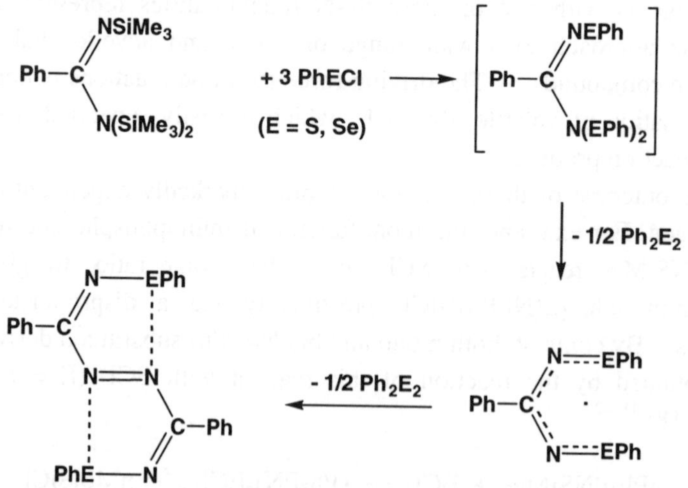

Scheme 2.3   Reactions of trisilylated benzamidines with benzenechalcogenyl halides

Phosphorus(V) can also be introduced into chalcogen–nitrogen rings via cyclocondensation reactions using $R_2PN_2(SiMe)_3$. In this case eight-membered rings of the type $R_2P(NEN)_2PR_2$ (E = S, Se) with folded structures (a cross-ring E•••E interaction) are formed. The reagents of choice are $SOCl_2$ (E = S)[47] or a mixture of $SeCl_4$ and Se_2Cl_2 (E = Se) (Section 13.2.1).[48]

In the case of tellurium four-membered PN_2Te rings of the type $Ph_2P(\mu\text{-}NSiMe_3)_2Te=NP(Ph_2)NSiMe_3$ are obtained from the reaction of $TeCl_4$ with $Li[Ph_2P(NSiMe_3)_2]$ (Eq. 2.19).[49]

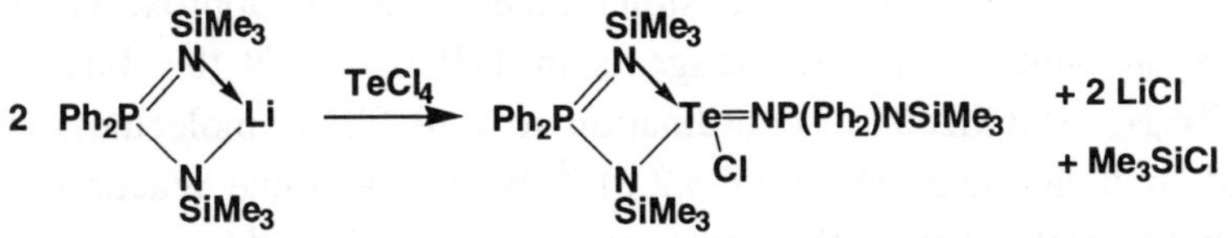

$$2 \; Ph_2P \quad \xrightarrow{TeCl_4} \quad Ph_2P \quad Te=NP(Ph_2)NSiMe_3 \qquad + 2 \; LiCl \qquad + Me_3SiCl \qquad (2.19)$$

The trifunctional amines $N(MMe_3)_3$ (M = Si, Sn) are attractive alternatives to ammonia for the generation of chalcogen–nitrogen bonds by reaction with chalcogen halides. The use of $(Me_3Sn)_3N$ instead of ammonia allows kinetic control of the reaction with S_2Cl_2 and results in the formation of a five-membered metallathiazene ring (Eq. 2.20),[15] which can be converted into other S–N ring systems by metathesis with element halides (Section 7.3.2).

$$4S_2Cl_2 + 4(Me_3Sn)_3N \rightarrow (Me_2SnS_2N_2)_2 + 2Me_4Sn +$$

$$8Me_3SnCl + 1/2S_8 \qquad (2.20)$$

Acyclic Se–N chlorides that are potential building blocks (as a source of the SeNSe unit) for cyclic Se–N systems are generated by the reactions of $[SeCl_3][AsF_6]$ or $SeCl_4$ with $(Me_3Si)_3N$ as illustrated by the sequence of transformations that produce the five-membered $[Se_3N_2Cl]^+$ cation (Scheme 2.4 and Section 8.4). [50-52]

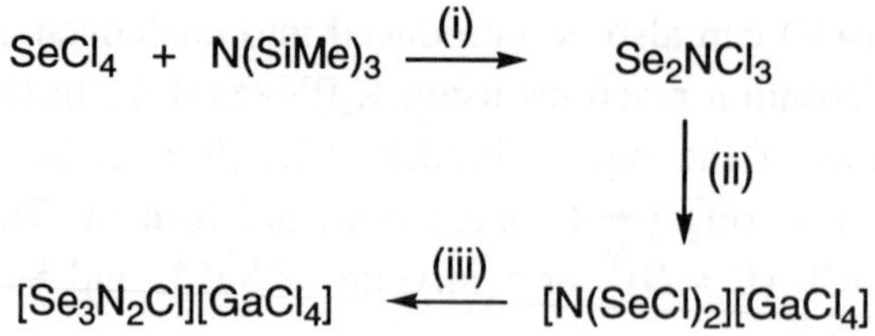

$$\text{SeCl}_4 \ + \ \text{N(SiMe)}_3 \ \xrightarrow{\text{(i)}} \ \text{Se}_2\text{NCl}_3$$

$$\Big\downarrow \text{(ii)}$$

$$[\text{Se}_3\text{N}_2\text{Cl}][\text{GaCl}_4] \ \xleftarrow{\text{(iii)}} \ [\text{N(SeCl)}_2][\text{GaCl}_4]$$

Scheme 2.4   Formation of acyclic and cyclic Se–N cations: (i) –3Me$_3$SiCl, –Cl$_2$ (ii) GaCl$_3$ (iii) Ph$_3$Sb

The major product obtained from the reaction of TeCl$_4$ with (Me$_3$Si)$_3$N is determined by the stoichiometry. When approximately equimolar amounts of the two reagents in THF are used the dimeric cluster Te$_6$N$_8$ (stabilized by coordination to four TeCl$_4$ molecules) is obtained in high yields (Section 5.2.8).[53] When the same reaction is carried out in acetonitrile with a molar ratio of 2:1, followed by treatment of the product with AsF$_5$ in SO$_2$, [Te$_4$N$_2$Cl$_8$][AsF$_6$]$_2$ is obtained Section 8.5).[54] The dication [Te$_4$N$_2$Cl$_8$]$^{2+}$ in this salt is the dimer of the hypothetical tellurium(IV) imide [Cl$_3$Te–N=TeCl]$^+$.

References

1. H. G. Heal, *The Inorganic Heterocyclic Chemistry of Sulfur, Nitrogen and Phosphorus*, Academic Press, London (1980).

2. T. Chivers, *Chem. Rev.*, **85**, 341 (1985).

3. R. T. Oakley, *Prog. Inorg. Chem.*, **36**, 1 (1988).

4. T. Chivers, Formation of S–N, Se–N and Te–N Bonds, in J. J. Zuckerman and A. P. Hagen (ed.) *"Inorganic Reactions and Methods"*, Vol. 5. VCH Publishers, Inc., New York, pp. 57-74 (1991).

5. T. Chivers, Sulfur–Nitrogen Compounds, in R. B. King (ed.) *Encyclopedia of Inorganic Chemistry*, Vol.7, John Wiley, pp. 3988-4009 (1994).

6. T. Torroba, *J. Prakt. Chem.*, **341**, 99 (1999).

7. S. Apler, M. Carruthers and L. H. Sutcliffe, *Inorg. Chim. Acta*, **31**, L455 (1978).

8. P. K. Gowik and T. M. Klapötke, *Spectrochim. Acta*, **46A**, 1371 (1990).

9. (a) W. L. Jolly and K. D. Maguire, *Inorg. Synth.*, **9**, 102 (1967); (b) T. Chivers, R. T. Oakley, O. J. Scherer and G. Wolmershäuser, *Inorg. Chem.*, **20**, 914 (1981).

10. H. G. Heal and J. Kane, *Inorg. Synth.*, **11**, 192 (1968).

11. A. Maaninen, J. Siivari, R. S. Laitinen and T. Chivers, *Inorg. Synth.*, **33**, 196 (2002).

12. M. Björgvinsson, H. W. Roesky, F. Pauer, D. Stalke and G. M. Sheldrick, *Inorg. Chem.*, **29**, 5140 (1990).

13. J. Siivari, T. Chivers and R. S. Laitinen, *Inorg. Chem.*, **32**, 1519 (1993).

14. T. Murai, K. Nonomura, K. Kimura and S. Kato, *Organometallics*, **10**, 1095 (1991).

15. C. P. Warrens and J. D. Woollins, *Inorg. Synth.*, **25**, 43 (1989).

16. F. Fockenberg and A. Haas, *Z. Naturforsch.*, **41B**, 413 (1986).

17. T. Chivers, N. Sandblom and G. Schatte, *Inorg. Synth.*, **34**, 42 (2004).

18. R. E. Allan, H. Gornitzka, J. Karcher, M. A. Paver, M-A. Rennie, C. A. Russell, P. R. Raithby, D. Stalke, A. Steiner and D. S. Wright, *J. Chem. Soc., Dalton Trans.*, 1727 (1996).

19. O. J. Scherer and G. Wolmershäuser, *Z. Anorg. Allg. Chem.*, **432**, 173 (1977).

20. R. Bussas, G. Kresze, H. Münsterer and A. Schwöbel, *Sulfur Rep.*, **2**, 215 (1983).

21. T. Maaninen, R. S. Laitinen and T. Chivers, *Chem. Commun.*, 1812 (2002).

22. Y. Inagakin, T. Hosogai, R. Okazaki and N. Inamoto, *Bull. Chem. Soc. Jpn.*, **53**, 205 (1980).

23. A. Maaninen, T. Chivers, M. Parvez, J. Pietikäinen and R. S. Laitinen, *Inorg. Chem.*, **38**, 4093 (1999).

24. T. Maaninen, T. Chivers, R. S. Laitinen, G. Schatte and M. Nissinen, *Inorg. Chem.*, **39**, 5241 (2000).

25. V. N. Kovtonyk, A. Yu. Makarov, M. M. Shakirov and A. V. Zibarev, *Chem. Commun.*, 1991 (1996).

26. R. T. Oakley, R. W. Reed, A. W. Cordes, S. L. Craig and J. B. Graham, *J. Am Chem. Soc.*, **109**, 7745 (1987).

27. I. Ernest, W. Holick, G. Rihs, D. Schomburg, G. Shoham, D. Wenkert and R. B. Woodward, *J. Am. Chem. Soc.*, **103**, 1540 (1981).

28. K. Seppelt, *Inorg. Chem.*, **12**, 2837 (1973).

29. J. Bojes, T. Chivers and I. Drummond, *Inorg. Synth.*, **18**, 203 (1978).

30. F. A. Kennett, G. K. MacLean, J. Passmore and M. N. S. Rao, *J. Chem. Soc., Dalton Trans.*, 851 (1982).

31. J. Siivari, T. Chivers and R. S. Laitinen, *Angew. Chem., Int. Ed.*, **31**, 1518 (1992).

32. A. J. Banister, R. G. Hey, G. K. MacLean and J. Passmore, *Inorg. Chem.*, **21**, 1679 (1982).

33. J. P. Johnson, G. K. MacLean, J. Passmore and P.S. White, *Can. J. Chem.*, **67**, 1687 (1989).

34. D. Bielefeld and A. Haas, *Chem. Ber.*, **116**, 1257 (1983).

35. T. M. Klapötke, B. Krumm and K. Polborn, *J. Am. Chem. Soc.*, **126**, 710 (2004).

36. (a) T. M. Klapötke, B. Krumm, P. Mayer, H. Piotrowski, O. P Ruscitti and A. Schiller, *Inorg. Chem.*, **41**, 1184 (2002); (b) T. M. Klapötke, B. Krumm, P. Mayer and I. Schwab, *Angew. Chem., Int. Ed. Engl.*, **42**, 5843 (2003); (c) R. Hiages, J. A. Boatz, A. Vij, M. Gerken, S. Schneider, T. Schroer and K. O. Christe, *Angew. Chem., Int. Ed. Engl.*, **42**, 5847 (2003).

37. G. Wolmershäuser, M. Schnauber and T. Wilhelm, *J. Chem. Soc., Chem. Commun.*, 573 (1984).

38. (a) H. Klüver and O. Glemser, *Z. Naturforsch.*, **32B**, 1209 (1977); (b) Z. Xiaoqing, L. Fengyi, S. Qiao, G. Maofa, Z. Jianping, A. Xicheng, M. Lingpeng, Z. Shijun and W. Dianxun, *Inorg. Chem.*, **43**, 4799 (2004).

39. T. J. Maricich, *J. Am. Chem. Soc.*, **90**, 7179 (1968).

40. J. Bojes, T. Chivers, A. W. Cordes, G. K. MacLean and R. T. Oakley, *Inorg. Chem.*, **20**, 16 (1981).

41. H. W. Roesky, K. L. Weber, U. Seseke, W. Pinkert, M. Noltemeyer, W. Clegg and G. M. Sheldrick, *J. Chem. Soc., Dalton Trans.*, 565 (1985).

42. (a) J. Münzenberg, M. Noltemeyer and H. W. Roesky, *Chem. Ber.*, **122**, 1915 (1989); (b) H. W. Roesky, J. Münzenberg, R. Bohra and M. Noltemeyer, *J. Organomet. Chem.*, **418**, 339 (1991).

43. H-U. Höfs, J. W. Bats, R. Gleiter, G. Hartman, R. Mews, M. Eckert-Maksić, H. Oberhammer and G. M. Sheldrick, *Chem. Ber.*, **118**, 3781 (1985).

44. P. Del Bel Belluz, A. W. Cordes, E. M. Kristof, P. V. Kristof, S. W. Liblong and R. T. Oakley, *J. Am. Chem. Soc.*, **111**, 6147 (1989).

45. E. Hey, C. Ergezinger and K. Dehnicke, *Z. Naturforsch.*, **44B**, 205 (1989).

46. (a) V. Chandrasekhar, T. Chivers, J. F. Fait and S. S. Kumaravel, *J. Am. Chem. Soc.*, **112**, 5373 (1990); (b) V. Chandrasekhar, T. Chivers, S. S. Kumaravel, M. Parvez and M. N. S. Rao, *Inorg. Chem.*, **30**, 4125 (1991).

47. T. Chivers, D. D. Doxsee and R. W. Hilts, Preparation and Some Reactions of the Folded $P_2N_4S_2$ Ring, in J. D. Woollins (ed.) *Inorganic Experiments*, 2nd Edn., Wiley-VCH, pp. 287-290 (2003).

48. T. Chivers, D. D. Doxsee and J. F. Fait, *J. Chem. Soc., Chem. Commun.*, 1703 (1989).

49. T. Chivers, D. D. Doxsee, X. Gao and M. Parvez, *Inorg. Chem.*, **33**, 5678 (1994).

50. M. Broschag, T. M. Klapötke, I. C. Tornieporth-Oetting and P. S. White, *J. Chem. Soc., Chem. Commun.*, 1390 (1992).

51. R. Wollert, A. H. Ilwarth, G. Frenking, D. Fenske, H. Goesman and K. Dehnicke, *Angew. Chem., Int. Ed. Engl.*, **31**, 1251 (1992).

52. R. Wollert, B. Neumüller and K. Dehnicke, *Z. Anorg. Allg. Chem.*, **616**, 191 (1992).

53. W. Massa, C. Lau, M. Möhlen, B. Neumüller and K. Dehnicke, *Angew., Chem., Int. Ed. Engl.*, **37**, 2840 (1998).

54. J. Passmore, G. Schatte and T. S. Cameron, *J. Chem. Soc., Chem. Commun.*, 2311 (1995).

Chapter 3

Applications of Physical Methods

This chapter will provide an overview, illustrated with recent examples, of some applications of the most commonly used physical methods for the characterization of chalcogen–nitrogen compounds.

3.1 Diffraction Techniques

The structures of chalcogen–nitrogen compounds are frequently unpredictable. For example, the reactions of heterocyclic systems often result in substantial reorganization of their structural frameworks, *e.g.* ring expansion or contraction. The formation of acyclic products from ring systems (or vice versa) is also observed.

Spectroscopic techniques alone are rarely sufficient to provide decisive structural information. Consequently, the ability to establish atomic arrangements in the solid state by X-ray crystallography has been of paramount importance in the development of chalcogen-nitrogen chemistry. Although there are no reviews dedicated exclusively to structural studies of S–N compounds, an account that focuses on the structures of compounds containing Se–N or Te–N bonds was published in 1991.[1] Since that time, the amount of structural information that is available for chalcogen–nitrogen compounds has increased dramatically, especially with the advent of CCD diffractometers that allow structure determinations to be completed in less than one day in many cases. X-ray crystallography also provides the details of intermolecular interactions, which are often significant in chalcogen–nitrogen compounds because of the polarity of the E–N bonds. In such compounds structural information is vitally important for understanding the physical properties, *e.g.*, conductivity and magnetic behaviour. Dithia- and diselena-diazolyl radicals, $[RCE_2N_2]^{\bullet}$ (E = S, Se), are an especially important class of chalcogen–nitrogen compound in this regard (Section 11.3.1). In a search for materials with novel electronic and magnetic properties compounds with two or three RCE_2N_2 ring systems attached to an organic framework have been prepared. The extended structures of these multifunctional derivatives adopt a wide variety of packing arrangements in the solid state which, in turn, give rise to unique properties.[2]

In general these intermolecular interactions become stronger along the series S•••N < Se•••N < Te•••N. As an example, the structural trends in the series $E(NMe_2)_2$ (E = S, Se, Te) can be considered. Whereas the sulfur and selenium derivatives are liquids that do not exhibit significant intermolecular interactions, the telluride $Te(NMe_2)_2$ (**3.1**) is a polymeric solid as a result of Te•••N contacts.[3a] The Te-N distances in the monomer units of **3.1** are 2.05 Å, while the intermolecular Te•••N contacts are 2.96 Å, *cf.* 2.76 Å for 1,2,5-telluradiazole (**3.2**).[3b] The sum of the van der Waals radii for Te and N is 3.61 Å. In this connection the properties of the unknown ditelluradiazolyls $[RCTe_2N_2]^{\bullet}$ will be of considerable interest.[4]

3.1   3.2

Electron diffraction studies provide valuable information about structures in the gas phase. Consequently, this method is important for chalcogen-nitrogen compounds that are liquids or gases at room temperature. The application of this technique has provided evidence for the monomeric structures of the 1,2,3,5-dithiadiazolyl radical $[CF_3CNSSN]^{\bullet}$ (**3.3**)[5] and the 1,3,2-dithiazolyl $[CF_3CSNSCCF_3]^{\bullet}$ (**3.4**), a paramagnetic liquid.[6] An electron diffraction study of the dithiatriazine $CF_3CN_3S_2F_2$ (3.5) showed that both S–F bonds are in axial positions.[7a] The structure of the trithiatriazine $(NSF)_3$ has also been determined in the gas phase by a combination of electron diffraction and microwave spectroscopy, which shows that all three fluorine atoms are axial (Section 8.7).[7b]

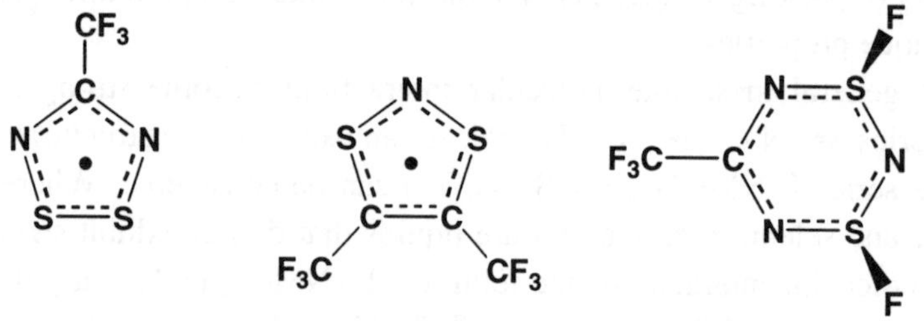

3.3　　　　　　　　3.4　　　　　　　　3.5

Gas-phase structures are not influenced by packing effects. As a result a comparison of the electron diffraction and X-ray diffraction structures for a certain compound allows an evaluation of the influence of packing effects on structural parameters to be made. For example, the elongation of the transannular S•••S distance in S_4N_4 by about 0.08 Å in the gas phase is attributed to the lack of intermolecular contacts that impart packing effects.[8] The gas-phase structures of the twelve π-electron systems 1,3,2,4-benzodithiadiazines $X_4C_6S_2N_2$ (3.6a, X = H; 3.6b, X = F) were determined by electron diffraction.[9] It was concluded that the parent system 3.6a is non-planar in the gas phase whereas the tetrafluoro derivative 3.6b is planar, in agreement with *ab initio* molecular orbital calculations. In the solid state, however, the crystal structure of 3.6a is planar,[10] whereas that of 3.6b is non-planar, presumably as a result of packing forces.[11] The structures of the sulfur diimide $S(N^tBu)_2$ (3.7) and the sulfur triimide $S(N^tBu)_3$ (3.8) have also been determined in the gas phase and in the solid state at low temperature.[12] Although a small, but significant, lengthening of the S=N bonds (by 0.001-0.002 Å) is observed in the gas phase, the structures of both compounds were found to be similar in both phases.

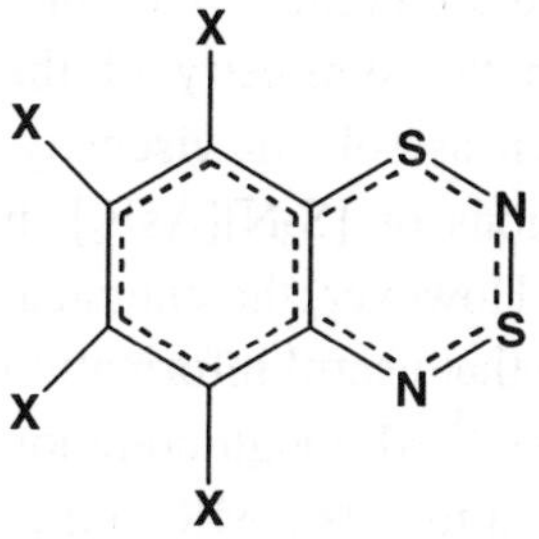

| 3.6a, X = H | 3.7 | 3.8 |
| 3.6b, X = F | | |

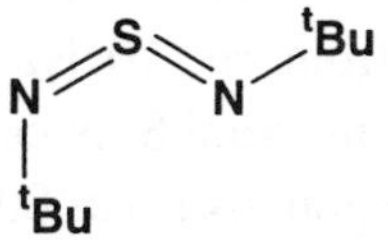

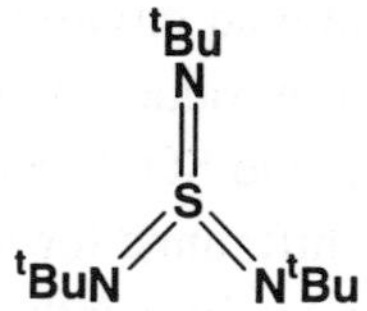

Specific structural information for various classes of chalcogen–nitrogen compounds will be discussed in the appropriate chapters.

3.2 ^{14}N and ^{15}N NMR Spectroscopy

The availability of high-field pulsed NMR spectrometers has facilitated nitrogen NMR investigations of chalcogen–nitrogen compounds. Nitrogen has two isotopes ^{14}N ($I = 1$, 99.6%) and ^{15}N ($I = \frac{1}{2}$, 0.4%) that are amenable to NMR studies. The advantages of ^{15}N NMR are narrow, well-resolved lines and the potential ability to observe spin-spin coupling patterns. The disadvantages include the lengthy acquisition times due to long spin-lattice relaxation times and the necessity to prepare isotopically enriched materials in order to observe the effects of spin-spin coupling. Methods for the introduction of ^{15}N into S_4N_4 and Se_4N_4 from ^{15}N-labelled ammonium chloride or ammonia, respectively, have been developed.[13,14] On the other hand, the ^{14}N nucleus gives rise to relatively broad lines, poorer resolution, and loss of coupling information. In order to offset these disadvantages very short delays between pulses are possible in view of the short spin-lattice relaxation times; good signal-to-noise ratios can be achieved in relatively short acquisition times when appropriate experimental conditions are used .[15]

Some important examples of the application of ^{14}N NMR spectroscopy in sulfur–nitrogen chemistry include (a) studies of the (NSCl)$_3$ ↔ 3NSCl equilibrium in solution[16] and (b) identification of the S-N species present in solutions of sulfur in liquid ammonia.[17] The linewidth of ^{14}N NMR resonances is dependent on the symmetry of the atomic environment as well as other factors, such as solvent viscosity. The values may vary from *ca.* 10 Hz for solutions of [S$_2$N][AsF$_6$] in liquid SO$_2$ to >1000 Hz in metal S–N complexes. However, the chemical shift range for S–N compounds is *ca.* 800 ppm so that useful information may frequently be obtained despite the broad lines.[15] Although common organic solvents, *e.g.*, toluene, CH$_2$Cl$_2$, may be used for ^{14}N NMR studies of many chalcogen–nitrogen compounds, solubility considerations or the high reactivity of the species under investigation may require the use of less common solvents for some species. Liquid SO$_2$ is a particularly good solvent for ^{14}N NMR studies of cationic S-N species,[16] while liquid ammonia is well suited to investigations of inorganic S–N anions.[15,17]

The first application of nitrogen NMR spectroscopy in S–N chemistry was reported in 1965.[18] The ^{15}N NMR spectrum of the [S$_4$*N$_3$]$^+$ cation (**3.9**, *N = 97% ^{15}N-enriched) exhibited the expected doublet and triplet (1:2:1) patterns [$^2J(^{15}N-^{15}N) \sim 7$ Hz] with relative intensities of 2:1. More recently, ^{15}N NMR spectroscopy has been used for the detection of new species in solution. For example, the acyclic [S$_2$N$_2$H]$^-$ ion (**3.10**), which is well known in metal complexes (Section 7.3.2), was identified by the observation of two characteristic resonances, one of which is a doublet [$^1J(^{15}N-^1H) = 55$ Hz].[17b] The ^{15}N–^{15}N coupling constant between the two inequivalent nitrogen atoms is *ca.* 2 Hz. ^{15}N NMR spectroscopy has also been applied to the elucidation of the mechanism of molecular rearrangements in S–N chemistry. For example, the bicyclic compound PhCN$_5$S$_3$ (**3.11**) has been shown to rearrange via 1,3-nitrogen shifts (Section 4.8.3).[19]

3.9   **3.10**   **3.11**   **3.12**

In S–N compounds containing ^1H bonded directly to nitrogen, inverse detection 2D $^1H\{^{15}N\}$ NMR spectroscopy can be applied to give ^{15}N chemical shifts and coupling constants involving ^{15}N nuclei using natural abundance ^{15}N molecules. The first application of this technique involved complexes of the type **3.12**.[20] This method has also been used for the analysis of a mixture of cyclic sulfur imides involving the diimide isomers 1,3-, 1,4-, and 1,5-$S_6(NH)_2$ and the triimide isomers 1,3,5- and 1,3,6-$S_5(NH)_3$ (Section 6.2.1); the ^{15}N NMR chemical shifts are determined by the environment of the NH group in these cyclic systems (Figure 3.1).[21]

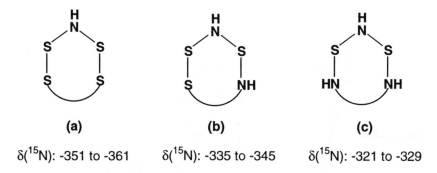

(a)   (b)   (c)

$\delta(^{15}N)$: -351 to -361 $\delta(^{15}N)$: -335 to -345 $\delta(^{15}N)$: -321 to -329

Fig. 3.1 ^{15}N NMR chemical shifts for NH groups in cyclic sulfur imides

3.3 ^{33}S, ^{77}Se and ^{125}Te NMR Spectroscopy

The only sulfur isotope with a nuclear spin is ^{33}S, which is quadrupolar ($I = 3/2$) and of low natural abundance (0.76%). In view of these inherent difficulties and the low symmetry around the sulfur nuclei in most S–N compounds, ^{33}S NMR spectroscopy has found very limited application in S–N chemistry. However, it is likely that reasonably narrow resonances could be obtained for sulfur in a tetrahedral environment, *e.g.* $[S(N^tBu)_4]^{2-}$,[22] *cf.* $[SO_4]^{2-}$. On the other hand both selenium and tellurium have isotopes with $I = \frac{1}{2}$ with significant natural abundances (^{77}Se, 7.6% and ^{125}Te, 7.0%). Consequently, NMR studies using these nuclei can provide useful information for Se–N and Te–N systems.

The ^{77}Se NMR chemical shifts of Se–N compounds cover a range of >1500 ppm and the value of the shift is characteristic of the local environment of the selenium atom. As a result, ^{77}Se NMR spectra can be used to analyse the composition of a complex mixture of Se–N compounds. For example, ^{77}Se NMR provides a convenient probe for analyzing the decomposition of selenium(IV) diimides RN=Se=NR (*e.g.*, R = [t]Bu). [23] By this method it was shown that the major decomposition products are the six-membered ring (SeN[t]Bu)$_3$, the five-membered ring Se$_3$(N[t]Bu)$_2$ and fifteen-membered ring Se$_9$(N[t]Bu)$_6$ (Figure 3.2 and Section 6.3).

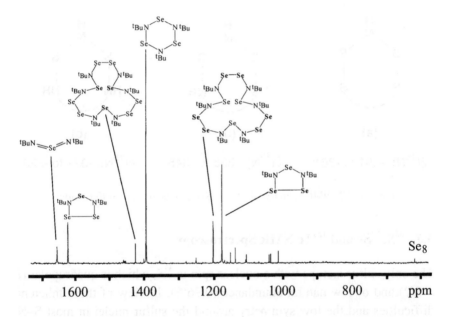

Fig. 3.2 ^{77}Se NMR spectrum for the products of the decomposition of [t]BuN=Se=N[t]Bu in toluene at 20°C after 2 days [Reproduced with permission from *Inorg. Chem.*, **39**, 5341 (2000)]

An elegant application of ^{77}Se NMR spectroscopy, in conjunction with ^{15}N NMR spectroscopy, involves the detection of the thermally unstable eight-membered rings (RSeN)$_4$ (**3.13**) from the reactions of a mixture of seleninic anhydrides (RSeO)$_2$O (R=Ph,[i]Pr) and ^{15}N-enriched hexamethyldisilazane $HN(SiMe_3)_2$.[24] On the basis of the number of signals observed in the ^{77}Se NMR spectrum, it was inferred that there are alternating single and double selenium–nitrogen bonds in **3.13**.

The ^{77}Se NMR spectrum of the cyclic dication $[Se_3N_2]^{2+}$ (**3.14**) exhibits a single resonance with no satellites due to $^{77}Se - {^{77}Se}$ coupling, even at $-70°C$.[25] This surprising observation is indicative of an intramolecular fluxional process, possibly via a trigonal bipyramidal intermediate of the type **3.15** in which the nitrogen atoms adopt a μ_3 bonding mode.

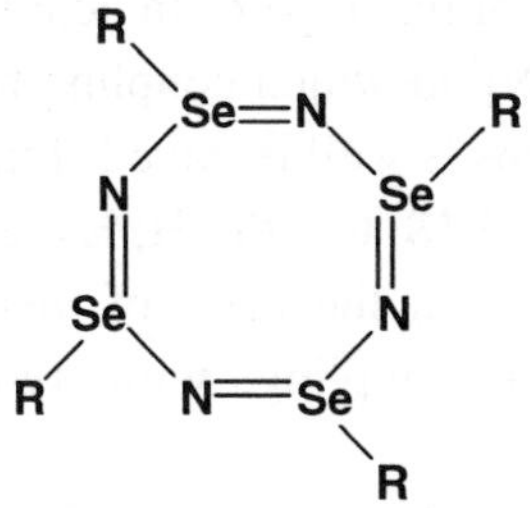

3.13 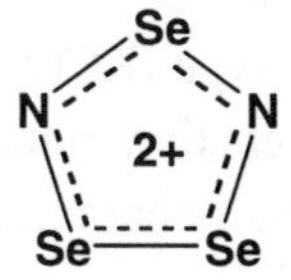   **3.14** 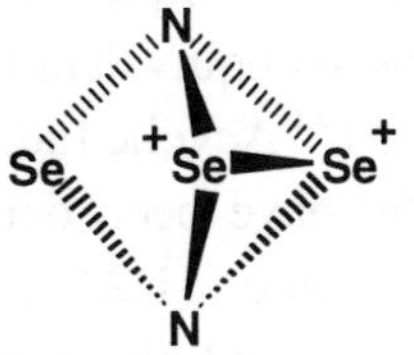   **3.15**

3.4 EPR Spectroscopy

EPR studies of S–N radicals were reviewed in 1990.[26] Many radicals containing the S–N linkage are persistent for more than several hours in solution at room temperature. Perhaps the best known example is the nitrosodisulfonate dianion $[ON(SO_3)]^{2-•}$, named as Fremy's salt. In the solid state this radical dianion dimerizes through weak N•••O interactions, but it forms a paramagnetic blue-violet monomer in solution. Although most chalcogen–nitrogen radicals dimerize in the solid state, a few heterocyclic C–S–N systems can be isolated as monomers (Section 11.3).

The ^{14}N nucleus ($I = 1$, 99%) has a moderately large magnetic moment and hyperfine splittings from this nucleus are a distinctive feature of the EPR spectra of chalcogen-nitrogen radicals. *N*-Arylthio-2,4,6-triphenylanilino radicals (**3.16**) are exceptionally persistent and oxygen-insensitive in solution.[27] They exhibit a characteristic 1:1:1

triplet in the EPR spectrum as a result of coupling of the unpaired electron with the single nitrogen centre. By using a combination of a bulky aryl group attached to nitrogen and an electron-withdrawing group on sulfur, dark green or purple crystals of these radicals may be isolated. The EPR spectra of *N,N'*-(bisarylthio)aminyls (**3.17**), which were first produced in 1925, have been studied in detail (Section 10.9).[28] They also exhibit 1:1:1 triplets and the [33]S hyperfine splittings could be observed for these radicals when the EPR spectra were recorded at very high amplification.[29] However, isotopic labeling is usually necessary to obtain information on spin density at sulfur ([33]S, I = 3/2, 0.76%).[30,31] A particularly pleasing example of the EPR spectra of this type of radical is the 1:1:1 triplet of septets observed for [(CF$_3$S)$_2$N]$^{•}$ in which coupling to the six equivalent fluorine atoms of two CF$_3$ groups is well resolved (Fig. 3.3).[32] Acyclic radicals of the type [RC(NEAr)$_2$]$^{•}$ (**3.18**, R = Ph, H; E = S, Se) have been detected as reaction intermediates on the basis of their five-line (1:2:3:2:1) patterns characteristic of coupling with two equivalent nitrogen atoms.[33, 34]

Many of the persistent chalcogen–nitrogen radicals are cyclic C–N–S systems (Chapters 11 and 12). For these heterocycles the unpaired electron often occupies a delocalized π-orbital, which may contribute to the stability of the species. In conjunction with molecular orbital calculations, EPR spectra can provide unique information about the electronic structures of these ring systems. For example, the EPR spectra of 1,2,4,6-thiatriazinyls (**3.19**) exhibit significantly larger hyperfine couplings to the unique nitrogen, except for Ar = 4-MeOC$_6$H$_4$, for which accidental equivalence of all three nitrogens is observed (Section 12.2.1).[35]

3.16   3.17   3.18 (E = S, Se)

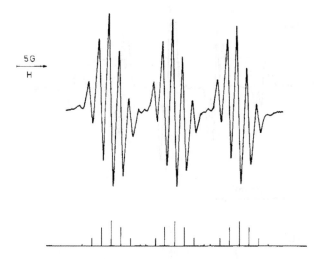

Fig. 3.3 EPR spectrum of [(CF$_3$S)$_2$N]$^{\bullet}$ [Reproduced with permission from. *J. Am. Chem. Soc.*, **105**, 1504 (1983)]

The EPR spectra of the related 1,2,4,6,3,5-thiatriazadiphosphinyl radicals (**3.20**) reveal a distinctly different electronic structure.[36] The observed spectrum consists of a quintet of triplets consistent with coupling of the unpaired electron with two equivalent nitrogen atoms and two equivalent phosphorus atoms [Fig. 3.4(a)]. This interpretation was confirmed by the observation that the quintet collapses to a 1:2:1 triplet when the nitrogen atoms in the ring are 99% ^{15}N-enriched [Fig 3.4(b)]. Thus the spin delocalization does not extend to the unique nitrogen atom in the phosphorus-containing system **3.20**.

3.19

3.20

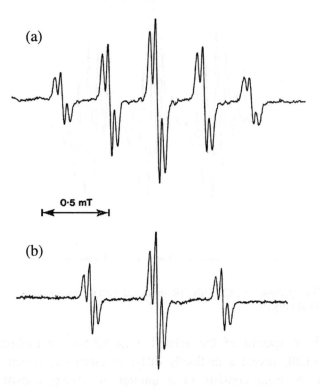

Fig. 3.4   EPR spectra of (a) $[Ph_4P_2N_3S]^{\bullet}$ and (b) $[Ph_4P_2{}^{15}N_3S]^{\bullet}$ in CH_2Cl_2 [Reproduced with permission from *J. Chem. Soc., Chem. Commun.*, 596 (1986)]

An intriguing class of persistent radicals are those formed by the one-electron oxidation of the hexagonal prismatic clusters $\{Li_2[E(N^tBu)_3]\}_2$ (**3.21**, E = S, Se). The air oxidation of **3.21** produces deep blue (E = S) or green (E = Se) solutions in toluene. The EPR spectra of these solutions consist of a septet (1:3:6:7:6:3:1) of decets (Fig. 3.5). This pattern results from interaction of the unpaired electron with three equivalent $I = 1$ nuclei, *i.e.*, ^{14}N, and three equivalent $I = 3/2$ nuclei, *i.e.*, ^7Li.[37, 38] It has been proposed that the one-electron oxidation of **3.21** is accompanied by the removal of an Li^+ cation from the cluster to give the neutral radical **3.22** in which the dianion $[S(N^tBu)_3]^{2-}$ and the radical monoanion $[S(N^tBu)_3]^{\bullet-}$ are bridged by three Li^+ cations.

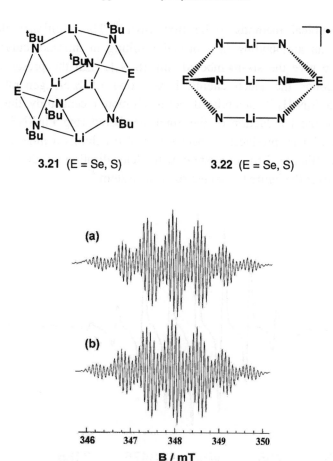

3.21 (E = Se, S) **3.22** (E = Se, S)

Fig. 3.5 (a) EPR spectrum of $\{Li_3[S(N^tBu)_3]_2\}^{\bullet}$ (**3.22**, E = S) in toluene at 20°C. (b) Simulated EPR spectrum [Reproduced with permission from *Inorg. Chem.*, **37**, 4633 (1998)]

3.5 Electrochemical Studies

The combination of electrochemical and EPR studies can provide valuable information about unstable S–N radical species. A classic early experiment involved the electrochemical reduction of S_4N_4 to the anion radical $[S_4N_4]^{\bullet-}$, which was characterized by a nine-line EPR spectrum.[39] The decay of the radical anion was shown by a combination of EPR and electrochemical methods to be first order.[40] The additional electron occupies an antibonding (π^*) orbital resulting in intramolecular bond rupture to give the six-membered ring $[S_3N_3]^-$.[41a] In the presence of a proton donor the electrochemical reduction of S_4N_4 generates the tetraimide $S_4N_4H_4$.[41b] In a related, but much more recent study, the *in situ* electrochemical reduction of the eight-membered ring $PhC(NSN)_2CPh$ (Section 12.4.1), produces a species that also exhibits a nine-line EPR spectrum (Figure 3.6), consistent with delocalization of the unpaired electron over the entire ten π-electron ring system.[42]

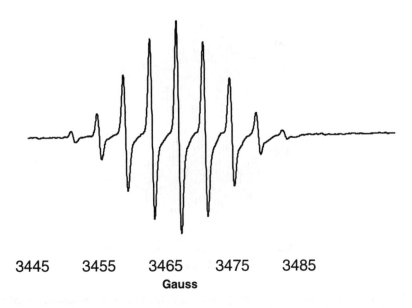

| 3445 | 3455 | 3465 | 3475 | 3485 |

Gauss

Fig. 3.6   EPR spectrum of the radical anion $[PhC(NSN)_2CPh]^{\cdot-}$ generated by electrochemical reduction.[42]

The electrochemistry of S–N and Se–N heterocycles has been reviewed comprehensively.[43] The emphasis is on the information that electrochemical studies provide about the redox properties of potential neutral conductors. To be useful as a molecular conductor the +1, 0, and −1 redox states should be accessible and the neutral radical should lie close to the centre of the redox spectrum. The chalcogen–nitrogen heterocycles that have been studied in most detail from this viewpoint are the 1,2,3,5-dithia- and diselena-diazoles (Scheme 3.1) (Section 11.3.1). Cyclic voltammetric studies of a series of 1,3,2-dithiazolyl systems have also been reported (Section 11.3.5).[44]

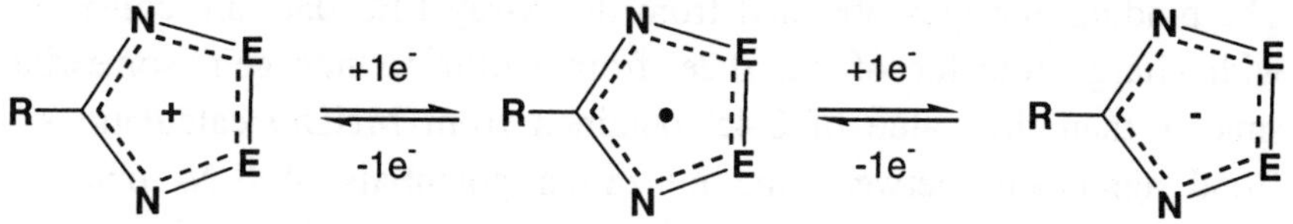

Scheme 3.1 Redox states for 1,2,3,5-dichalcogenadiazoles

Paradoxically, although they are electron-rich, S–N compounds are good electron acceptors because the lowest unoccupied molecular orbitals (LUMOs) are low-lying relative to those in the analogous carbon systems. For example, the ten π-electron $[S_3N_3]^-$ anion undergoes a two-electron electrochemical reduction to form the trianion $[S_3N_3]^{3-}$ whereas benzene, the aromatic hydrocarbon analogue of $[S_3N_3]^-$, forms the monoanion radical $[C_6H_6]^{-\bullet}$ upon reduction.[41a]

Electrochemical methods may also be used in the synthesis of chalcogen–nitrogen compounds. For example, the electrochemical reduction of salts of the $[S_5N_5]^+$ cation (Section 5.3.9) in SO_2 or CH_2Cl_2 at low temperatures produces microcrystals of the superconducting polymer $(SN)_x$.[45]

3.6 Photoelectron Spectroscopy

Photoelectron spectra (PES) supply information about the binding energies of either inner core (X-ray PES) or valence-level (UV–PES) electrons. The data obtained provide experimental support for molecular orbital calculations of electronic energy levels and atomic charges in various molecules. UV–PES are most readily obtained for reasonably volatile compounds and studies of S–N compounds have included S_2N_2,[46] S_4N_4,[46,47] and the eight-membered rings $RC(NSN)_2CR$ (R = Ph, Me_2N, ^tBu).[48] A recent application of PES involves the *in situ* characterization of the thermally unstable gas $SO(N_3)_2$.[49]

The UV–PES for S_2N_2 show that the two upper π levels are nearly degenerate with the π_S orbital (primarily nonbonding electrons located on the two sulfur atoms) being of slightly higher energy than π_N.[46] This ordering is supported by recent density functional theory calculations.[50] The binding energies obtained from the X-ray PES data are consistent with charge transfer of *ca.* 0.2e⁻ from sulfur to nitrogen, somewhat smaller than the value of 0.4e⁻ obtained from MNDO calculations.[48] Good agreement between the ionization potentials of S_4N_4 obtained from PES measurements and the Hartree–Fock–Slater eigenvalues has been obtained.[51] The X-ray PES data indicate a slightly greater charge transfer from sulfur to nitrogen compared to S_2N_2, consistent with molecular orbital calculations. The UV and X-ray PES data are also compatible with a modest S•••S bonding interaction in S_4N_4.[48] The UV–PES of the eight-membered ring ${}^tBuC(NSN)_2C^tBu$ shows four well-resolved peaks with ionization potentials less than 11 eV. The calculated orbital energies for the highest five molecular orbitals of the model compound $HC(NSN)_2CH$ corroborate the experimental results.[51]

PES also provides information about the binding energies of sulfur or nitrogen atoms that are in different environments in a given molecule. For example, the X-ray PES of $S_4N_4O_2$ (**3.23**) exhibits distinct binding energy values for the three types of sulfur atoms and two pairs of inequivalent nitrogen atoms consistent with the known structure.[52a] Similarly, the X-ray PES of S_3N_2 derivatives, *e.g.*, S_3N_2O (**3.24**), show three sulfur ($2p$) and two nitrogen ($1s$) binding energies as expected.[52b]

3.23   **3.24**

3.7 UV–Visible and MCD Spectroscopy

In contrast to unsaturated organic systems, most chalcogen-nitrogen compounds exhibit intense colours that are due to low energy ($\pi^* \to \pi^*$ or $n \to \pi^*$) transitions. In many cases the experimental values (from electronic spectra) can be correlated with the calculated excitation energies for these transitions.[53-57] Unambiguous assignments of UV–visible absorption bands to the appropriate electronic transitions can usually be made by comparing the calculated transition moments for the various alternatives.

In a few instances the technique of magnetic circular dichroism (MCD) spectroscopy has been used to corroborate assignments based on UV–visible spectroscopy. For example, the assignment of the intense 360 nm band for $[S_3N_3]^-$ to a $\pi^*(2e'') \to \pi^*(2a_2'')$ (HOMO $\to$ LUMO) excitation has been confirmed by the measurement of the MCD spectrum of $[S_3N_3]^-$.[58a] The MCD spectrum of $[S_4N_3]^+$ indicates that each of the two electronic absorption bands at 340 and 263 nm consists of nearly degenerate transitions,[58b] a conclusion that is consistent with Hartree–Fock–Slater calculations of the electronic transitions.[56] Monosubstituted derivatives of the type S_3N_3X (X = O⁻, NPPh₃, NAsPh₃) exhibit remarkably similar electronic spectra and molecular structures. The attachment of the substituent X to a sulfur atom of the S_3N_3 ring results in the loss of degeneracy for the HOMOs ($2e''$). Consequently, the strong absorptions observed at 480-510 and 330-340 nm have been assigned to the HOMO (π^*) $\to$ LUMO (π^*) and HOMO-1 (π^*) $\to$ LUMO (π^*) transitions, respectively. This assignment is supported by the MCD spectra of $Ph_3E=N-S_3N_3$ (E = P, As).[59]

The similarity in the electronic structures of the eight π-electron systems $R_2PN_3S_2$ (**3.25**) and $[S_3N_3O_2]^-$ (**3.26**) is also reflected in their UV–visible spectra. Both these heterocycles have an intense purple colour due to a visible band at *ca.* 560 nm attributed to the HOMO (π^*) $\to$ LUMO (π^*) transition.[55, 60] The 1,3- and 1,5- isomers of the eight-membered $R_2P(NSN)_2PR_2$ (R = Me, Ph) rings, **3.27** and **3.28**, respectively, exhibit characteristically different colours. The 1,3-isomer **3.27** is dark orange with an absorption maximum at *ca.* 460 nm, whereas the 1,5 isomer has a very pale yellow colour. In the absence of a cross-ring S•••S bond, the former are π-electron rich (ten π-electrons) and the strong visible absorption band is ascribed to the HOMO (π*) → LUMO (π*) transition.[61]

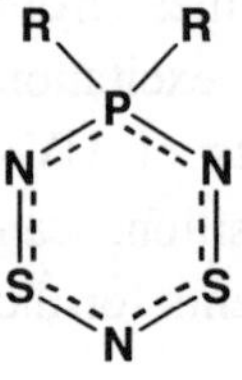

3.25 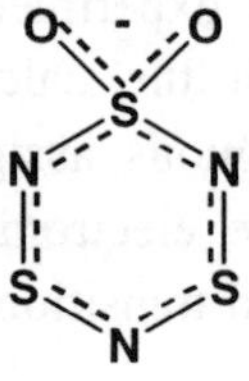   3.26 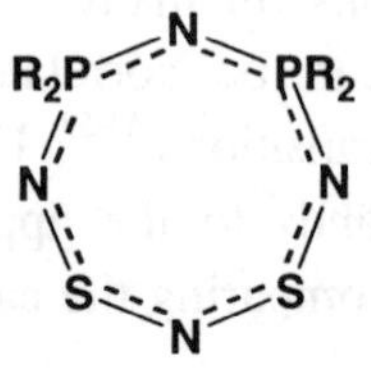   3.27 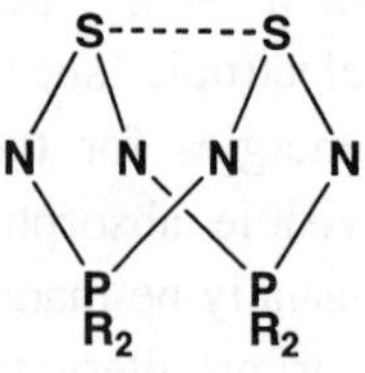   3.28

Some acyclic sulfur–nitrogen compounds also exhibit intense colours. For example, S-nitrosothiols RSNO are either green, red or pink (Section 9.7). Their UV-visible spectra show an intense band in the 330-350 nm region ($n_O → π*$) and a weaker band in the visible region at 550-600 nm ($n_N → π*$).[62]

3.8 Infrared and Raman Spectroscopy

Vibrational spectra have been used to identify small chalcogen–nitrogen species that are unstable under ambient conditions.[63,64] The experimental technique involves subjecting nitrogen gas and the chalcogen vapour to a microwave discharge in an argon atmosphere. The products are trapped in the argon matrix at 12 K and, with the aid of isotopic substitution (^{15}N, ^{34}S, ^{76}Se and ^{80}Se), the IR and Raman spectra can be assigned to specific molecules. In this way the species NS, SNS (and the less stable isomer NSS), NNS, NSe, SeNSe and [SeNSe]$^+$ have been identified (Sections 5.2.1-5.2.3). The strongest absorptions are observed for the antisymmetric stretching vibration of the ENE species (E = S, 1225 cm^{-1}; E = Se, 1021 cm^{-1}). The <ENE bond angles are estimated to be 153.5° and 146.5°, respectively. The diatomic species SN exhibits an absorption at 1209 cm^{-1} and the nitrogen–nitrogen stretching fundamental of N$_2$S is observed at 2040 cm^{-1}.[63] A band at 1253 cm^{-1} is assigned to the [SeNSe]$^+$

cation, *cf.* 1300 cm^{-1} observed for the isoelectronic, neutral molecule SeCSe.[64] Stable salts of this cation have not been isolated (Section 5.3.2).

The technique of ^{15}N-enrichment has been used in several cases to distinguish S–N from S–S vibrations. The IR and Raman spectra of S$_4$^{14}N$_4$ and S$_4$^{15}N$_4$ have been assigned in accordance with the D_{2d} structure.[65] Force constants were calculated and good agreement between observed and calculated wave numbers was obtained. A strong Raman band at 218 cm^{-1} for solid S$_4$N$_4$ was assigned to the stretching vibration of the weak cross-ring S•••S bonds. Indeed Raman spectroscopy is a useful probe for detecting transannular S•••S interactions in bicyclic or cage S-N molecules or ions.[66] The strongly Raman active vibrations occur at frequencies in the range 180-300 cm^{-1} for S•••S bond lengths in the range 2.4-2.7 Å. These bonds are substantially longer than a S–S single bond (ca. 2.06 Å) for which vibrational frequencies of 400-420 cm^{-1} are observed. Valence force field calculations for the fundamental vibrations of Se$_4$N$_4$ are in excellent agreement with the experimental values.[67] On the basis of symmetry considerations the Raman spectrum of the mixed sulfur–selenium nitride S$_2$Se$_2$N$_4$ was assigned to the 1,5- rather than the 1,3- isomer.[67]

In cases where information about atomic arrangements cannot be obtained by X-ray crystallography owing to the insolubility or instability of a compound, vibrational spectroscopy may provide valuable insights. For example, the explosive and insoluble black solid Se$_3$N$_2$Cl$_2$ was shown to contain the five-membered cyclic cation [Se$_3$N$_2$Cl]$^+$ by comparing the calculated fundamental vibrations with the experimental IR spectrum.[68]

3.9 Mass Spectrometry

Low ionizing potentials or soft ionization methods are necessary to observe the parent ions in the mass spectra of many S–N compounds because of their facile thermal decomposition. Mass spectrometry has been used to investigate the thermal breakdown of S$_4$N$_4$ in connection with the formation of the polymer (SN)$_x$. On the basis of the appearance potentials of various S$_x$N$_y$ fragments, two important steps were identified:

(a) the fragmentation of S_4N_4 into S_2N_2 and (b) the transformation of cyclic S_2N_2 into an open form of $(SN)_2$.[69] By contrast, the 12 eV mass spectrum of $Se_2S_2N_4$ shows $Se_2SN_2^+$ as the predominant fragment ion;[67a] the molecular ion $[Se_2S_2N_4]^+$ could be observed with modern instrumentation.[67b] The $[S_3N_3]^{\cdot}$ radical has been identified by mass spectrometry (in conjunction with PES) as the major product of the vaporization of $(SN)_x$ polymer.[70]

Mass spectrometry played an important role in the recent characterization of small cyclic sulfur imides that are formally derived from the unstable cyclic sulfur allotropes S_6 and S_7 by the replacement of one sulfur atom by an NR group. The compounds S_5NOct and S_6NOct (Section 6.2.2), which are yellow oils, exhibit molecular ions of medium intensity in their mass spectra.[71]

References

1. M. Björgvinsson and H. W. Roesky, *Polyhedron*, **10**, 2353 (1991).

2. (a) R. T. Oakley, *Can. J. Chem.*, **71**, 1775 (1993); (b) J. M. Rawson, A. J. Banister and I. Lavender, *Adv. Heterocycl. Chem.*, **62**, 137 (1995).

3. (a) R. E. Allan, H. Gornitzka, J. Kärcher, M. A. Parver, M-A. Rennie, C. A. Russell, P. R. Raithby, D. Stalke, A. Steiner and D. S. Wright, *J. Chem. Soc., Dalton Trans.*, 1727 (1996); (b) V. Bertini, P. Dapporto, F. Lucchesini, A. Sega and A. De Munno, *Acta Crystallogr.*, **C40**, 653 (1984).

4. W. M. Davis and J. D. Goddard, *Can. J. Chem.*, **74**, 810 (1996).

5. H. U. Höfs, J. W. Bats, R. Gleiter, G. Hartman, R. Mews, M. Eckert-Maksić, H. Oberhammer and G. M. Sheldrick, *Chem. Ber.*, **118**, 3781 (1985).

6. E. G. Awere, N. Burford, C. Mailer, J. Passmore, M. J. Schriver, P. S. White, A. J. Banister, H. Oberhammer and L. H. Sutcliffe, *J. Chem. Soc., Chem. Commun.*, 66 (1987).

7. (a) E. Jaudas-Prezel, R. Maggiulli, R. Mews, H. Oberhammer, T. Paust and W-D. Stohrer, *Chem. Ber.*, **123**, 2123 (1990); (b) E. Jaudas-Prezel, R. Maggiulli, R. Mews, H. Oberhammer and W-D. Stohrer, *Chem. Ber.*, **123**, 2117 (1990).

8. M. J. Almond, G. A. Forsyth, D. A. Rice, A. J. Downs, T. L. Jeffery and K. Hagen, *Polyhedron*, **8**, 2631 (1989).

9. F. Blockhuys, S. L. Hinchley, A. Y. Makarov, Y. V. Gatilov, A. V. Zibarev, J. D. Woollins and D. W. H. Rankin, *Chem. Eur. J.* **7**, 3592 (2001).

10. A. W. Cordes, M. Hojo, H. Koenig, M. C. Noble, R. T. Oakley and W.T. Pennington, *Inorg. Chem.*, **25**, 1137 (1986).

11. A. V. Zibarev, Y. V. Gatilov and A. O. Miller, *Polyhedron*, **11**, 1137 (1992).

12. S. L. Hinchley, P. Trickey, H. E. Robertson, B. A. Smart, D. W. H. Rankin, D. Leusser, B. Walfort, D. Stalke, M. Buhl and S. J. Obrey, *J. Chem. Soc., Dalton Trans.*, 4607 (2002).

13. T. Chivers, R. T. Oakley, O. J. Scherer and G. Wolmerhäuser, *Inorg. Chem.*, **20**, 914 (1981).

14. V. C. Ginn, P. F. Kelly and J. D. Woollins, *J. Chem. Soc., Dalton Trans.*, 2129 (1992).

15. I. P. Parkin, J. D. Woollins and P. S. Belton, *J. Chem. Soc., Dalton Trans.*, 511 (1990).

16. J. Passmore and M. Schriver, *Inorg. Chem.*, **27**, 2751 (1988).

17. (a) T. Chivers, D. D. McIntyre, K. J. Schmidt and H. J. Vogel, *J. Chem. Soc., Chem. Commun.*, 1341 (1990); (b) T. Chivers and K. J. Schmidt, *J. Chem. Soc., Chem. Commun.*, 1342 (1990).

18. N. Logan and W. L. Jolly, *Inorg. Chem.*, **4**, 1508 (1965).

19. R. T. Boeré, A. W. Cordes and R. T. Oakley, *J. Am. Chem. Soc.*, **109**, 7781 (1987).

20. B. Wrackmeyer, K. Schamel, K. Guldner and M. Herberhold, *Z. Naturforsch.*, **42B**, 702 (1987).

21. T. Chivers, M. Edwards, D, D. McIntyre, K. J. Schmidt and H. J. Vogel, *Mag. Reson. Chem.*, **30**, 177 (1992).

22. R. Fleischer, A. Rothenberger and D. Stalke, *Angew. Chem., Int. Ed. Engl.*, **36**, 1105 (1997).

23. T. Maaninen, T. Chivers, R. S. Laitinen, G. Schatte and M. Nissinen, *Inorg. Chem.*, **39**, 5341 (2000).

24. D. H. R. Barton and S. I. Parekh, *J. Am. Chem. Soc.*, **115**, 948 (1993).

25. E. G. Awere, J. Passmore and P. S. White, *J. Chem. Soc., Dalton Trans.*, 299 (1993).

26. K. F. Preston and L. H. Sutcliffe, *Mag. Reson. Chem.*, **28**, 189 (1990).

27. Y. Miura and A. Tanaka, *J. Chem. Soc., Chem. Commun.*, 441 (1990).

28. Y. Miura, N. Makita and M. Kinoshita, *Bull. Chem. Soc. Jpn.*, **50**, 482 (1977).

29. Y. Miura and M. Kinoshita, *Bull. Chem. Soc. Jpn.*, **53**, 2395 (1980).

30. S. A. Fairhurst, K. M. Johnson, L. H. Sutcliffe, K. F. Preston, A.J. Banister, Z. V. Hauptman and J. Passmore, *J. Chem. Soc., Dalton Trans.*, 1465 (1986).

31. K. F. Preston, J. P. B. Sandall and L. H. Sutcliffe, *Mag. Reson. Chem.*, **26**, 755 (1988).

32. K. Schlosser and S. Steenken, *J. Am. Chem. Soc.*, **105**, 1504 (1983).

33. V. Chandrasekhar, T. Chivers, S. S. Kumaravel, M. Parvez and M. N. S. Rao, *Inorg. Chem.*, **30**, 4125 (1991).

34. T. Chivers, B. McGarvey, M. Parvez, I. Vargas-Baca and T. Ziegler, *Inorg. Chem.*, **35**, 3839 (1996).

35. (a) R. T. Boeré, R. T. Oakley, R. W. Reed and N. P. C. Westwood, *J. Am. Chem. Soc.*, **111**, 1180 (1989); (b) R. T. Boeré and T. L. Roemmele, *Phosphorus, Sulfur and Silicon*, **179**, 875 (2004).

36. R. T. Oakley, *J. Chem. Soc., Chem. Commun.*, 596 (1986).

37. R. Fleischer, S. Freitag and D. Stalke, *J. Chem Soc., Dalton Trans.*, 193 (1998).

38. J. K. Brask, T. Chivers, B. McGarvey, G. Schatte, R. Sung and R. T. Boeré, *Inorg. Chem.*, **37**, 4633 (1998).

39. R. A. Meinzer, D. W. Pratt and R. J. Myers, *J. Am. Chem. Soc.*, **91**, 6623 (1969).

40. J. D. Williford, R. E. Van Reet, M. P. Eastman and K. B. Prater, *J. Electrochem. Soc.*, **120**, 1498 (1973).

41. (a) T. Chivers and M. Hojo, *Inorg. Chem.*, **23**, 1526 (1984); (b) M. Hojo, *Bull. Chem. Soc. Jpn.*, **53**, 2856 (1980).

42. R. T. Boeré, personal communication, 2003.

43. R. T. Boeré and T. L. Roemmele, *Coord. Chem. Rev.*, **210**, 369 (2000).

44. J. L. Brusso, O. P. Clements, R. C. Haddon, M. E. Itkis, A. A. Leitch, R. T. Oakley, R. W. Reed and J. F. Richardson, *J. Am. Chem. Soc.*, **121**, published on-line June 15 (2004).

45. (a) H. P. Fritz and R. Bruchhaus, *Z. Naturforsch.*, **38B**, 1375 (1983); (b) A. J. Banister, Z. V. Hauptman, A. G. Kendrick and R. W. H. Small, *J. Chem. Soc., Dalton Trans.*, 915 (1987).

46. (a) R. H. Findlay, M. H. Palmer, A. J. Downs, R. G. Egdell and R. Evans, *Inorg. Chem.*, **19**, 1307 (1980); (b) F. L. Skrezenek and R. D. Harcourt, *J. Am. Chem. Soc.*, **106**, 3934 (1984).

47. P. Brant, D. C. Weber, C. T. Ewing, F. L. Carter and J. A. Hashmall, *Inorg. Chem.*, **19**, 2829 (1980).

48. R. Gleiter, R. Bartetzko and D. Cremer, *J. Am. Chem. Soc.*, **106**, 3427 (1984).

49. Z. Xiaoqing, L. Fengyi, S. Qiao, G. Maofa, Z. Jianping, A. Xicheng, M. Lingpeng, Z. Shijun and W. Dianxun, *Inorg. Chem.*, **43**, 4799 (2004).

50. Y. Jung, T. Heine, P. von R. Schleyer and M. Head-Gordon, *J. Am. Chem. Soc.*, **126**, 3132 (2004).

51. M. Trsic and W. G. Laidlaw, *Int. J. Quantum Chem., Quantum Chem. Symp.*, **17**, 357 (1983).

52. (a) M. V. Andreocci, M. Bossa, V. Di Castro, C. Furlani, G. Maltogno and H. W. Roesky, *Z. Phys. Chem.*, **118**, 137 (1979); (b) M. V. Andreocci, M. Bossa, V. Di Castro, C. Furlani, G. Maltogno and H. W. Roesky, *Gazz. Chim. Ital.*, **109**, 1 (1979).

53. T. Chivers, P. W. Codding, W. G. Laidlaw, S. W. Liblong, R. T. Oakley and M. Trsic, *J. Am. Chem. Soc.*, **105**, 1186 (1983).

54. J. Bojes, T. Chivers, W. G. Laidlaw and M. Trsic, *J. Am. Chem. Soc.*, **101**, 4517 (1979).

55. N. Burford, T. Chivers, A. W. Cordes, W. G. Laidlaw, M.C. Noble, R. T. Oakley and P. N. Swepston, *J. Am. Chem. Soc.*, **104**, 1282 (1982).

56. M. Trsic and W. G. Laidlaw, *Inorg. Chem.*, **23**, 1981 (1984).

57. M. Trsic, W. G. Laidlaw and R. T. Oakley, *Can. J. Chem.*, **60**, 2281 (1982).

58. (a) J. W. Waluk and J. Michl, *Inorg. Chem.*, **20**, 963 (1981); (b) J. W. Waluk and J. Michl, *Inorg. Chem.*, **21**, 556 (1982).

59. J. W. Waluk, T. Chivers, R. T. Oakley and J. Michl, *Inorg. Chem.*, **21**, 832 (1982).

60. T, Chivers, A. W. Cordes, R. T. Oakley and W. T. Pennington, *Inorg. Chem.*, **22**, 2429 (1983).

61. N. Burford, T. Chivers and J. F. Richardson, *Inorg. Chem.*, **22**, 1482 (1983).

62. D. L. H. Williams, *Acc. Chem. Res.*, **32**, 869 (1999).

63. P. Hassanzadeh and L. Andrews, *J. Am. Chem. Soc.*, **114**, 83 (1992).

64. L. Andrews and P. Hassanzadeh, *J. Chem. Soc., Chem. Commun.*, 1523 (1994).

65. R. Steudel, *Z. Naturforsch.*, **36A**, 850 (1981).

66. N. Burford, T. Chivers, P. W. Codding and R. T. Oakley, *Inorg. Chem.*, **21**, 982 (1982).

67. (a) A. Maaninen, R. S. Laitinen, T. Chivers and T. A. Pakkanen, *Inorg. Chem.*, **38**, 3450 (1999); A. Maaninen, J. Konu, R. S. Laitinen, T. Chivers, G. Schatte, J. Pietikäinen and M. Ahlgrén, *Inorg. Chem.*, **40**, 3539 (2001).

68. J. Siivari, T. Chivers and R. S, Laitinen, *Inorg. Chem.*, **32**, 4391 (1993).

69. E. Besenyei, G. K. Eigendorf and D. C. Frost, *Inorg. Chem.*, **25**, 4404 (1986).

70. W. M. Lau, N. P. C. Westwood and M. H. Palmer, *J. Am. Chem. Soc.*, **108**, 3229 (1986).

71. R. Steudel, O. Schumann, J. Buschmann and P. Luger, *Angew. Chem., Int. Ed. Engl.*, **37**, 492 (1998).

Chapter 4

Electronic Structures and Reactivity Patterns

4.1 Introduction

Interest in the electronic structures of chalcogen–nitrogen compounds was stimulated initially by the synthesis and structural determination of various binary sulfur nitrides. The unusual cage structure of S_4N_4 (**4.1**)[1] and the planar, cyclic structures of S_2N_2 (**4.2**),[2] $[S_4N_3]^+$ (**4.3**)[3] and $[S_5N_5]^+$ (**4.4**)[4] were prominent among these early discoveries. The discovery of the unique solid-state properties of the polymer $(SN)_x$ (**4.5**)[5] added an important incentive for the need to gain an understanding of the electronic structures of these electron-rich species. In 1972 Banister proposed that planar S–N heterocycles belong to a class of "electron-rich aromatics" which conform to the well known Hückel $(4n + 2)$ π-electron rule of organic chemistry.[6] On the reasonable assumption that each sulfur contributes two and each nitrogen one electron to the π-system, the known species **4.2** (6π), **4.3** (10π) and **4.4** (14π) were cited as examples in support of this contention. This empirical proposal stimulated synthetic chemists to discover other members of this series. Notable examples include $[S_4N_4]^{2+}$ (**4.6**)[7] and $[S_3N_3]^-$ (**4.7**)[8] both of which are ten π-electron systems with planar, cyclic structures. More recently, the first example of a nitrogen-rich species $[S_2N_3]^+$ (**4.8**) that conforms to the Hückel rule was reported.[9]

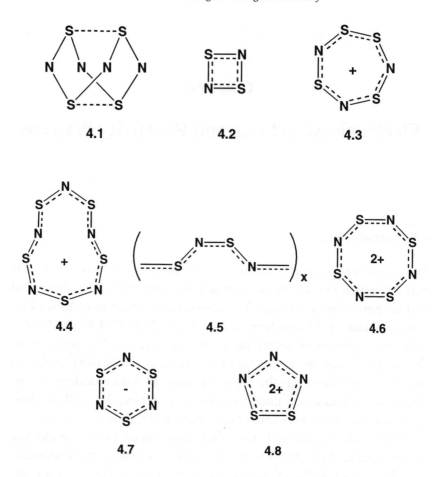

4.2   Bonding in Sulfur–Nitrogen Compounds: Comparison with Organic Systems

In order to understand the physical properties and reactivity patterns of S–N compounds it is particularly instructive to compare their electronic structures with those of the analogous organic systems.[10] On a qualitative level, the simplest comparison is that between the hypothetical HSNH radical and the ethylene molecule; each of these units can be considered as the building blocks from which conjugated –S=N– or –CH=CH– systems can be constructed. To a first approximation the σ-framework of both molecules can be described in terms of sp^2 hybridization of the non-hydrogen atoms. In HSNH only two of the sp^2 hybrid orbitals (and two valence electrons) are involved in bonding to neighbouring atoms, while a non-bonding pair of electrons occupies the third sp^2 hybrid orbital. Consequently, the sulfur atom has two electrons in a $3p$ orbital and nitrogen has one electron in a $2p$ orbital to contribute to the π system. It is for this reason that unsaturated S–N compounds with two-coordinate sulfur are referred to as "electron-rich".

The π-systems of HSNH and $H_2C=CH_2$ are compared in Figure 4.1. Several important conclusions can be derived from consideration of this

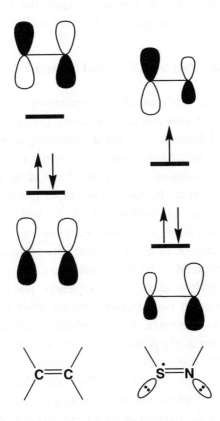

Fig. 4.1   Qualitative comparison of the π-molecular orbitals of $H_2C=CH_2$ and HSNH

simple diagram. First, the extra π-electron in HSNH occupies an antibonding level. As a result the π-bond order for the S=N bond is approximately 0.5 and this bond is predictably substantially weaker than a C=C bond with a π-bond order of 1.0. In addition to their thermal instability, the consequences of the partial occupation of the π^* levels will be apparent in the subsequent discussions of the physical properties and reactivity of S–N compounds. It should also be noted that the energies of the π-orbitals in HSNH are lower than those in ethylene as a result of the higher atom electronegativities of sulfur and, particularly nitrogen, compared to that of carbon. Finally, the small difference in the electronegativities of S and N results in a π-bonding orbital in HSNH that is polarized towards the nitrogen atom while the singly occupied π^* orbital has more electron density on sulfur than on nitrogen.

4.3 Bonding in the Polymer (SN)$_X$ and in Sulfur–Nitrogen Chains

The polymer (SN)$_x$ has remarkable properties for a material that is composed of two non-metallic elements (Section 14.2).[5] It has a metallic lustre and behaves as a highly anisotropic conductor at room temperature. The conductivity along the polymer chains is of a similar order of magnitude as that of mercury metal and about fifty times greater than that perpendicular to the chain. The conductivity increases by three orders of magnitude on cooling to 4 K and (SN)$_x$ becomes a superconductor at 0.3 K. The metallic behaviour can be understood from the molecular description of the π-bonding illustrated in Figure 4.2. The repeat unit is considered to be the hypothetical acyclic S$_2$N$_2$ molecule, which is a six π-electron species. Consequently, one of the antibonding π-orbitals is occupied by two electrons and the other is unoccupied. The polymer (SN)$_x$ is comprised of an infinite number of S$_2$N$_2$ units juxtaposed in an essentially planar arrangement. In the molecular orbital description of (SN)$_x$ this gives rise to a completely filled valence band and a half-filled conducting band. Thus the electronic structure is reminiscent of that of alkali metals, for which the half-filled conducting band is the result of the overlap of an infinite number of singly occupied ns orbitals, *e.g.*, $2s^1$ for Li. The electrons in the partially filled band in

$(SN)_x$ are free to move under the influence of an applied potential difference and thus conduction occurs along the polymer chain. The S–N distances in the chain are essentially equal, consistent with a delocalized structure. The increase in conductivity with decreasing temperature is characteristic of a metallic conductor. The predicted Peierls distortion is apparently inhibited by weak interactions between the polymer chains ($S{\cdots}S = 3.47\text{-}3.70$ Å; $S{\cdots}N = 3.26\text{-}3.38$ Å.).[5b]

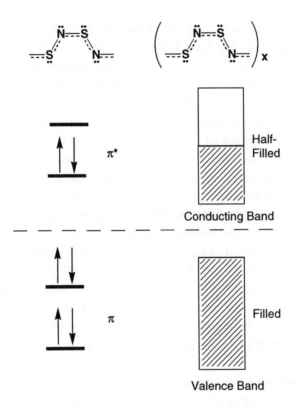

Fig. 4.2 Molecular orbital description of the bonding in acyclic S_2N_2 and $(SN)_x$

Materials that are comprised of small fragments of $(SN)_x$ with organic terminal groups, *e.g.*, ArS_3N_2Ar and ArS_5N_4Ar (Ar = aryl), are of potential interest as molecular wires in the development of nanoscale technology.[11] Consistent with simple band theory, the energy gap between π-orbitals decreases as the chain length (number of SN units) increases. This trend has a noticeable influence on the optical properties of sulfur–nitrogen chains. Thus, the five-atom ArS_3N_2Ar chains are yellow or orange whereas the longer ArS_5N_4Ar chains exhibit a deep green, royal blue or purple colour (Section 14.3).

4.4 Aromaticity in Planar Sulfur–Nitrogen Rings

At first sight the discovery of numerous examples of cyclic, binary sulfur–nitrogen species **4.2-4.8** that conform to the Hückel $(4n + 2)$ π-electron rule would appear to provide ample justification for Banister's proposal that these species can be regarded as "electron-rich aromatics".[6] In organic chemistry the criteria for aromaticity include physical data such as heats of formation and diamagnetic ring currents, as well as structural data (planarity and equality of bond lengths). While the geometrical parameters for S–N systems are consistent with delocalized structures, the absence of other physical data make it difficult to assess whether the term aromaticity is appropriate for these inorganic systems. The electronic origin of aromaticity in the case of cyclic polyenes lies in the degeneracy of the frontier molecular orbital levels for systems of high symmetry [Fig. 4.3(a)]. For a system with n atoms (and n orbitals) with n-fold symmetry any 4n-electron count will give rise to an open shell species [Fig. 4.3(b)]. For example, the four π-electron system cyclobutadiene should, as a consequence of Hund's rule, be a diradical. The degeneracy can be removed if the molecule is distorted to a structure of lower symmetry, *e.g.*, a rectangle in the case of cyclobutadiene [Fig. 4.3(c)].

It is informative to compare the π-orbital manifolds of some common sulfur-nitrogen ring systems with those of their aromatic counterparts on the basis of simple Hückel molecular orbital calculations. In Figure 4.4 this comparison is made for the following pairs of molecules: S_2N_2 and cyclobutadiene, $[S_3N_3]^-$ and benzene, $[S_4N_4]^{2+}$ and cyclooctatetraene. In the case of S_2N_2 the lowering of symmetry form D_{4h} in C_4H_4 to C_{2v} results in a loss of degeneracy. Consequently, there is no fundamental reason for the four π-electron system, *i.e.*, $[S_2N_2]^{2+}$, to be unstable. The

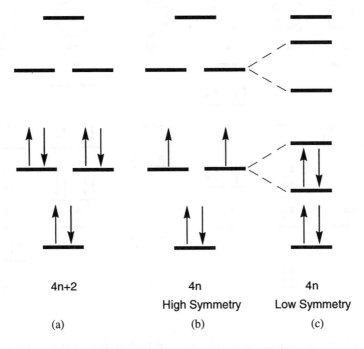

4n+2	4n	4n
	High Symmetry	Low Symmetry
(a)	(b)	(c)

Fig. 4.3 π-Orbital occupations of 4n + 2 and 4n systems

formation of the neutral six π-electron system S_2N_2 illustrates the ability of S–N systems to accommodate more π-electrons than their hydrocarbon counterparts. This is a reflection of the higher electronegativities of sulfur and nitrogen compared to that of carbon, which give rise to lower energy molecular orbitals for the S–N systems. However, four of the six π-electrons in S_2N_2 occupy two non-bonding π-orbitals located on the nitrogen and sulfur atoms, respectively. Consequently, the bond order in this four-membered ring is 1.25.[12a] Recent *ab initio* calculations reveal a small amount (7%) of diradical character for S_2N_2, which increases significantly for the heavier chalcogens, *i.e.*, in E_2N_2 (E = Se, Te).[12b]

This feature of the electronic structure of these inorganic rings is especially dramatic for the six-atom system $[S_3N_3]^-$, which is able to accommodate ten π-electrons, *cf.* six π-electrons for C_6H_6. The electron-richness of $[S_3N_3]^-$ results in a significantly lower π-bond order of 0.17

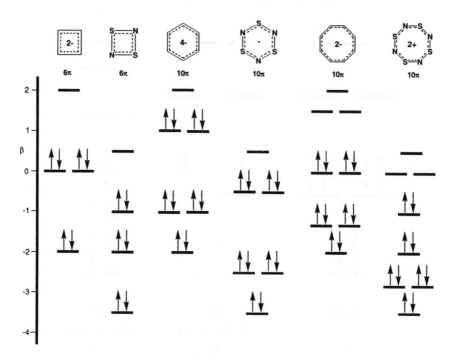

Fig. 4.4   The π-manifolds for $[C_4H_4]^{2-}$, S_2N_2, $[C_6H_6]^{4+}$, $[S_3N_3]^-$, $[C_8H_8]^{2-}$ and $[S_4N_4]^{2+}$

compared to the corresponding value of 0.5 for benzene. In the case of eight-atom systems, planar C_8H_8 is unstable because of Hund's rule, but the corresponding dianion $[C_8H_8]^{2-}$, a ten π-electron system which is isoelectronic with $[S_4N_4]^{2+}$, is stable.

The criterion of ipsocentric ring current has been used to assess aromaticity in S–N heterocycles (and related inorganic ring systems).[12] Current density maps indicate that the ten π-electron systems $[S_3N_3]^-$, $[S_4N_3]^+$ and $[S_4N_4]^{2+}$, and the fourteen π-electron system $[S_5N_5]^+$ support diatropic π currents, reinforced by σ circulations.[12b]

4.5   Bonding in Heterocyclothiazenes: The Isolobal Analogy

Consideration of S–N heterocycles in which one (or more) of the sulfur atoms is replaced by another group leads to further insights into the bonding in these electron-rich systems. Six-membered rings derived from an S_3N_3 framework provide an informative example. From the π-manifold for $[S_3N_3]^-$ shown in Fig. 4.4 it is apparent that the corresponding cation $[S_3N_3]^+$ (eight π-electrons) is expected to be a diradical unless the hexagonal (D_{3h}) structure distorts to remove the degeneracy of the singly occupied antibonding orbitals. This cation is not known, but it has been stabilized as a cycloaddition complex with norbornadiene.[13] However, heterocycles that are isoelectronic with $[S_3N_3]^+$ (**4.9**), *e.g.*, $PhCN_3S_2$ (**4.10**)[14a] and $Ph_2PN_3S_2$ (**4.11**),[14b] have been isolated as stable species. As illustrated in Fig. 4.5, the groups S^+, CR and PR_2 can be considered as isolobal; each of them contributes one electron to the π-system in an orbital that has similar symmetry properties, but with different energies. In S^+ this electron occupies the $3p_z$ orbital, whereas for the three-coordinate carbon in a CR group it occupies the $2p_z$ orbital and for the four-coordinate phosphorus in a PR_2 group it occupies a $3d_{xz}$ orbital. Thus the heterocycles **4.9-4.11** are all eight π-electron "antiaromatic" systems.

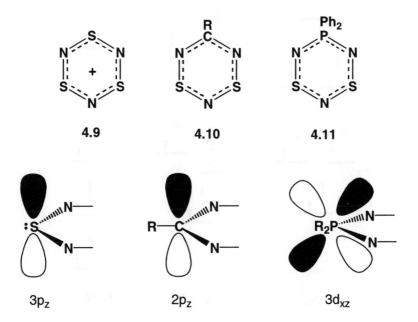

Fig. 4.5 The isolobal groups S^+, CR and PR_2

The formal replacement of S^+ in the hexagonal (D_{3h}) structure of **4.9** by a CR or PR_2 group lowers the symmetry of the ring to C_{2v}, if the ring retains planarity, or C_s if the substituent lies out of the N_3S_2 plane, as occurs for **4.11**. Under these conditions, the degeneracies in the π-levels are removed and there is no fundamental difference between a 4n + 2 and a 4n system, since Hund's rule no longer applies. Consequently, the stability of these heterocycles depends on the ability of the system to accommodate additional electrons in the π-system. Since the molecular orbitals in S–N rings are of lower energies than those of their organic analogues, the eight π-electron species **4.10** and **4.11** can be isolated. In the case of **4.11** (R = Ph) a monomeric structure is stable, whereas dithiatriazines of the type **4.10** form dimers with weak S•••S intermolecular interactions in the solid state (Section 12.2.2). This difference between **4.10** and **4.11** arises because the CR group induces a smaller perturbation of the degenerate π^* orbitals than the PR_2 group does (Fig. 4.6).[14d]

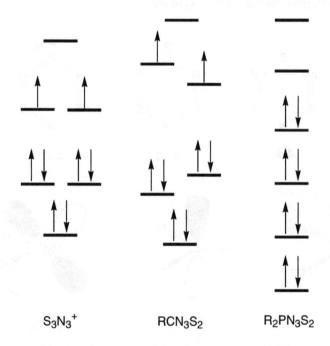

$S_3N_3^+$ RCN_3S_2 $R_2PN_3S_2$

Fig. 4.6 π-Manifolds of $[S_3N_3]^+$, RCN_3S_2 and $R_2PN_3S_2$

The preceding discussion shows that simple Hückel or extended Hückel molecular orbital calculations can give a useful indication of the π-electron structures of π-electronic-rich S–N rings. It is important, however, to recognize their limitations. For example, the choice of Coulomb parameters α_N and α_S is particularly difficult. The assumption that nitrogen is more electronegative than sulfur is appropriate for the frontier orbitals, but is not likely correct for the deeper π-orbitals. *Ab initio* MO calculations are to be preferred for a deeper understanding of the properties of S–N compounds, *e.g.*, excitation energies, because of their less arbitrary input.

The most recent theoretical studies for individual binary chalcogen-nitrogen species are cited in the following list: SN,[15] [NS]$^+$,[16] [NS$_2$]$^+$,[16] SNN,[17,18] SNS,[19] [NSN]$^{2-}$,[20] [SNSS]$^-$,[21] [SSNSS]$^-$,[22] *cyclo*-S$_2$N$_2$,[12,23,24] *cyclo*-[S$_2$N$_3$]$^+$,[9] *cyclo*-[S$_3$N$_2$]$^+$,[25] *cyclo*-[S$_3$N$_2$]$^{2+}$,[16,25] *cyclo*-S$_3$N$_3$,[26] *cyclo*-[S$_3$N$_3$]$^-$,[12b,27] *cyclo*-S$_4$N$_2$,[28a,b] *cyclo*-S$_{4-x}$Se$_x$N$_2$ (x = 1-4),[28c] *cyclo*-[S$_4$N$_3$]$^+$,[12b,29] S$_4$N$_4$,[30] Se$_4$N$_4$,[30] *cyclo*-[S$_4$N$_4$]$^{2+}$,[12b,31] [S$_4$N$_5$]$^+$,[32] [S$_4$N$_5$]$^-$,[32] [S$_5$N$_5$]$^+$,[33] S$_5$N$_6$.[32,34] The usefulness of the Mayer bond order in the context of the role of ring size and electron count in S–N rings has been evaluated.[35] The chemistry of binary chalcogen–nitrogen species is discussed in Chapter 5.

4.6 Weak Intramolecular Chalcogen–Chalcogen Interactions

A characteristic feature of larger chalcogen–nitrogen rings is the formation of folded structures with weak transannular E•••E interactions. For sulfur systems the S•••S distances are typically in the range 2.50-2.60 Å, about 0.5 Å longer than the S–S bonds in *cyclo*-S$_8$. The classic example of this phenomenon is the cage structure of S$_4$N$_4$ (**4.1**); the selenium analogue Se$_4$N$_4$ has a similar structure.[36] Although the details of the electronic structure of the bonding in **4.1** are still a matter of debate,[37a] a simple and useful way of rationalizing the molecular structure was provided by Gleiter in 1970.[37b] Symmetry arguments predict that a planar S$_4$N$_4$ molecule (a twelve π-electron system) with D_{4h} symmetry would possess a triplet ground state and hence be susceptible to Jahn–Teller distortion. The distortion of the planar molecule into a D_{2d}

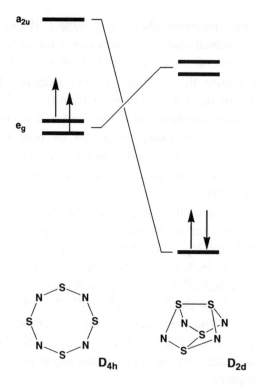

Fig. 4.7   Frontier orbitals for the planar and folded forms of S_4N_4

structure inverts the ordering of the a_{2u} LUMO and e_g HOMOs thereby affording a singlet ground state (Fig. 4.7) and, at the same time, allowing the development of cross-ring S•••S bonding.

The dithiatetrazocines $(RC)_2N_4S_2$ (**4.12**)[38] and $(R_2P)_2N_4S_2$ (**4.13**)[39] provide further insights into this interesting structural feature. These eight-membered rings are formally derived from $[S_4N_4]^{2+}$ by the replacement of two S^+ atoms by RC or R_2P groups. Thus they are both 4n + 2 (10 π) systems. The structures of **4.12** are markedly dependent on the nature of the exocyclic substituent R. When R = Ph the $C_2N_4S_2$ adopts a planar, delocalized structure, whereas the dimethylamino substituents (**4.12**, R = NMe_2) give rise to a folded structure that is also observed for the phosphorus-containing systems **4.13**. This dichotomy is explained by the destabilizing influence of the π-donor Me_2N substituents on the HOMO of the $C_2N_4S_2$ ring and subsequent second-order Jahn Teller distortion. The replacement of two S^+ units by the relatively electropositive PR_2 substituents enforces a similar destabilization of the HOMO. This orbital is primarily antibonding with respect to the two NSN units in these eight-membered rings. Ring folding allows an intramolecular π^*–π^* interaction, which is bonding with respect to the sulfur atoms on opposite sides of the ring, to occur (Fig. 4.8). This bonding arrangement is reminiscent of that found in the dithionite dianion $[S_2O_4]^{2-}$ [d(S–S) = 2.39 Å] in which a dimeric structure is formed by a similar interaction of the π^* orbitals of two sulfur dioxide radical anions $[SO_2]^{-\cdot}$.

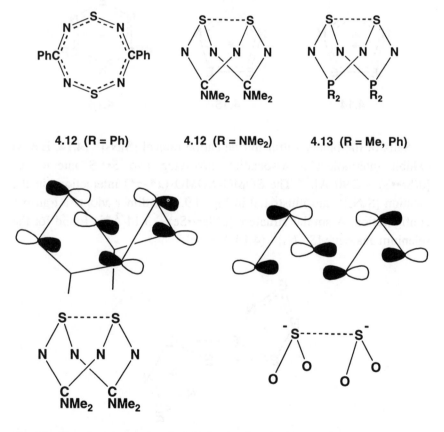

4.12 (R = Ph)   **4.12 (R = NMe₂)**   **4.13 (R = Me, Ph)**

Fig. 4.8 The π^*–π^* interactions in $E_2N_4S_2$ (E = $CNMe_2$, PR_2) and the dithionite dianion $[S_2O_4]^{2-}$

4.7 Intermolecular Association: Radical Dimerization

Weak interactions may occur between molecules (intermolecular association) as well as within a molecule (intramolecular) for chalcogen–nitrogen ring systems. This behaviour is especially significant for odd electron species, *e.g.*, $[E_3N_2]^{+\bullet}$ (**4.14**, E = S, Se) and $[PhCN_2E_2]^{\bullet}$ (**4.15**, E = S, Se), both of which are seven π-electron molecules. As mentioned in the previous section, it also occurs for the eight π-electron dithiatriazines **4.10**.

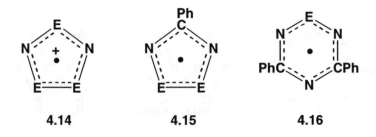

| **4.14** | **4.15** | **4.16** |

Salts of the binary sulfur nitride cation radical $[S_3N_2]^{+\bullet}$ (**4.14**, E = S) exhibit intermolecular association involving two S•••S interactions $[d(S•••S) \sim 2.90 \text{ Å}]$.[40a] The SOMO–SOMO $(\pi^*-\pi^*)$ interactions in the dication $[S_6N_4]^{2+}$ are illustrated in Fig. 4.9. The dimer adopts a transoid configuration. A similar structure $[d(Se•••Se) \sim 3.14 \text{ Å}]$ is found for the selenium analogue $[Se_6N_4]^{2+}$ (**4.14**, E = Se).[40b]

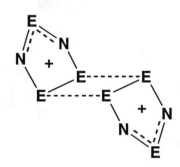

Fig. 4.9   Structure of $E_6N_4^{2+}$ (E = S, Se)

The formal replacement of an S^+ unit in **4.14** by an RC group generates an isoelectronic 1,2,3,5-dithiadiazolyl $[RCN_2E_2]^{\bullet}$ (**4.15**, E = S). These neutral radicals represent very important building blocks for the construction of solid-state materials with novel conducting or magnetic properties (Section 11.3.1).[41] The SOMO of **4.15** has a similar composition to that of **4.14**, but the substituent attached to carbon influences the dimerization mode. With a flat substituent, *e.g.*, an aryl group, association occurs in a cisoid cofacial manner with two weak S•••S interactions (Fig. 4.10). The selenium analogue **4.15** (E = Se) adopts a similar arrangement. In the case of non-planar substituents, *e.g.*, $[CF_3CN_2S_2]^{\bullet}$ and $[Me_2NCN_2S_2]^{\bullet}$, the monomer units are held together through one S•••S interaction (~3.10 Å) and the rings are twisted *ca.* 90° to one another.[42]

1,3,2-Dithiazolyl radicals have also attracted considerable attention recently as potential molecular conductors (Section 11.3.6).[43] The advantage of these systems over 1,2,3,5-dithiadiazolyls lies in their relatively low disproportionation energies (Section 11.3.1).

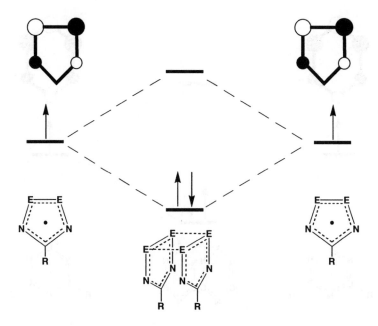

Fig. 4.10   Structure and bonding in $(PhCN_2E_2)_2$ (E = S, Se)

In the solid state interactions occur between dimeric units that influence the physical properties of these materials, *e.g.*, conducting and magnetic behaviour. As a result these chalcogen–nitrogen ring systems have been the subject of intensive study in the design of molecular conductors. Different packing arrangements can be induced by introducing substituents such as CN on the aryl group attached to carbon or by the synthesis of multifunctional materials with two or three CN_2E_2 units attached to an organic scaffold (Section 11.3.1).

The S•••S distance in the dimer of the thiatrazinyl radical **4.16** (E = S) is 2.67 Å,[44a] similar to the values observed for the intramolecular S•••S interactions in **4.12** (R = NMe_2) and **4.1**. Indeed this dimerization involves a $\pi^*-\pi^*$ interaction analogous to those depicted in Figure 4.8 for $E_2N_4S_2$ (E = $CNMe_2$, PR_2) and $[S_2O_4]^{2-}$. The electron density in **4.16** is primarily located on the NEN unit and the SOMO is antibonding with respect to the two chalcogen–nitrogen bonds. A face-to-face arrangement of two radicals results in the in-phase overlap of the two SOMOs, which generates a weak E•••E bond (Fig. 4.11). Consequently, ring systems of

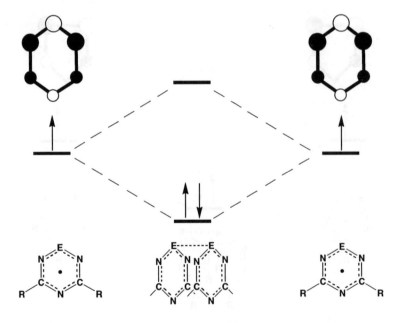

Fig. 4.11   Cofacial overlap of two $[(PhC)_2N_3E]^{\bullet}$ (E = S, Se) radicals

type **4.16** are diamagnetic in the solid state. On the basis of the experimentally determined dissociation constants, the overlap in the selenium derivative **4.16** (E = Se) is stronger than that in the sulfur analogue, *i.e.*, $4p–4p$ overlap is more effective than $3p–3p$ overlap.[44b]

4.8 Reactivity Patterns: The Role of Frontier Orbital Theory

The knowledge of the composition of the frontier orbitals gleaned from molecular orbital theory has been crucial to the development of an understanding of the various transformations that are observed for chalcogen-nitrogen systems. These processes include: cycloaddition reactions, polar and radical oxidations, intermolecular rearrangements, ring contraction, ring expansion, and polymerization.

4.8.1 *Cycloaddition Reactions*

Early work established that S_4N_4 forms di-adducts with alkenes such as norbornene or norbornadiene.[45] Subsequently, structural and spectroscopic studies established that cycloaddition occurs in a 1,3-*S,S′*-fashion.[46,47] The regiochemistry of addition can be rationalized in frontier orbital terms; the interaction of the alkene HOMO with the low-lying LUMO of S_4N_4 exerts kinetic control.[48] Consistently, only electron-rich alkenes add to S_4N_4.

More recent investigations of the 1:1 adducts of alkenes with the eight π-electron six-membered rings **4.9-4.11** have provided further insight into the nature of this cycloaddition process. All three of these heterocycles undergo an *S,S′* mode of addition with norbornadiene.[12-14] This regiochemistry is readily explained from frontier orbital considerations as shown in Figure 4.12 for $R_2PN_3S_2$ (**4.11**). Both the HOMO and LUMO of **4.11** are sulfur-based and of the correct symmetry to overlap with the LUMO and HOMO, respectively, of an alkene. Energy considerations indicate that the HOMO (olefin)–LUMO (S–N heterocycle) controls the kinetics of these reactions. The formation of these norbornadiene adducts occurs rapidly and is of practical value in characterizing and storing unstable derivatives of **4.11** (R = Me, F, CF$_3$),

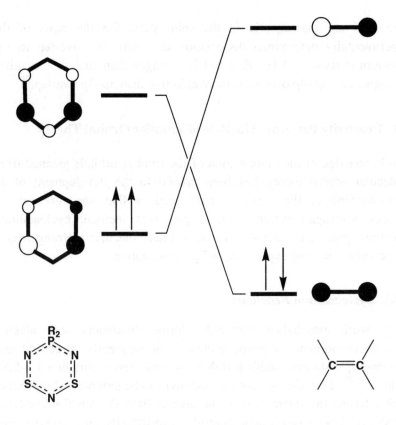

Fig. 4.12   Cycloaddition of $R_2PN_3S_2$ and alkenes

since the free heterocycle can be regenerated by mild heating of the adduct.[49] The formation of a norbornadiene adduct has been used to trap the unstable $[S_3N_3]^+$ cation **4.9**[13] or monomeric dithiatriazines **4.10**.[14a,c]

Cycloaddition reactions also have important applications for acyclic chalcogen-nitrogen species. Extensive studies have been carried out on the cycloaddition chemistry of $[NS_2]^+$ which, unlike $[NO_2]^+$, undergoes quantitative, cycloaddition reactions with unsaturated molecules such as alkenes, alkynes and nitriles (Section 5.3.2).[50] The frontier orbital interactions involved in the cycloaddition of $[NS_2]^+$ and alkynes are illustrated in Fig. 4.13. The HOMO (π_n) and LUMO (π^*) of the sulfur-nitrogen species are of the correct symmetry to interact with the LUMO (π^*) and HOMO (π) of a typical alkyne, respectively. Although both interactions are symmetry-allowed, in practice the HOMO (alkyne) – LUMO (NS_2^+) interaction will dominate, *i.e.*, $[NS_2]^+$ behaves as a Lewis acid, because they are much closer in energy than the other pair of orbitals. Electron-attracting substituents in the alkyne, *e.g.*, CF_3, which decrease the energy of the HOMO will increase the HOMO (alkyne) – LUMO (NS_2^+) energy gap and, hence, decrease the rate of cycloaddition.

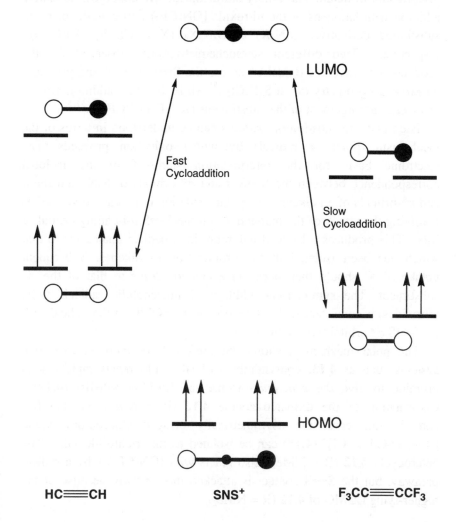

Fig. 4.13 The frontier orbitals of $[NS_2]^+$, HC≡CH and CF_3C≡CCF_3

4.8.2 *Polar and Radical Oxidations*

The unusual nature of the transannular S•••S "bond" in S_4N_4 (**4.1**) and the bicyclic dithiatetrazocines **4.12** (R = NMe_2) and **4.13** poses interesting questions about the reactivity of this functionality. The reactions of **4.1**, **4.12** and **4.13** with oxidizing agents have been investigated in detail. The binary sulfur nitride **4.1** undergoes oxidative addition with halogens or the nitroxide [ON(CF$_3$)$_2$]• to give di- or tetra-substituted derivatives $S_4N_4X_2$ or $S_4N_4X_4$ (X = Cl, F, ON(CF$_3$)$_2$), respectively. Two different stereochemistries are observed for the products; one involves oxidation across the S•••S contact and produces an *exo,endo* geometry (as in $S_4N_4Cl_2$),[51] while the other addition affords an *exo,exo* arrangement of the substituents (X = F, ON(CF$_3$)$_2$).[52, 53]

Each of these substitution patterns can be understood in terms of the mechanism (polar or radical) by which oxidation proceeds. The electronic basis for the interpretation stems from the isolobal correspondence between the S•••S σ and σ^* orbitals of S_4N_4 and the π and π^*-orbitals of an alkene. The polar oxidation route can be viewed as involving the attack of the halogen X_2 on the S•••S σ-bonding orbital of S_4N_4. This produces a bridged halonium intermediate analogous to that which has been isolated in the bromination of alkenes. Subsequent uptake of X⁻ should then occur in a position *trans* to that of the X⁺ substituent. The symmetrical addition of nucleophilic radicals (*e.g.*, nitroxides) is consistent with interaction of the SOMO of the radical with the S•••S σ^* orbital (see Figure 4.14).

The polar mechanism should be preferred for more electron-rich systems such as **4.13**; consistently **4.13** (R = Ph) reacts rapidly with bromine to give the *exo, endo* isomer of 1,5-Ph$_4$P$_2$N$_4$S$_2$Br$_2$ (**4.17**).[54] Chlorination of the dithiatetrazocine **4.12** (R = NMe_2) is also fast and, in this case, the asymmetrically bridged chloronium cation [(Me$_2$N)$_2$C$_2$N$_4$S$_2$Cl]⁺ (**4.18**) can be isolated as the trichloride salt.[55] The heterocycle **4.12** (R = NMe_2) also reacts with [ON(CF$_3$)$_2$]• by a radical pathway, but the S•••S contact is attacked more slowly because of the higher-lying LUMO of **4.12** (R = NMe_2).

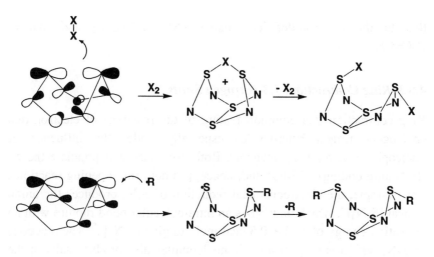

Fig. 4.14 Polar and radical oxidations of S_4N_4

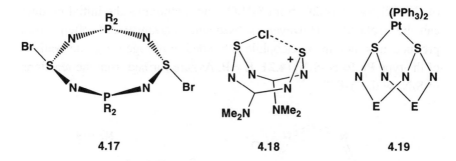

4.17　　　　　　4.18　　　　　　4.19

The isolobal analogy between the S•••S σ and $\sigma*$ orbitals in **4.12** (R = NMe₂) and **4.13** and the π and $\pi*$ orbitals of an alkene instigated the notion that these heterocycles should form η^2-metal complexes by reaction with zerovalent Pt (or Pd) reagents, *cf.* alkene–Pt complexes. Indeed complexes of the type Pt(1,5-E₂N₄S₂)(PPh₃)₂ (E = CNMe₂, PPh₂) (**4.19**) are readily obtained by displacement of the alkene ligand in Pt(CH₂=CH₂)(PPh₃)₂.[56] Relativistic DFT calculations support the notion that the bonding in **4.19** is analogous to that in classic alkene–Pt complexes. The strongest contributor to the stability of **4.19** is back donation to the $\sigma*$ (S•••S) orbital of **4.12** (R = NMe₂). Calculations show that the extent of back donation to the heterocyclic ligand lies between that to the electron-deficient alkene $(NC)_2C=C(CN)_2$ and that to $H_2C=CH_2$.

4.8.3 Ring Contraction: 1,3-Nitrogen Shifts

Ring contraction is a common feature of the reactions of electron-rich chalcogen-nitrogen heterocycles especially under the influence of nucleophilic or reducing reagents. Both processes will populate the π* (E–N antibonding) LUMOs and, hence, promote ring-opening reactions. For example, the electrochemical reduction of S_4N_4 produces the anion radical $[S_4N_4]^{\cdot-}$, which undergoes an intramolecular bond rupture with an activation energy of 11.1 ± 0.5 kcal mol^{-1} to give $[S_3N_3]^{\cdot}$.[57] The reactions of S_4N_4 with a variety of anionic nucleophiles also produce salts of the $[S_3N_3]^-$ anion.[58] The initial nucleophilic attack occurs at sulfur because of the polarity of the S–N bond, $S^{\delta+}$–$N^{\delta-}$. In some cases, *e.g.*, the formation of $[MeOS_4N_4O_2]^-$ (**4.20**) from $S_4N_4O_2$ and methoxide, the initial product can be isolated.[59] However, subsequent rearrangement often gives products in which the nucleophile is bonded to nitrogen, *e.g.*, derivatives of the type $Ph_3E=N–S_3N_3$ (**4.21**, E = P, As) are formed from the reactions of S_4N_4 with Ph_3E.[60]

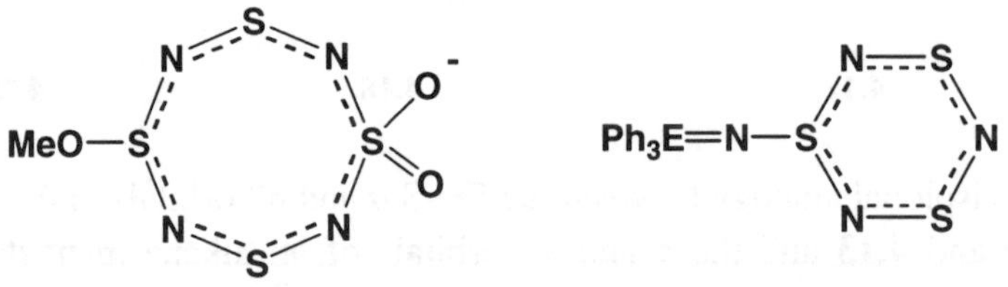

4.20   **4.21**

Mechanistic investigations of the reaction of $PhCN_5S_3$ with tertiary phosphines by a combination of ^{15}N and ^{31}P NMR spectroscopy indicate that ring contraction occurs via a 1,3-nitrogen shift.[61] This process is illustrated in Figure 4.15 for the formation of the six-membered ring $[S_3N_3O_2]^-$ from the eight-membered ring $S_4N_4O_2$ with the elimination of the unstable NSN fragment.

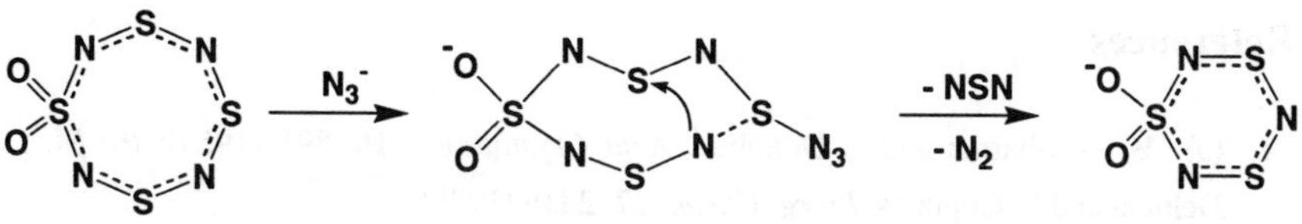

Fig. 4.15 1,3- Nitrogen shift promoted by a nucleophilic reagent

4.8.4 *Fluxional Processes*

Several different classes of chalcogen–nitrogen compounds undergo degenerate rearrangements. In the case of bicyclic systems such as $[S_4N_5]^+$ it has been suggested on theoretical grounds that this fluxional behaviour can be rationalized via a series of 1,3-shifts.[32] Mechanistic support for this proposal was provided by an ^{15}N NMR investigation of $PhCN_2*N_3S_3$ ($*N = 99\%$ [15]N).[63] When this bicyclic compound, which is isoelectronic with $[S_4N_5]^+$, is dissolved in chloroform a slow scrambling process, involving a series of 1,3-nitrogen shifts, exchanges the labelled and unlabelled nitrogen atoms (Fig. 4.16).

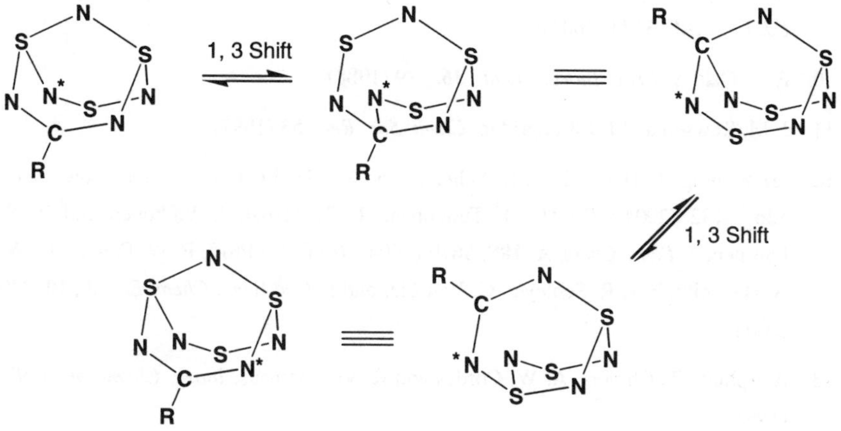

Fig. 4.16 Exchange of nitrogen sites in $PhCN_5S_3$ via 1,3-shifts

References

1. (a) B. D. Sharma and J. Donohue, *Acta Crystallogr.*, **16**, 891 (1963); (b) M. L. Delucia and P. Coppens, *Inorg. Chem.*, **17**, 2336 (1978).

2. C. M. Mikulski, P. J. Russo, M. S. Saran, A.G. MacDiarmid, A. F. Garito and A. J. Heeger, *J. Am. Chem. Soc.*, **97**, 6358 (1975).

3. (a) J. Weiss, *Z. Anorg. Allg. Chem.*, **333**, 314 (1964); (b) A. W. Cordes, R. F. Kruh and E. K. Gordon, *Inorg. Chem.*, **4**, 681 (1965).

4. A. J. Banister, P. J. Dainty, A. C. Hazell, R. G. Hazell and J. G. Lomborg, *Chem. Commun.*, 1187 (1969).

5. (a) M. M. Labes, P. Love and L. F. Nichols, *Chem. Rev.*, **79**, 1 (1979); (b) A. J. Banister and I. B. Gorrell, *Adv. Mater.*, **10**, 1415 (1998).

6. A. J. Banister, *Nature Physical Science*, **237**, 92 (1972).

7. R. J. Gillespie, D. R. Slim and J. D. Tyrer, *J. Chem., Soc., Chem. Commun.*, 253 (1977).

8. J. Bojes and T. Chivers, *J. Chem. Soc., Chem Commun.*, 391 (1978).

9. S. Herler, P. Mayer, H. Nöth, A. Schulz, M. Suter and M. Vogt, *Angew. Chem., Int. Ed. Engl.*, **40**, 3173 (2001).

10. R. T. Oakley, *Prog. Inorg. Chem.*, **36**, 299 (1988).

11. J. M. Rawson and J. J. Longridge, *Chem. Soc. Rev.*, 53 (1997).

12. (a) Y. Jung, T. Heine, P. v. R. Schleyer and M. Head-Gordon, *J. Am. Chem. Soc.*, **126**, 3132 (2004); (b) H. M. Tuononen, R. Suontamo, J. Valkonen and R. S. Laitinen, *J. Phys. Chem. A*, **108**, 5670 (2004); (c) F. De Proft, P. W. Fowler, R. W. A. Havenith, P. v. R. Schleyer, G. Van Lier and P. Geerlings, *Chem. Eur. J.*, **10**, 940 (2004).

13. A Apblett, T. Chivers, A. W. Cordes and R. Vollmerhaus, *Inorg. Chem.*, **30**, 1392, (1991).

14. (a) R. T. Boeré, C. L. French, R. T. Oakley, A. W. Cordes, J. A. J. Privett, S. L. Craig and J. B. Graham, *J. Am. Chem. Soc.*, **107**, 7710 (1985); (b) N. Burford, T. Chivers, A. W. Cordes, W. G. laidlaw, M. Noble, R. T. Oakley, and P. N. Swepston, *J. Am. Chem. Soc.*, **104**, 1282 (1982); (c) T. Chivers, F. Edelmann,

J. F. Richardson, N. R. M. Smith, O. Treu, Jr., and M. Trsic, *Inorg. Chem.*, **25**, 2119 (1986); (d) R. E. Hoffmeyer, W-T. Chan, J. D. Goddard and R. T. Oakley, *Can. J. Chem.*, **66**, 2279 (1988).

15. D. R. Salahub and R. P. Messmer, *J. Chem. Phys.*, **64**, 2039 (1976).

16. F. Grein, *Can. J. Chem.*, **71**, 335 (1993).

17. R. D. Davy and H. F. Schaefer III, *J. Am. Chem. Soc.*, **113**, 1917 (1991).

18. L. Behera, T. Kar and Sannigrahi, *J. Mol. Struct. Theochem.*, **209**, 111 (1990).

19. Y. Yamaguchi, Y. Xie, R. S. Grev and H. F. Schaefer III, *J. Chem. Phys.*, **92**, 3683 (1990).

20. T. Borrman, E. Lork, R. Mews, M. M. Shakirov and A. V. Zibarev, *Eur. J. Inorg. Chem.*, 2452 (2004).

21. J. Bojes, T. Chivers, W. G. Laidlaw and M. Trsic, *J. Am. Chem. Soc.*, **104**, 4837 (1982).

22. (a) R. Gleiter and R. Bartetzko, *Z. Naturforsch.*, **36B**, 492 (1981); (b) T. Chivers, W. G. Laidlaw, R. T. Oakley and M. Trsic, *J. Am. Chem. Soc.*, **102**, 5773 (1980).

23. J. Gerratt, S. J. McNicholas, P. B. Kardakov, M. Sironi, M. Raimondi and D. L. Cooper, *J. Am. Chem. Soc.*, **118**, 6472 (1996).

24. D. S. Warren. M. Zhao and B. M. Gimarc, *J. Am. Chem. Soc.*, **117**, 10345 (1995).

25. B. M. Gimarc and D. S. Warren, *Inorg. Chem.*, **30**, 3276 (1991).

26. W. M. Lau, N. P. C. Westwood and M. H. Palmer, *J. Am. Chem. Soc.*, **108**, 3229 (1986).

27. M. Bénard, W. G. Laidlaw and J. Paldus, *Can. J. Chem.*, **63**, 1797 (1985).

28. (a) D. Jayatilaka and M. Wajrak, *Chem. Phys.*, **198**, 169 (1995) ; (b) K. Somasundram and N. C. Handy, *J. Phys Chem.*, **100**, 17485 (1996); (c) J. Siivari, R. J. Suontamo, J. Konu, R. S. Laitinen and T. Chivers, *Inorg. Chem.*, **36**, 2170 (1997).

29. H. Johansen, *J. Am. Chem. Soc.*, **110**, 5322 (1988).

30. G. Chung and D. Lee, *J. Mol. Struct. Theochem.*, **582**, 85 (2002).

31. M. Trsic, W. G. Laidlaw and R. T. Oakley, *Can. J. Chem.*, **60**, 2281 (1982).

32. R. Bartetzko and R. Gleiter, *Chem. Ber.*, **113**, 1138 (1980).

33. A. J. Banister, Z. V. Hauptman, A. G. Kendrick and R. W. H. Small, *J. Chem. Soc., Dalton Trans.*, 915 (1987).

34. M. Trsic, K. Wagstaff and W. G. Laidlaw, *Int. J. Quantum Chem.*, **22**, 903 (1982).

35. A. J. Bridgeman, G. Cavigliasso, L. R. Ireland and J. Rothery, *J. Chem. Soc., Dalton Trans.*, 2095 (2001).

36. H. Folkerts, B. Neumüller and K. Dehnicke, *Z. Anorg. Allg. Chem.*, **620**, 1011 (1994).

37. (a) W. Scherer, M. Spiegler, B. Pedersen, M. Tafipolsky, W. Hieringer, B. Reinhard, A. J. Downs and G. S. McGrady, *Chem. Commun.*, 635 (2000); (b) R. Gleiter, *J. Chem. Soc. A*, 3174 (1970).

38. I. Ernest, W. Holick, G. Rihs, G. Schomburg, G. Shoham. D. Wenkert and R. B. Woodward, *J. Am. Chem. Soc.*, **103**, 1540 (1981).

39. N. Burford, T. Chivers, P. W. Codding and R. T. Oakley, *Inorg. Chem.*, **21**, 982 (1982).

40. (a) R. J. Gillespie, J. P. Kent and J. F. Sawyer, *Inorg. Chem.*, **20**, 3784 (1981); (b) E. G. Awere, J. Passmore, P. S. White and T. Klapötke, *J. Chem. Soc., Chem. Commun.*, 1415 (1989).

41. J. M. Rawson, A. J. Banister and I. Lavender, *Adv. Heterocycl. Chem.*, **62**, 137 (1995).

42. E. G. Awere, J. Passmore and P. S. White, *J. Chem. Soc., Dalton Trans.*, 299 (1993).

43. J. L. Brusso, O. P. Clements, R. C. Haddon, M. E. Itkis, A. A. Leitch, R. T. Oakley, R. W. Reed and J. F. Richardson, *J. Am. Chem. Soc.*, **126**, 8256 (2004).

44. (a) P. J. Hayes, R. T. Oakley, A. W. Cordes and W. T. Pennington, *J. Am. Chem. Soc.*, **107**, 1346 (1985); (b) R. T. Oakley, R. W. Reed, A. W. Cordes, S. L. Craig and J. B. Graham, *J. Am. Chem. Soc.*, **109**, 7745 (1987).

45. M. Becke-Goehring and D. Schlafer, *Z. Anorg. Allg. Chem.*, **356**, 234 (1968).

46. G. Ertl and J. Weiss, *Z. Anorg. Allg. Chem.*, **420**, 155 (1976).

47. A. M. Griffin and G. M. Sheldrick, *Acta Crystallogr.*, **31B**, 895 (1975).

48. T. Yamabe, K. Tanaka, A. Tachibana, K. Fukui and H. Kato, *J. Phys. Chem.*, **83**, 767 (1979).

49. N. Burford, T. Chivers, R. T. Oakley and T. Oswald, *Can. J. Chem.*, **62**, 712 (1984).

50. S. Parsons and J. Passmore, *Acc. Chem. Res.*, **27**, 101 (1994).

51. Z. Zak, *Acta Crystallogr.*, **37B**, 23 (1981).

52. R. A. Forder and G. M. Sheldrick, *J. Fluorine Chem.* **1**, 23 (1971/72).

53. G. A. Weigers and A. Vos, *Acta Crystallogr.*, **16**, 152 (1963).

54. T. Chivers, M. N. S. Rao and J. F. Richardson, *Inorg. Chem.*, **24**, 2237 (1985).

55. R. T. Boeré, A. W. Cordes, S. L. Craig, R. T. Oakley and R. W. Reed, *J. Am. Chem. Soc.*, **109**, 868 (1987).

56. T. Chivers, K. S. Dhathathreyan and T. Ziegler, *J. Chem. Soc., Chem. Commun.*, 86 (1989).

57. J. D. Williford, R. E. VanReet, M. P. Eastman and K. B. Prater, *J. Electrochem. Soc.*, **120**, 1498 (1973).

58. T. Chivers and R. T. Oakley, *Topics Curr. Chem.*, **102**, 117 (1982).

59. H. W. Roesky, M. Witt, B. Krebs and H. J. Korte, *Angew. Chem., Int. Ed. Engl.*, **18**, 415 (1979).

60. T. Chivers, A. W. Cordes, R. T. Oakley and P. N. Swepston, *Inorg. Chem.*, **20**, 2376 (1981).

61. R. T. Boeré, A. W. Cordes and R. T. Oakley, *J. Am. Chem. Soc.*, **109**, 7781 (1987).

62. (a) T. Chivers, A. W. Cordes, R. T. Oakley and W. T. Pennington, *Inorg. Chem.*, **22**, 2429 (1983); (b) M. Witt and H. W. Roesky, *Z. Anorg. Allg. Chem.*, **515**, 51 (1984).

63. (a) R. T. Boeré, R. T. Oakley and M. Shevalier, *J. Chem. Soc., Chem. Commun.*, 110 (1987); (b) K. T. Bestari, R. T. Boeré and R. T. Oakley, *J. Am. Chem. Soc.*, **111**, 1579 (1989).

Chapter 5

Binary Systems

5.1 Introduction

This chapter will deal with species that are comprised of only chalcogen and nitrogen atoms. In contrast to their N–O analogues, for which small molecules such as NO, N_2O and NO_2 are well known, stable chalcogen–nitrogen molecules of this type often aggregate to form ring or cage structures. The reasons for these differences are considered in Chapter 1. These species can be conveniently classified into neutral species, cations and anions and this classification will be adopted in the ensuing discussion.

5.2 Neutral Molecules

Sulfur or selenium analogues of NO or NO_2 are thermally unstable; the most common binary systems are S_4N_4 and Se_4N_4. This behaviour reflects the tendency of multiply bonded species containing heavier main group elements to oligomerize in the absence of steric protection. The neutral binary systems are thermodynamically unstable with respect to the formation of the elements and care must be taken in handling solid samples of these compounds owing to their tendency to explode when subjected to heat or friction.

5.2.1 *Thiazyl and selenazyl monomer, NS and NSe[1,2]*

Thiazyl monomer is a radical with one unpaired electron. In contrast to NO, it polymerizes so readily that it cannot be isolated as a monomeric solid or liquid and has only a transient existence in the gas phase. Thiazyl monomer may be generated in a number of ways, *e.g.*, the pyrolysis of S_4N_4 over quartz wool above 300°C.[3] The emission spectrum and EPR spectrum have been reported.[4] The bond length is calculated to be 1.497 Å from the spectroscopic moment of inertia.[5] This value indicates a bond order between two and three. The bond energy of this strong sulfur–nitrogen bond is estimated to be 111 kcal mol^{-1} from spectroscopic data.[6] Nevertheless, like other sulfur nitrides, NS is endothermic and unstable with respect to its elements. Thiazyl monomer exhibits an IR band at 1209 cm^{-1}.[7] The experimental dipole moment is 1.83 ± 0.03 D reflecting the difference in the electronegativities of sulfur and nitrogen (Section 1.1). The direction of the dipole is opposite to that in NO for which the dipole moment is 0.16 D. The first ionization potential of NS is 9.85 eV.[8] This ionization involves loss of an electron from the singly occupied π^* orbital. Much less is known about selenazyl monomer, SeN, but it has been characterized by infrared spectroscopy by using a combination of matrix isolation techniques and isotopic enrichment (^{15}N, ^{76}Se and ^{80}Se).[9]

Thiazyl monomer can be stabilized by coordination to a metal and many thionitrosyl complexes are known (Section 7.2). Comparison of the spectroscopic properties and electronic structures of M–NS and M–NO complexes indicates that NS is a better σ-donor and π-acceptor ligand than NO.[10] The only selenonitrosyl metal complex is TpOs(NSe)Cl$_2$ (Tp = hydrotris(1-pyrazolyl)borate).[11] Comparison with the corresponding NS complex indicates similar bonding and reactivity. Oxygen transfer from an NO$_2$ to an NS ligand on the same metal centre occurs in ruthenium porphyrin complexes.[12]

5.2.2 Dinitrogen sulfide, N_2S

A variety of acyclic and cyclic S–N compounds decompose at moderate temperatures (100-150°C) with the formal loss of a symmetrical NSN fragment, but this molecule has never been detected. The lowest energy isomer, linear NNS, is generated by flash vacuum pyrolysis of 5-phenyl-1,2,3,4-thiatriazole (Eq. 5.1).[13]

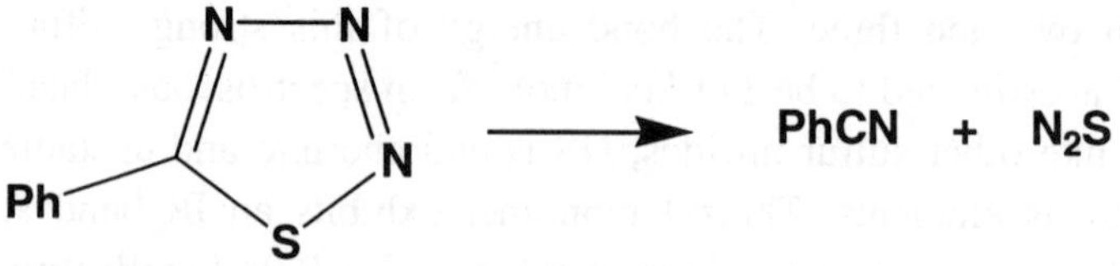

$$\text{(structure)} \longrightarrow \text{PhCN} + \text{N}_2\text{S} \qquad (5.1)$$

Unlike the stable molecule N_2O, the sulfur analogue N_2S decomposes above 160 K. In the vapour phase N_2S has been detected by high-resolution mass spectrometry. The IR spectrum is dominated by a very strong band at 2040 cm^{-1} [$\nu(NN)$].[9] The first ionization potential has been determined by photoelectron spectroscopy to be 10.6 eV.[14] These data indicate that N_2S resembles diazomethane, CH_2N_2, rather than N_2O. It decomposes to give N_2 and diatomic sulfur, S_2, and, hence, elemental sulfur, rather than monoatomic sulfur. *Ab initio* molecular orbital calculations of bond lengths and bond energies for linear N_2S indicate that the resonance structure $N \equiv N^+ - S^-$ is dominant.[15]

5.2.3 Nitrogen disulfide and diselenide, NS_2 and NSe_2

The unstable molecules NS_2 and NSe_2 can be produced in an argon/nitrogen/chalcogen microwave discharge and trapped in solid argon at 12 K.[7,9] Isotopic substitution (^{15}N, ^{34}S, ^{76}Se and ^{80}Se) facilitates the assignment of the IR spectrum. The more stable isomers have symmetrical bent structures with <ENE bond angles of 153 ± 5° (E = S) and 146 ± 5° (E = Se). The antisymmetric stretching vibrations are observed at 1225 and 1021 cm^{-1}, respectively.

5.2.4 *Disulfur and diselenium dinitride, S_2N_2 and Se_2N_2*

S_2N_2 forms large colourless crystals with an iodine-like smell. It detonates with friction or on heating above 30°C, but can be sublimed at 10^{-2} Torr at 20°C. It has a square planar structure with S–N bond distances of 1.654 Å and bond angles of $90.0 \pm 0.4°$.[16] It is prepared by the thermolysis of other cyclic S–N compounds, *e.g.*, S_4N_4 (over silver wool at 220°C),[16] $[S_4N_3]Cl$,[17] or $Ph_3AsNS_3N_3$[18] at *ca.* 130°C. The four-membered ring S_2N_2 is an important precursor for the synthesis of the polymer $(SN)_x$ (Section 14.2). S_2N_2 is a six π-electron system, but four of these electrons occupy non-bonding orbitals (Section 4.4). Consequently, it has been suggested that S_2N_2 should be regarded as a two π–electron system.[19a] Recent *ab initio* calculations reveal that the four-membered rings E_2N_2 (E = S, Se, Te) exhibit significant diradical character, which increase along the series S<Se<Te.[19b] Both theoretical and experimental evidence suggest the existence of the acyclic isomer *trans*-N≡S–S≡N.[19c]

S_2N_2 forms both mono- and di-adducts, $S_2N_2 \cdot L$ and $S_2N_2 \cdot 2L$, with Lewis acids such as $AlCl_3$, BCl_3 and $SbCl_5$,[20, 21] and with a variety of transition-metal halides.[22] The S_2N_2 ligand is attached to the Lewis acid through nitrogen in these complexes, *e.g.*, **5.1**, and coordination has very little effect on the geometry of the four-membered ring. Although the selenium analogue Se_2N_2 is unknown as the free ligand both main group and transition-metal halide adducts have been structurally characterized. The bis-adduct $Se_2N_2 \cdot 2AlBr_3$ (**5.2**)[23] is prepared by the reaction of Se_4N_4 with $AlBr_3$, while the dianion $[Cl_3Pd(\mu\text{-}Se_2N_2)PdCl_3]^{2-}$ (**5.3**)[24] is similarly obtained by the reaction of Se_4N_4 with $[PPh_4]_2[Pd_2Cl_6]$ in CH_2Cl_2. The mean Se–N bond lengths in these adducts is *ca.* 1.79 Å.

| 5.1 | 5.2 | 5.3 |

The formation of π-complexes, *e.g.*, $(\eta^4\text{-}S_2N_2)M(CO)_3$ (M = Cr, Mo, W) (**5.4**), in which S_2N_2 acts as a six π-electron ligand has been proposed on the basis of frontier orbital considerations,[25] but only N-bonded adducts are observed. Cyclometallathiazenes of the type $L_nMS_2N_2$ in which a metal has formally inserted into an S–N bond of the S_2N_2 ring are also known (Section 7.3.2).

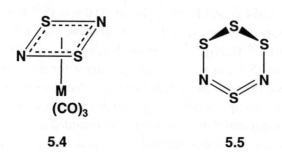

5.4   5.5

5.2.5 Tetrasulfur dinitride, S_4N_2

S_4N_2 forms dark red needles (mp 23°C) upon recrystallization from diethyl ether. It sublimes readily at room temperature, but must be stored at −20°C to avoid decomposition. Several routes are available for the preparation of S_4N_2. The decomposition of $Hg(S_7N)_2$ at room temperature gives the best yield, while the reaction of S_2Cl_2 with aqueous ammonia is quick, cheap and provides a purer product.[26]

The S_4N_2 molecule consists of a six-membered ring in a half-chair conformation (5.5).[26,27] There are long S–N bonds (1.68 Å) connecting the –SSS– and –N=S=N– units, which have S–N distances of 1.56 Å. Although few chemical reactions of S_4N_2 have been studied owing to its thermal instability, the reaction with norbornadiene produces the 1:1 adduct $S_4N_2.C_7H_8$ in which a >C=C< unit is inserted into one of the S–S bonds.[28a] A minor product of this reaction is $C_7H_8(SNSS)_2$.[28b]

The selenium analogue Se_4N_2 is unknown. *Ab initio* molecular orbital calculations indicate that six-membered ring systems of the type $S_xSe_{4-x}N_2$ (x = 0 - 4) containing the –N=Se=N– fragment are less stable than those containing the –N=S=N– group.[29]

5.2.6 Tetrasulfur and tetraselenium tetranitride, S_4N_4 and Se_4N_4

The standard synthesis of S_4N_4 involves the treatment of S_2Cl_2 with chlorine, followed by ammonia gas in carbon tetrachloride at 20-50°C.[30] Alternatively, S_4N_4 can be prepared by the reaction of $[(Me_3Si)_2N]_2S$ with an equimolar mixture of SCl_2 and SO_2Cl_2 (Eq. 5.2).[31] Several methods are available for the synthesis of Se_4N_4. The older methods involve the reaction of $(EtO)_2SeO$ or SeX_4 (X = Br, Cl) with ammonia under high pressure. The former procedure has been adapted to the preparation of $Se_4{}^{15}N_4$ by using a stoichiometric amount of $^{15}NH_3$.[32] A more recent alternative to the established methods of synthesis of Se_4N_4 is shown in Eq. 5.3.[33]

$$[(Me_3Si)_2N]_2S + SCl_2 + SO_2Cl_2 \rightarrow 1/2S_4N_4 + 4Me_3SiCl + SO_2 \qquad (5.2)$$

$$12(Me_3Si)_2NLi + 2Se_2Cl_2 + 8SeCl_4$$
$$\rightarrow 3Se_4N_4 + 24Me_3SiCl + 12LiCl \qquad (5.3)$$

The heat of formation of Se_4N_4 was determined to be $\Delta H^{\circ}{}_f = +163$ kcal mol^{-1}.[34] Thus Se_4N_4 is even more endothermic than S_4N_4 ($\Delta H^{\circ}{}_f = +110$ kcal mol^{-1}). The mean E–N bond energies in E_4N_4 were estimated to be 59 kcal mol^{-1} (E = Se)[34] and 72 kcal mol^{-1} (E = S)[35] from the enthalpies of formation.

Tetrasulfur tetranitride (**5.6a**) adopts a cage structure with equal S–N bond lengths (1.62 Å) and two weak transannular S•••S interactions of *ca.* 2.60 Å at room temperature. Tetrasulfur tetranitride is orange-yellow at room temperature, but becomes almost colourless at 80 K. Despite this thermochromism there is no significant change in the S•••S distance at 120 K.[36] Tetraselenium tetranitride has a cage-like structure (**5.6b**) similar to that of S_4N_4 with Se–N bond lengths of 1.79 Å (*cf.* 1.86 Å for Se–N single bonds) and transannular Se•••Se interactions of 2.75 Å (*cf.* 2.34 Å for Se–Se single bonds).[37] The hybrid tetrachalcogen tetranitride 1,5-$S_2Se_2N_4$ (**5.7**) also has a cage structure with transannular chalcogen•••chalcogen contacts of 2.70 Å.[38]

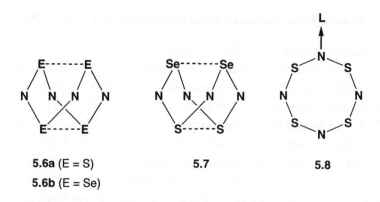

5.6a (E = S)   **5.7**   **5.8**
5.6b (E = Se)

Rietveld analysis indicates that S_4N_4 undergoes a transition to a new orthorhombic phase at 397 K.[39] Theoretical and experimental charge densities reveal intermolecular attractions that correspond to molecular recognition in the solid state.[40] The intermolecular forces in **5.6b** are stronger than those in **5.6a** and this probably accounts for the insolubility of Se_4N_4 in common organic solvents. The unusual structures of **5.6a** and **5.6b**, in which the chalcogen atoms are three coordinate and the nitrogen atoms are two coordinate, is the inverse of that found for P_4E_4 (E = S, Se), in which the more electronegative atoms occupy the lowest coordination sites. A planar (D_{4h}) S_4N_4 molecule would be a twelve π-electron system (Section 4.6) with an open-shell configuration, unstable with respect to Jahn-Teller distortion. Correlation of the molecular orbitals of the hypothetical planar molecule with those of the experimental (D_{2d}) conformation shows that the orbital degeneracy of the ground state is lost in the latter and four of the previously π^* electrons are accommodated in S•••S σ-bonds.[41] The DFT method gives a good estimate of the transannular chalcogen–chalcogen bonds in the series $Se_nS_{4-n}N_4$ (n = 0-4).[42] The IR and Raman spectra of S_4N_4 and Se_4N_4 have been assigned on the basis of slightly deformed D_{2d} symmetry, consistent with the X-ray structures.[43, 44]

The S_4N_4 molecule exhibits a very versatile chemical behaviour and is a source of many other S-N compounds, as shown in Scheme 5.1. The synthesis of other S–N rings from S_4N_4 can be classified as follows: reactions with (a) halogens or other oxidizing agents (b) nucleophiles or reducing agents, and (c) metal halides or organometallic reagents. The eight-membered ring is retained in $1,5\text{-}S_4N_4Cl_2$[45] and $[S_4N_4]^{2+}$,[46,47] the initial products of chlorination or oxidation of S_4N_4, respectively. Further chlorination or oxidation results in ring contraction to give $(NSCl)_3$ or the $[S_6N_4]^{2+}$ cation.[48]

The reaction of S_4N_4 with nucleophiles leads to ring contraction. For example, the degradation of S_4N_4 by anionic nucleophiles produces $[S_3N_3]^-$, while $Ph_3EN\text{-}S_3N_3$ (E = P, As) is obtained from S_4N_4 and triphenyl-phosphine or -arsine in benzene.[49, 50] In acetonitrile, however, the major product is $1,5\text{-}(Ph_3P=N)_2S_4N_4$.[51] *N*-Bonded adducts of the type S_4N_4.L (**5.8**), in which the eight-membered ring is flattened into a puckered boat with no cross-ring interactions, are formed with a variety of Lewis or Brønsted acids and some transition-metal halides, *e.g.*, $FeCl_3$ and VCl_5. However, the reactions of S_4N_4 with metal halides or organometallic reagents more often give rise to cyclometallathiazenes in which the metal becomes part of the S–N ring (Section 7.3). Molecular orbital calculations of the protonation of S_4N_4 reveal a localization of π^* electrons into the skeletal bonds to the protonated nitrogen and a strengthening of the almost planar SNSNSNS fragment.[52]

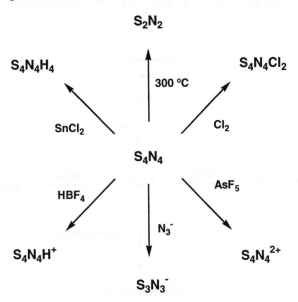

Scheme 5.1 Preparation of S–N rings from S_4N_4

The reactions of S_4N_4 with alkynes have also been investigated in considerable detail. These reactions produce 1,3,5,2,4-trithiadiazepines (**5.9**) and 1,3,5,2,4,6-trithiatriazepines (**5.10**), both of which are planar, ten π-electron seven-membered rings (Section 12.3), but the major products are 1,2,5-thiadiazoles.[53]

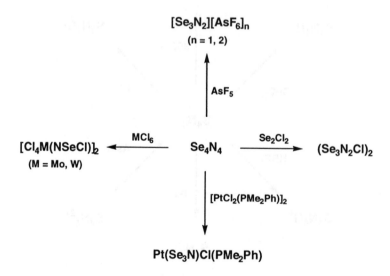

5.9   **5.10**

In spite of the hazardous nature of Se_4N_4, this binary selenium nitride has been used for the synthesis of other Se-N compounds, all of which have sulfur analogues (Scheme 5.2).[54-57] However, safer alternatives to the use of Se_4N_4, *e.g.*, selenium-nitrogen halides and silicon–nitrogen–selenium reagents, are available for the development of Se–N chemistry.[58]

$$[Se_3N_2][AsF_6]_n$$
$$(n = 1, 2)$$

$\uparrow$ AsF_5

$[Cl_4M(NSeCl)]_2$ $\xleftarrow{\;MCl_6\;}$ Se_4N_4 $\xrightarrow{\;Se_2Cl_2\;}$ $(Se_3N_2Cl)_2$

(M = Mo, W)

$\downarrow$ $[PtCl_2(PMe_2Ph)]_2$

$Pt(Se_3N)Cl(PMe_2Ph)$

Scheme 5.2 Preparation of Se–N compounds from Se_4N_4

5.2.7 *Pentasulfur hexanitride, S_5N_6*

Pentasulfur hexanitride (**5.11**) is an explosive, air-sensitive orange solid, which is best prepared from $S_4N_4Cl_2$ (Eq. 5.4).[59] It may also be obtained by the oxidation of $[S_4N_5]^-$.[60] The structure of S_5N_6 resembles a cradle in which an –N=S=N– unit bridges two antipodal sulfur atoms of an S_4N_4 cage via S–N single bonds. It decomposes in warm solvents to give S_4N_4. Because of its hazardous nature, very few reactions of S_5N_6 have been reported. An exception is the reaction with $[Ph_4P]_2[Pd_2Cl_6]$ to give $[Ph_4P]_2[Pd_2Cl_4(S_2N_3)]$.[61]

$$S_4N_4Cl_2 + Me_3SiNSNSiMe_3 \rightarrow S_5N_6 + 2Me_3SiCl \quad (5.4)$$

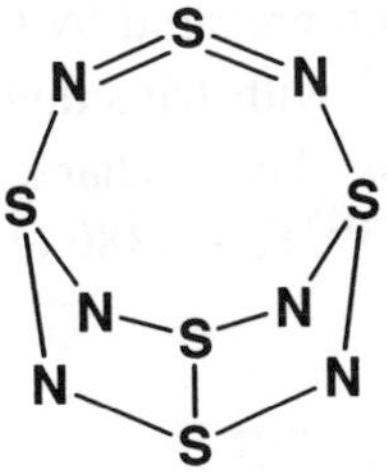

5.11 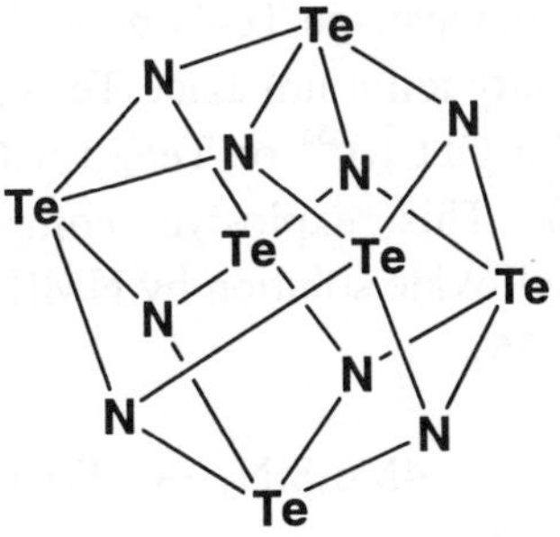   5.12

5.2.8 *Tellurium nitrides*

Tellurium nitride was first obtained by the reaction of $TeBr_4$ with liquid ammonia more than 100 years ago. The empirical formula TeN was assigned to this yellow, highly insoluble and explosive substance. However, subsequent analytical data indicated the composition is Te_3N_4[62] which, in contrast to **5.6a** and **5.6b**, would involve tetravalent tellurium. This conclusion is supported by the recent preparation and structural determination of $Te_6N_8(TeCl_4)_4$ from tellurium tetrachloride and tris(trimethylsilyl)amine (Eq. 5.5).[63] The Te_6N_8 molecule (**5.12**), which is a dimer of Te_3N_4, forms a rhombic dodecahedron in which the tellurium atoms form the corners of a distorted octahedron and the eight nitrogen atoms bridge each face of the octahedron as μ_3 ligands.

$$10\text{TeCl}_4 + 8\text{N(SiMe)}_3 \rightarrow [\text{Te}_6\text{N}_8(\text{TeCl}_4)_4] + 24\text{Me}_3\text{SiCl} \quad (5.5)$$

5.2.9 Tellurium azides

Several species in which three or more azido groups are bonded to tellurium have been well characterized.[64a] The first example was the salt $[\text{Te(N}_3)_3][\text{SbF}_6]$. The triazido tellurium cation was formed from the reaction of $\text{Te}_4[\text{SbF}_6]_2$ with potassium azide in liquid sulfur dioxide.[64b] The <NTeN bond angles in the pyramidal cation are in the range 91.9-97.3° and the average Te–N bond distance is 1.99 Å. More recently, the neutral binary tellurium azide $\text{Te(N}_3)_4$ has been prepared by the reaction of TeF_4 (in CFCl_3)[65a] or TeF_6 (in CH_3CN)[65b] with trimethylsilyl azide (Eq. 5.6). This explosive compound has been characterized in dimethylsulfoxide solution by NMR spectra ($\delta^{125}\text{Te} = 1380$ and $\delta^{14}\text{N} = -140$ and -235).

$$\text{TeF}_4 + 4\text{Me}_3\text{SiN}_3 \rightarrow \text{Te(N}_3)_4 + 4\text{Me}_3\text{SiF} \quad (5.6)$$

In a similar fashion $[\text{Me}_4\text{N}][\text{TeF}_5]$ reacts with trimethylsilyl azide to produce $[\text{Me}_4\text{N}][\text{Te(N}_3)_5]$ (Eq. 5.7), which has been structurally characterized in the solid state. The pentaazidotellurite anion is a distorted *pseudo*-octahedron with Te–N bond lengths in the range 2.07-2.26 Å.

$$[\text{Me}_4\text{N}][\text{TeF}_5] + 5\text{Me}_3\text{SiN}_3 \rightarrow [\text{Me}_4\text{N}][\text{Te(N}_3)_5] + 5\text{Me}_3\text{SiF} \quad (5.7)$$

The reaction of $\text{Te(N}_3)_4$ with ionic azides generates the $[\text{Te(N}_3)_6]^{2-}$ anion (Eq. 5.8).[65b] The structure of this anion is strongly distorted from octahedral by the stereochemically active lone pair on the tellurium atom, which gives rise to substantial differences in the Te–N bond lengths. Four of these bonds are in the range 2.09-2.24 Å, *cf.* $[\text{Te(N}_3)_5]^-$, while the other two are 2.42 Å and 2.53 Å, suggesting substantial ionic character in the latter two bonds.

$$Te(N_3)_4 + 2[PPh_4]N_3 \rightarrow [PPh_4]_2[Te(N_3)_6] \tag{5.8}$$

5.3 Cations

Investigations of binary S–N cations have played an important role in the development of S–N chemistry. The simple cations $[SN]^+$ and $[S_2N]^+$ are especially useful reagents for the synthesis of other S–N compounds. The selenium analogues $[SeN]^+$ and $[Se_2N]^+$ have not been isolated as simple salts, but they have been generated in the gas phase by laser ablation of a selenium target in the presence of N_2. The cyclic cations $[S_2N_3]^+$, $[S_3N_2]^{2+}$, $[S_4N_3]^+$, $[S_4N_4]^{2+}$, and $[S_5N_5]^+$ are of interest in the context of Banister's proposal that planar S-N rings obey the Hückel $(4n + 2)\pi$–electron rule (Section 4.4). The only selenium analogue of this series is $[Se_3N_2]^{2+}$ and there are no known tellurium congeners. The radical cation $[S_3N_2]^{+\bullet}$, which dimerizes to $[S_6N_4]^{2+}$ in the solid state, and the bicyclic cation $[S_4N_5]^+$ are also important cyclic species.

5.3.1 *The thiazyl cation, $[SN]^+$*

Thiazyl salts were first prepared in 1971 by the reaction of NSF with AsF_5 or SbF_5.[66a] They may also be obtained from $(NSCl)_3$ by reaction with (a) $Ag[AsF_6]$ in liquid SO_2 (Eq. 5.9)[66b] or (b) $AlCl_3$ in CH_2Cl_2 under the influence of heat or ultrasound.[66c]

$$(NSCl)_3 + 3Ag[AsF_6] \rightarrow 3 [SN][AsF_6] + 3AgCl \tag{5.9}$$

The sulfur–nitrogen bond length in thiazyl salts is about 1.42 Å and the vibrational frequency [ν(SN) occurs at 1437 cm^{-1} in $[SN][AsF_6]$. The $[SN]^+$ cation exhibits an ^{14}N NMR resonance at *ca.* 200 ppm and this technique is useful for monitoring reactions of $[SN]^+$.[67]

The thiazyl cation is used for the preparation of other important S–N compounds (Scheme 5.3). For example, the insertion reactions with S_4N_4

or SCl_2 produce $[S_5N_5]^+$ or the acyclic $[ClSNSCl]^+$ cation,[68] respectively. Insertion of thiazyl cations into the metal–halogen bond of $Re(CO)_5X$ (X = Cl, Br) provides a route to thiazyl halide complexes (Section 7.5). Intensely coloured, thermally labile charge-transfer complexes are formed between $[SN]^+$ and arenes such as mesitylene or hexamethylbenzene.[69]

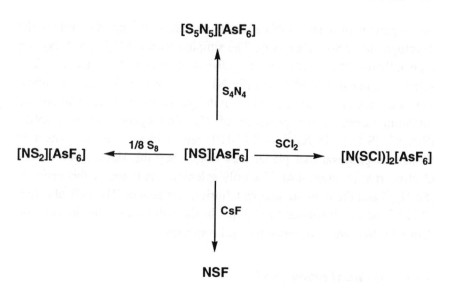

Scheme 5.3   Some reactions of $[SN][AsF_6]$

5.3.2 *The dithianitronium cation, $[S_2N]^+$*

The $[S_2N]^+$ cation was first obtained in low yield in 1978 by the oxidation of S_7NH with $SbCl_5$.[70] The reaction of *in situ* generated $[SN]^+$ salts with sulfur (Scheme 5.3) or the reduction of the $[ClSNSCl]^+$ cation with anhydrous tin(II) chloride in SO_2 or CH_2Cl_2 can be used to prepare $[S_2N]^+$ salts.[71a] The best large scale synthesis involves the reaction of AsF_5 with a mixture of S_4N_4 and sulfur in the presence of a trace amount of bromine (as a catalyst) (Eq. 5.10).[71b]

$$S_8 + 2S_4N_4 + 12AsF_5 \rightarrow 8[SNS][AsF_6] + 4AsF_3 \qquad (5.10)$$

The [SNS]$^+$ cation is a linear species isoelectronic with CS_2. The S–N bond distances are in the range 1.46-1.48 Å indicating a bond order of 2, *i.e.*, S=N$^+$=S.[70a,72] Consistently, the estimated SN bond energy of 113 kcal mol^{-1} reflects strong bonds.[35] The asymmetric stretching vibration is observed at 1498 cm^{-1} in the IR spectrum. The corresponding value for [SeNSe]$^+$ in an argon matrix at 12 K is 1253 cm^{-1} (*cf.* 1300 cm^{-1} for the isoelectronic, neutral molecule SeCSe).[9] The [S$_2$N]$^+$ cation exhibits a very narrow ^{14}N NMR resonance at –91 ppm ($v_{1/2}$ = 8 Hz) in SO_2. As in the case of [SN]$^+$, this technique is also useful for monitoring reactions of the [S$_2$N]$^+$ cation.

Although they are isovalent, the chemistry of [S$_2$N]$^+$ is very different from that of [NO$_2$]$^+$. The electrophilic reactions of [NO$_2$]$^+$ with aromatics are important in organic chemistry. By contrast, the [S$_2$N]$^+$ cation forms highly coloured charge-transfer complexes with arenes, which may involve the sulfur-protonated substitution product [C$_6$H$_5$(S$_2$N)H]$^+$.[73a] However, the [S$_2$N]$^+$ cation is an important reagent in S–N chemistry, especially in thermally allowed cycloaddition reactions with organic nitriles and alkynes, which give quantitative yields of heterocyclic cations (Scheme 5.4).[73b] The dominant orbital interaction in these cycloadditions is between the LUMO of [S$_2$N]$^+$ and the HOMO of the alkyne or nitrile (Section 4.8.1). Cycloaddition reactions also occur with alkenes.

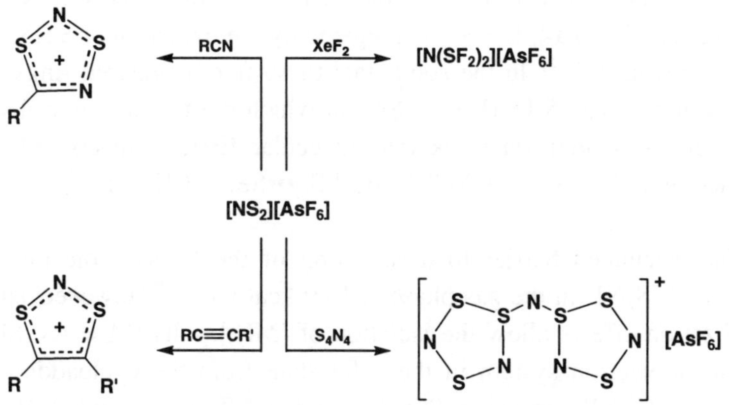

Scheme 5.4 Some cycloaddition reactions of [NS$_2$][AsF$_6$]

5.3.3 *The dithiatriazyl cation,* $[S_2N_3]^+$

The cyclic $[S_2N_3]^+$ cation (**5.13**), a six π-electron system, is the only example of a nitrogen-rich, binary S-N cation. The thermally stable salt $[S_2N_3]_2[Hg_2Cl_6]$ is obtained from the reaction of $(NSCl)_3$ with $HgCl_2$ in CH_2Cl_2.[74] The bond lengths in the five-membered ring indicate delocalized π-bonding that is attenuated across the S–S bond.

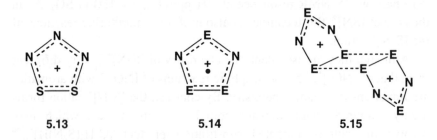

| 5.13 | 5.14 | 5.15 |

5.3.4 *The trithia and triselena-diazyl cations* $[E_3N_2]^{+\bullet}$, $[E_6N_4]^{2+}$ *and* $[E_3N_2]^{2+}$ *(E = S, Se)*

The $[S_3N_2]^{+\bullet}$ radical cation (**5.14**, E = S) has been well characterized in solution by its five-line (1:2:3:2:1) EPR signal.[75] It is prepared by oxidation of S_4N_4 with reagents such as AsF_5 or HSO_3F.[76]

All the possible selenathiadiazolyl radicals $[Se_{(3-n)}S_nN_2]^{+\bullet}$ (n = 0-3) have also been characterized by their EPR spectra.[77] The binary Se-N cation $[Se_3N_2]^{+\bullet}$ (**5.14**, E = Se) is prepared by the oxidation of Se_4N_4 with AsF_5 (Scheme 5.2).[54] In the solid state these five-membered rings form dimers of the type **5.15** (E = S, Se), in which the two seven π-electron rings are associated via weak intermolecular E•••E contacts ($\pi^* - \pi^*$ interactions) [d(S•••S) = 3.00-3.10 Å, d(Se•••Se) = 3.12-3.15 Å] (Section 4.7).[78]

The calculated barrier to dissociation of the $[S_3N_2]^{2+}$ dication into $[SN]^+$ and $[S_2N]^+$ in the gas phase is 10.9 kcal mol^{-1}.[79] However, lattice-stabilization effects allow the isolation of $[MF_6]^-$ salts (M = As, Sb) of this six π-electron system in the solid state from the cycloaddition of $[SN]^+$ and $[S_2N]^+$ cations in SO_2 (Eq. 5.11).[80] The S–S and S–N bond distances in the planar, monomeric dication are shorter than those in the dimeric radical cation dimer, as anticipated for the removal of an electron from a π^* orbital.

$$[NS][AsF_6] + [NS_2][AsF_6] \leftrightarrow [S_3N_2][AsF_6]_2 \qquad (5.11)$$

The selenium analogue $[Se_3N_2]^{2+}$ is prepared by the oxidation of Se_4N_4 with an excess of AsF_5 (Scheme 5.2).[54] In contrast to $[S_3N_2]^{2+}$, the selenium analogue does not dissociate significantly to $[SeN]^+$ and $[Se_2N]^+$ in solution. This behaviour is likely related to the relative magnitude of σ compared to π-bond strengths for chalcogen–nitrogen systems. The ^{77}Se NMR spectrum of $[Se_3N_2][AsF_6]_2$ at $-70°C$ in SO_2 solution shows one resonance with no satellites due to $^{77}Se-^{77}Se$ coupling, rather than the expected two resonances. This surprising result is indicative of a rapid exchange process. An intramolecular rearrangement involving structures in which the nitrogen atoms adopt a μ_3 bonding mode may be involved (Section 3.3).

5.3.5 *The cyclotrithiazyl cation, [S₃N₃]⁺*

The $[S_3N_3]^+$ cation is of interest as an example of an antiaromatic eight π-electron system (Section 4.4). *Ab initio* molecular orbital calculations indicate that a triplet cation, with a planar ring, is more stable than the singlet cation. The $[S_3N_3]^+$ cation has been obtained as the norbornene adduct **5.16,** but salts of the free cation have not been isolated.[82]

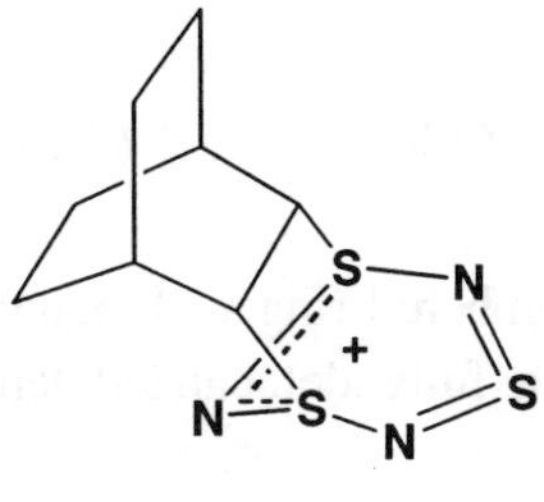

5.16 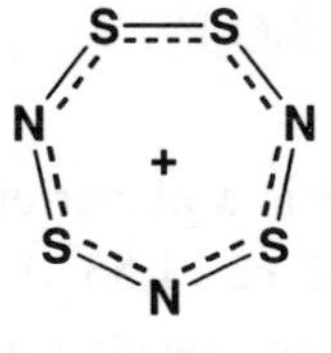   **5.17** 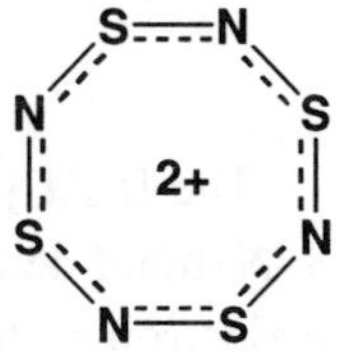   **5.18**

5.3.6 The thiotrithiazyl cation, $[S_4N_3]^+$

The thiotrithiazyl cation in $[S_4N_3]Cl$ was one of the first S–N heterocycles to be prepared and structurally characterized. It is obtained as a reasonably air-stable, yellow solid by the reaction of S_4N_4 or $[S_3N_2Cl]Cl$ with S_2Cl_2 in CCl_4 (Eq. 5.12).[83]

$$3[S_3N_2Cl]Cl \ + \ S_2Cl_2 \ \rightarrow \ 2[S_4N_3]Cl \ + \ 3SCl_2 \qquad (5.12)$$

The $[S_4N_3]^+$ cation (**5.17**) is a planar seven-membered ring with approximately equal S–N bond lengths and an S–S bond. This cation is a ten π-electron system.[84] A deformation density study of $[S_4N_3]Cl$ indicates significant charge transfer with charges of +0.75 e⁻ for the ring and –0.75 e⁻ for the chloride ion as a result of the close contacts (2.81-2.82 Å) between Cl⁻ and the two sulfur atoms of the S–S bond.[85] In addition, there are π–facial interactions between Cl⁻ and the $[S_4N_3]^+$ ring involving primarily the other two sulfur atoms.[86] The ^{15}N NMR spectrum shows two resonances, a doublet and a 1:2:1 triplet.[87] The seven-membered ring **5.17** fragments upon heating at *ca.* 250°C in the presence of silver wool to give S_2N_2 and, hence, the polymer $(SN)_x$.[17]

5.3.7 The cyclotetrathiazyl dication, $[S_4N_4]^{2+}$

The $[S_4N_4]^{2+}$ cation was first reported in 1977. It is prepared by the oxidation of S_4N_4 with an excess of a Lewis acid, such as $SbCl_5$ or AsF_5, or with $S_2O_6F_2$ (Eq. 5.13).[88]

$$S_4N_4 \ + \ S_2O_6F_2 \ \rightarrow \ [S_4N_4][FSO_3]_2 \qquad (5.13)$$

The $[S_4N_4]^{2+}$ cation **5.18** is a planar eight-membered ring with equal S–N bond lengths (D_{4h}) of *ca.* 1.55 Å. It is a fully delocalized ten π-electron system with a strong π-network.[89]

5.3.8 The tetrasulfur pentanitride cation, [S₄N₅]⁺

The nitrogen-rich $[S_4N_5]Cl$ is a yellow-orange, hygroscopic solid which decomposes violently on heating. It is readily prepared from $(NSCl)_3$ and bis(trimethylsilyl)sulfur diimide (Eq. 5.14).[90] The treatment of $[S_4N_5]Cl$ with AgF_2, $SbCl_5$, or $AgAsF_6$ gives $[S_4N_5]F$, $[S_4N_5][SbCl_6]$, or $[S_4N_5][AsF_6]$, respectively.

$$(NSCl)_3 + Me_3SiNSNSiMe_3 \rightarrow [S_4N_5]Cl + 2Me_3SiCl \quad (5.14)$$

$[S_4N_5]Cl$ has a polymeric, predominantly ionic, structure in which bicyclic $[S_4N_5]^+$ cations (**5.19**) are bridged by Cl^- anions with $d(S–Cl) =$ 2.81 Å. The unbridged S•••S distance is 4.01 Å.[90] By contrast, the S–F distance of 1.67 Å in $[S_4N_5]F$ is only slightly longer than the value of 1.61 Å found for the covalent S–F bonds in $(NSF)_3$.[91] A number of covalent derivatives of the S_4N_5 cage have been prepared by the reaction of $[S_4N_5]Cl$ with various silylated reagents.[92]

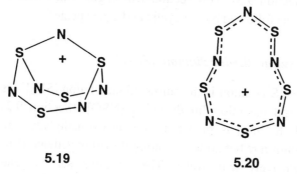

5.19   **5.20**

5.3.9 The cyclopentathiazyl cation, [S₅N₅]⁺

The first $[S_5N_5]^+$ salt was reported in 1969.[93] These salts are moisture sensitive, but they dissolve without decomposition in $SOCl_2$ or formic acid. The $[S_5N_5]^+$ cation is formed by the reaction of S_4N_4 with the $[SN]^+$ cation (Scheme 5.3).[94]

The $[S_5N_5]^+$ cation is a fourteen π-electron system.[95] With the exception of $[S_5N_5]Cl$, which is red, salts of $[S_5N_5]^+$ are yellow.[96] The planar, ten-membered ring usually has an azulene shape (**5.20**), with alternating sulfur and nitrogen atoms. Electrochemical reduction of $S_5N_5^+$ salts in acetonitrile produces the polymer $(SN)_x$.[96]

5.4 Anions

Sulfur and nitrogen are a versatile combination in the formation of binary anions with acyclic (including catenated systems), cyclic and cage structures. In addition, several S–N anions are known only in complexes with transition metals. The analogous Se–N anions are not known in ionic salts, but some of them are stabilized in metal complexes (Sections 7.3.2 and 7.3.3). In addition to their function as versatile ligands, S–N anions play an important role as intermediates in the formation of cyclic sulfur imides (Sections 6.2.1) and as constituents of solutions of sulfur in liquid ammonia (Section 5.4.4). Catenated anions that involve oxygen, in addition to S and N, are thought to be involved in the gunpowder reaction (Section 9.3). This section will begin with a discussion of the acyclic anions followed by the cyclic and cage species.

5.4.1 The sulfur diimide dianion, $[SN_2]^{2-}$

The simplest S-N anion is the sulfur diimide anion $[SN_2]^{2-}$, isoelectronic with SO_2 and the thionylimide anion $[NSO]^-$ (Section 9.2). The pale yellow salt $K_2[SN_2]$ is prepared from bis(trimethylsilyl)sulfur diimide and potassium *tert*-butoxide in boiling dimethoxyethane (Eq. 5.15).[97] It reacts explosively with water. The dilithium salt is generated from $Me_3SnNSNSnMe_3$ and MeLi in the same solvent.[98] The S=N bond distances in the $[K(18\text{-crown-}6)]^+$ salt are 1.484 Å, *ca.* 0.05 Å longer the the S=O bonds in SO_2. The <NSN bond angle of 129.9° is *ca.* 12° wider than the corresponding bond angle in SO_2, presumably reflecting electrostatic repulsion between the negatively charged nitrogen atoms.[99] The frequencies of the three fundamental vibrations of $[SN_2]^{2-}$ are 1198, 1001, and 528 cm^{-1}.

$$Me_3SiNSNSiMe_3 + 2KO^tBu \rightarrow K_2[SN_2] + 2Me_3SiO^tBu \qquad (5.15)$$

The salt $K_2[SN_2]$ is an important reagent for the preparation of other sulfur diimide derivatives when $Me_3SiNSNSiMe_3$ is not sufficiently reactive. For example, both acyclic and cyclic arsenic(III) compounds, $^tBu_2AsNSNAs^tBu_2$ and $^tBuAs(NSN)_2As^tBu$, respectively, have been obtained in this way.[100a]

Protonation of the anion $[SN_2]^{2-}$ by acetic acid in diethyl ether produces the thermally unstable sulfur diimide $S(NH)_2$.[100b] Like all sulfur diimides, the parent compound $S(NH)_2$ can exist as three isomers (Scheme 5.5). *Ab initio* molecular orbital calculations indicate that the (*cis,cis*) configuration is somewhat more stable than the (*cis,trans*) isomer, while the (*trans,trans*) isomer is expected to possess considerably higher energy. The alternative *syn,anti* or *E,Z* nomenclatures may also be used to describe these isomers. The structures of organic derivatives $S(NR)_2$ (R = alkyl, aryl) are discussed in Section 10.4.2.

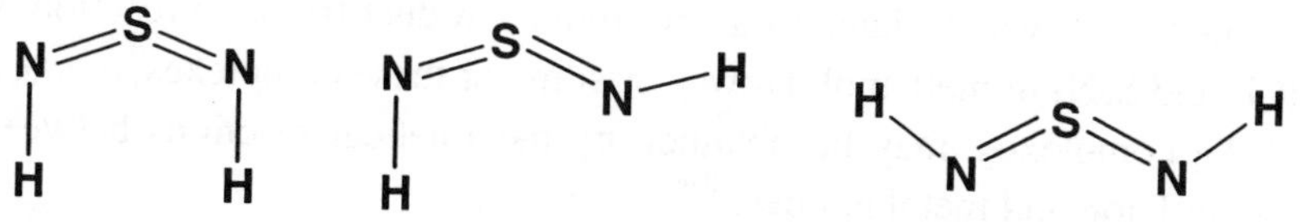

(*cis, cis*)   (*cis, trans*)   (*trans, trans*)

Scheme 5.5   Isomers of sulfur diimide, HNSNH

Salts of mono-alkylated or arylated sulfur diimide anion [RNSN]⁻ (R = aryl, ^tBu, $SiMe_3$) are prepared by Si–N cleavage of $RNSNSiMe_3$ with $[(Me_2N)_3S][Me_3SiF_2]$.[101,102] These anions adopt *cis* configurations with very short terminal S–N bond lengths (1.44 - 1.49 Å) indicative of a thiazylamide anion, [RNS≡N]⁻ (**5.21**).[99,101,102]

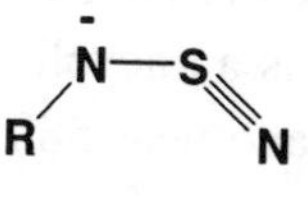

5.21 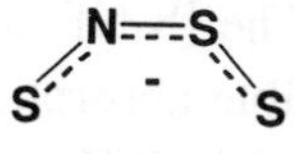   **5.22** 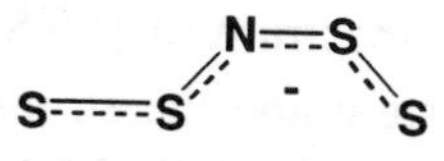   **5.23** 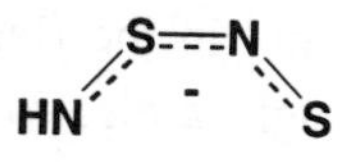   **5.24**

5.4.2 The [SSNS]⁻ anion

The orange-red [S_3N]⁻ anion (λ_{max} 465 nm) is obtained by the addition of triphenylphosphine to a solution of a [S_4N]⁻ salt in acetonitrile.[103] It can be isolated as a salt in combination with large counterions, *e.g.*, [Ph_4As]⁺ or [$N(PPh_3)_2$]⁺, but it is unstable with respect to the formation of the blue [S_4N]⁻ anion in solution or in the solid state under the influence of heat or pressure.

Although the structure of [S_3N]⁻ has not been established by X-ray crystallography, the vibrational spectra of 30% ¹⁵N-enriched [S_3N]⁻ suggest an unbranched [SNSS]⁻ (**5.22**) arrangement of atoms in contrast to the branched structure (D_{3h}) of the isoelectronic [CS_3]²⁻ and the isovalent [NO_3]⁻ ion (Section 1.2). Mass spectrometric experiments also support the SNSS connectivity in the gas phase.[104] Many metal complexes are known in which the [S_3N]⁻ ion is chelated to the metal by two sulfur atoms (Section 7.3.3). Indeed the first such complex, Ni(S$_3$N)$_2$, was reported more than twenty years before the discovery of the anion.[105a] It was isolated as a very minor product from the reaction of NiCl$_2$ and S$_4$N$_4$ in methanol. However, some of these complexes, *e.g.*, Cu and Ag complexes, may be obtained by metathetical reactions between the [S_3N]⁻ ion and metal halides.[105b]

5.4.3 The [SSNSS]⁻ anion

The dark blue [S_4N]⁻ anion (λ_{max} 580 nm) was first obtained in 1972 from the decomposition of [$^nBu_4N$][S_7N] produced by deprotonation of S$_7$NH with [$^nBu_4N$]OH in diethyl ether at –78°C.[106] It is more conveniently prepared by the thermolysis of salts of the [S_3N_3]⁻ ion with large cations in boiling acetonitrile.[107] The [S_4N]⁻ ion (**5.23**) is a planar (*cis,trans*) chain with nitrogen as the central atom and short, terminal S–S bonds (*ca.* 1.90 Å), which gives rise to strong bands at *ca.* 565 and 590 cm⁻¹ in the Raman spectrum. As expected, only one band is observed at *ca.* 570 cm⁻¹ in the Raman spectrum of [SNSS]⁻. The [S_4N]⁻ anion is an acyclic eight π-electron system and the strong visible absorption band near 580 nm has been assigned to a π*(HOMO) → π*(LUMO) transition.

5.4.4 *Sulfur–nitrogen anions in sulfur–liquid ammonia solutions*

Sulfur dissolves in liquid ammonia to give intensely coloured solutions. The colour is concentration-dependent and the solutions are photosensitive.[108] Extensive studies of this system by several groups using a variety of spectroscopic techniques, primarily Raman, ^{14}N/^{15}N NMR and visible spectra have established that a number of S-N anions are present in such solutions.[109-111] The primary reduction products are polysulfides S_x^{2-}, which dissociate to polysulfur radical anions, especially the deep blue S_3^{-} ion (λ_{max} ~620nm). In a 1M solution the major S–N anion detected by ^{14}N NMR spectroscopy is *cyclo*-[S$_7$N]$^-$ with smaller amounts of the [SSNSS]$^-$ ion and a trace of [SSNS]$^-$. The formation of the acyclic anion **5.23** from the decomposition of *cyclo*-S$_7$N$^-$ is well established from chemical investigations (Section 5.4.3). The acyclic anions **5.22** and **5.23** have been detected by their characteristic visible and Raman spectra. It has also been suggested that a Raman band at 858 cm^{-1} and a visible absorption band at 390 nm may be attributed to the [S$_2$N]$^-$ anion formed by cleavage of a S–S bond in [SSNS]$^-$.[108] However, this anion cannot be obtained as a stable species when [S$_3$N]$^-$ is treated with one equivalent of PPh$_3$.

5.4.5 *The [S$_2$N$_2$H]$^-$ anion*

The [S$_2$N$_2$H]$^-$ anion (**5.24**) (a monoprotonated form of [S$_2$N$_2$]$^{2-}$) has been characterized in solution by ^{14}N and ^{15}N NMR spectroscopy. It is formed by treatment of S$_4$N$_4$H$_4$ (or S$_7$NH) with potassium amide in liquid ammonia (Scheme 5.6).[109] In the ^{14}N NMR spectrum two well-separated resonances are observed for the inequivalent nitrogen atoms, one of which is a doublet [$^1J(^{14}$N–^{1}H) = 36 Hz]. In the ^{15}N NMR spectrum the coupling between the inequivalent nitrogen atoms is 2.2 Hz for the inequivalent and $^1J(^{15}$N–^{1}H) = 55 Hz. Although the [S$_2$N$_2$H]$^-$ anion has not been isolated as an ionic salt, numerous metal complexes in which **5.24** acts as a bidentate (*S,N*) ligand are known (Section 7.3.2). In solution **5.24** decomposes to give [S$_3$N$_3$]$^-$.

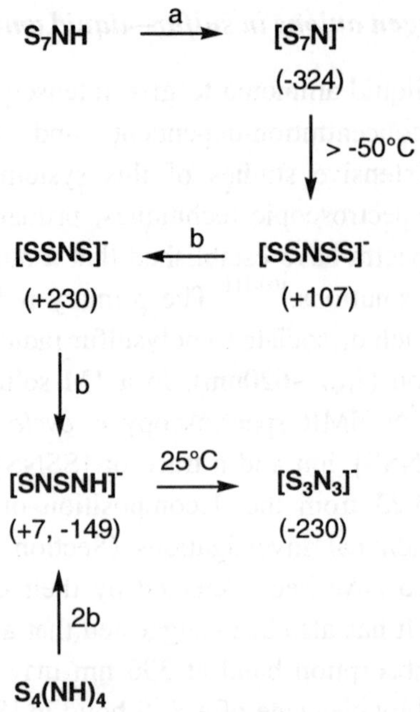

$$S_7NH \xrightarrow{\ a\ } [S_7N]^-$$
$$(-324)$$

$$\downarrow {\scriptstyle >\text{-}50°C}$$

$$[SSNS]^- \xleftarrow{\ b\ } [SSNSS]^-$$
$$(+230) \qquad\qquad (+107)$$

$$\downarrow b$$

$$[SNSNH]^- \xrightarrow{\ 25°C\ } [S_3N_3]^-$$
$$(+7, -149) \qquad\qquad (-230)$$

$$\uparrow 2b$$

$$S_4(NH)_4$$

Scheme 5.6 Sulfur–nitrogen anions formed in the deprotonation of S_7NH and $S_4N_4H_4$ (a = $NH_3(l)$; b = KNH_2). Nitrogen NMR chemical shifts (in ppm) are given in parentheses [ref. $MeNO_2$ (*l*)]

5.4.6 The trisulfur trinitride anion, [S₃N₃]⁻

The $[S_3N_3]^-$ anion (**5.25**), a ten π-electron system, is an important example of a π-electron-rich ring system (Section 4.4). It is formed, together with $[S_4N_5]^-$, from the reaction of S_4N_4 with a variety of nucleophilic reagents, *e.g.*, ionic azides, organolithium reagents, or secondary amines.[112] The best preparation involves the reaction of S_4N_4 with an azide of a large cation, *e.g.*, $[(Ph_3P)_2N]^+$. The large cation is necessary to stabilize the anion; alkali metal salts of $[S_3N_3]^-$ are explosive in the solid state.

The $[S_3N_3]^-$ anion is an essentially planar, six-membered ring with bond angles at nitrogen and sulfur of *ca.* 123° and 117°, respectively, and almost equal S–N bond lengths.[113] Two $[S_3N_3]^-$ salts with interesting solid-state structures are $[PhCN_2S_2][S_3N_3]$,[114] prepared from $[PhCN_2S_2]_2$

and [S$_5$N$_5$]Cl, and [Cp$_2$Co][S$_3$N$_3$],[115] obtained by reduction of S$_4$N$_4$ with cobaltocene. The electronic structure of [S$_3$N$_3$]⁻ is discussed in Section 4.4. The yellow colour (λ_{max} 360 nm) is assigned to a π*(HOMO) → π*(LUMO) transition and this assignment is confirmed by the MCD spectrum. The vibrational spectra of the alkali-metal salts are consistent with D_{3h} symmetry for the anion.[113]

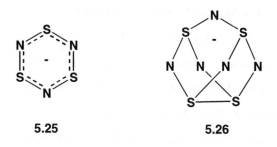

5.25   **5.26**

5.4.7 The tetrasulfur pentanitride anion, [S$_4$N$_5$]⁻

The yellow [S$_4$N$_5$]⁻ anion (**5.26**) was first reported in 1975 from the methanolysis of Me$_3$SiNSNSiMe$_3$.[116a] It can also be prepared by the treatment of S$_4$N$_4$ with certain nucleophiles, *e.g.*, secondary amines or azide salts of small alkali metal cations (Eq. 5.16).[112] The reaction of (NSCl)$_3$ with dry liquid ammonia at –78°C also generates [NH$_4$][S$_4$N$_5$] in *ca.* 50% yield.[117]

$$S_4N_4 + LiN_3 \rightarrow Li[S_4N_5] + N_2 \qquad (5.16)$$

The structure of **5.26** is closely related to that of S$_4$N$_4$.[116b] In [S$_4$N$_5$]⁻, five of the six edges of the S$_4$ tetrahedron are bridged by nitrogen atoms. The unbridged S•••S distance is *ca.* 2.70 Å. Alkali metal salts of [S$_4$N$_5$]⁻ are explosive when subjected to heat or mechanical shock, but salts with large counterions, *e.g.*, [Ph$_4$As]⁺ or [(Ph$_3$P)$_2$N]⁺, are safe to handle in the solid state. The carefully controlled thermolysis of solutions of [S$_4$N$_5$]⁻ in boiling acetonitrile generates [S$_3$N$_3$]⁻ and, subsequently, the [S$_4$N]⁻ anion.[107] The reaction of [S$_4$N$_5$]⁻ with bromine or iodine produces pentasulfur hexanitride, S$_5$N$_6$, whereas oxidation with chlorine yields [S$_4$N$_5$]Cl.[60]

References

1. K. K. Pandey, D. K. M. Raju. H. L. Nigam and U. C. Agarwala, *Proc. Indian Nat. Sci. Acad.*, **48A**, 16 (1982).

2. K. K. Pandey, *Prog. Inorg. Chem.*, **40**, 445 (1992).

3. R. D. Smith, *J. Chem. Soc., Dalton Trans.*, 478 (1979).

4. H. Uehara and Y. Morino, *Mol. Phys.*, **17**, 239 (1969).

5. P. B. Zeeman, *Can. J. Phys.*, **29**, 174 (1951).

6. D. L. Hildebrand and E. Murad, *J. Chem. Phys.*, **51**, 807 (1969).

7. P. Hassanzadeh and L. Andrews, *J. Am Chem. Soc.*, **114**, 83 (1992).

8. P. A. G. O'Hare, *J. Chem. Phys.*, **52**, 2992 (1970).

9. L. Andrews and P. Hassanzadeh, *J. Chem. Soc., Chem. Commun.*, 1523 (1994).

10. D. L. Lichtenberger and J. L. Hubbard, *Inorg. Chem.*, **24**, 3835 (1985).

11. T. J. Crevier, S. Lovell and J. M. Mayer, *J. Am. Chem. Soc.*, **120**, 6607 (1998).

12. D. S. Bohle, C-H. Hung, A. K. Powell, B. D. Smith and S. Wocadlo, *Inorg. Chem.*, **36**, 1992 (1997).

13. C. Wentrup and P. Kambouris, *Chem. Rev.*, **91**, 363 (1991).

14. H. Bender, F. Carnovale, J. B. Peel and C. Wentrup, *J. Am. Chem. Soc.*, **110**, 3458 (1988).

15. R. D. Davy and H. F. Schaefer, III, *J. Am. Chem. Soc.*, **113**, 1917 (1991).

16. C. M. Mikulski, P. J. Russo, M. S. Soran, A. G. MacDiarmid, A. F. Garito and A. J. Heeger, *J. Am. Chem. Soc.*, **97**, 6358 (1975).

17. A. J. Banister and Z. V. Hauptman, *J. Chem. Soc., Dalton Trans.*, 731 (1980).

18. T. Chivers, A. W. Cordes, R. T. Oakley and P. N. Swepston, *Inorg. Chem.*, **20**, 2376 (1981).

19. (a) Y. Jung, T. Heine, P. v. R. Schleyer and M. Head-Gordon, *J. Am. Chem. Soc.*, **126**, 3132 (2004); (b) H. M. Tuononen, R. Suontamo, J. Valkonen, and R. S. Laitinen, *J. Phys. Chem. A*, **108**, 5670 (2004); (c) R. C. Mawhinney and J. D. Goddard, *Inorg. Chem.*, **42**, 6323 (2003).

20. R. L. Patton and K. N. Raymond, *Inorg. Chem.*, **8**, 2426 (1969).

21. U. Thewalt and M. Burger, *Angew. Chem., Int. Ed. Engl.*, **21**, 634 (1982).

22. K. Dehnicke and U. Müller, *Transition Metal Chem.*, **10**, 361 (1985).

23. P. F. Kelly and A. M. Z. Slawin, *J. Chem. Soc., Dalton Trans.*, 4029 (1996).

24. P. F. Kelly and A. M. A. Slawin, *Angew. Chem., Int. Ed. Engl.*, **34**, 1758 (1995).

25. M. Bénard, *New J. Chem.*, **10**, 529 (1986).

26. T. Chivers, P. W. Codding, W. G. Laidlaw, S. W. Liblong, R. T. Oakley and M. Trsic, *J. Am. Chem. Soc.*, **105**, 1186 (1983).

27. R. W. H. Small, A. J. Banister and Z. V. Hauptman, *J. Chem. Soc., Dalton Trans.*, 2188 (1981).

28. (a) H. Koenig, R. T. Oakley. A. W. Cordes and M. C. Noble, *Can. J. Chem.*, **61**, 1185 (1983); (b) A. W. Cordes, H. Koenig, M. C. Noble and R. T. Oakley, *Inorg. Chem.*, **22**, 3375 (1983).

29. A. Maaninen, J. Siivari, R. J. Suontamo, J. Konu, R. S. Laitinen and T. Chivers, *Inorg. Chem.*, **36**, 2170 (1997).

30. M. Villena-Blanco and W. L. Jolly, *Inorg. Synth.*, **9**, 98 (1967).

31. A. Maaninen, J. Siivari, R. S. Laitinen and T. Chivers, *Inorg. Synth.*, **33**, 196 (2002).

32. V. C. Ginn, P. F. Kelly and J. D. Woollins, *J. Chem. Soc., Dalton Trans.*, 2129 (1992).

33. J. Siivari, T. Chivers and R. S. Laitinen, *Inorg. Chem.*, **32**, 1519 (1993).

34. C. K. Barker, A. W. Cordes and J. L. Margrave, *J. Phys. Chem.*, **69**, 334 (1965).

35. S. Parsons and J. Passmore, *Inorg. Chem.*, **31**, 526 (1992).

36. M. L Delucia and P. Coppens, *Inorg. Chem.*, **17**, 2336 (1978).

37. H. Barnighausen, T. v. Volkmann and J. Jander, *Acta Crystallogr.*, **21**, 751 (1966).

38. J. Konu, A. Maaninen, K. Paananen, P. Ingman, R. S. Laitinen, T. Chivers and J. Valkonen, *Inorg. Chem.*, **41**, 1430 (2002).

39. S. H. Irsen, P. Jacobs and R. Dronskowski, *Z. Anorg. Allg. Chem.*, **627**, 321 (2001).

40. W. Scherer, M. Spiegler, B. Pedersen, M. Tafipolsky, W. Hieringer, B. Reinhard, A. J. Downs and G. S. McGrady, *Chem. Commun.*, 635 (2000).

41. R. Gleiter, *J. Chem. Soc. A*, 3174 (1970).

42. G. Chung and D. Lee, *J. Mol. Struct.- Theochem.*, **582**, 85 (2002).

43. R. Steudel, *Z. Naturforsch.*, **36A**, 850 (1981).

44. (a) J. Adel, C. Ergezinger, R. Figge and K. Dehnicke, *Z. Naturforsch.*, **43B**, 639 (1988); (b) P. K. Gowik and T. M. Klapötke, *Spectrochim Acta*, **46A**, 1371 (1990).

45. Z. Zak, *Acta Crystallogr.*, **B37**, 23 (1981).

46. R. J. Gillespie, J. P. Kent, J. F. Sawyer, D. R. Slim and J. D. Tyrer, *Inorg. Chem.*, **20**, 3799 (1981).

47. R. D. Sharma, F. Aubke and N. L. Paddock, *Can. J. Chem.*, **59**, 3157 (1981).

48. R. J. Gillespie, J. P. Kent and J. F. Sawyer, *Inorg. Chem.*, **20**, 3784 (1981).

49. J. Bojes, T. Chivers, I. Drummond and G. MacLean, *Inorg. Chem.*, **17**, 3668 (1978).

50. T. Chivers and R. T. Oakley, *Topics Curr. Chem.*, **102**, 117 (1982).

51. J. Bojes, T. Chivers, A. W. Cordes, G. MacLean and R. T. Oakley, *Inorg. Chem.*, **20**, 16 (1981).

52. A. W. Cordes, C. G. Marcellus, M. C. Noble, R. T. Oakley and W. T. Pennington, *J. Am. Chem. Soc.*, **105**, 6008 (1983).

53. J. L. Morris and C. W. Rees, *Chem. Soc. Rev.*, **15**, 1 (1986).

54. E. G. Awere, J. Passmore and P. S. White, *J. Chem. Soc., Dalton Trans.*, 299 (1993).

55. J. Siivari, T. Chivers and R. Laitinen, *Inorg. Chem.*, **32**, 4391 (1993).

56. P. F. Kelly, A. M. Z. Slawin, D. J. Williams and J. D. Woollins, *J. Chem. Soc., Chem. Commun.*, 408 (1989).

57. J. Adel and K. Dehnicke, *Chimia*, **42**, 413 (1988).

58. T. Chivers, *Main Group Chem. News*, **1**, 6 (1993).

59. W. S. Sheldrick, M. N. S. Rao and H. W. Roesky, *Inorg. Chem.*, **19**, 538 (1980).

60. T. Chivers and J. Proctor, *Can. J. Chem.*, **57**, 1286 (1979).

61. P. F. Kelly, A. M. Z. Slawin, D. J. Williams and J. D. Woollins, *Angew. Chem., Int. Ed. Engl.*, **31**, 616 (1992).

62. O. Schmitz-Dumont and B. Ross, *Angew. Chem., Int. Ed. Engl.*, **6**, 1071 (1967).

63. W. Mosa, C. Lau, M. Möhlen, B. Neumüller and K. Dehnicke, *Angew. Chem., Int. Ed. Engl.*, **37**, 2840 (1998).

64. (a) C. Knapp and J. Passmore, *Angew. Chem., Int. Ed. Engl.*, **43**, in press (2004); (b) J. P. Johnson, G. K. MacLean, J. Passmore and P. S. White, *Can. J. Chem.*, **67**, 1687 (1989).

65. (a) T. M. Klapötke, B. Krumm, P. Mayer and I. Schwab, *Angew. Chem., Int. Ed. Engl.*, **42**, 5843 (2003); (b) R. Hiages, J. A. Boatz, A. Vij, M. Gerken, S. Schneider, T. Schroer and K. O. Christe, *Angew. Chem., Int. Ed. Engl.*, **42**, 5847 (2003).

66. (a) O. Glemser and W. Koch, *Angew. Chem.*, **83**, 145 (1971); (b) A. Apblett, A. J. Banister, D. Biron, A. G. Kendrick, J. Passmore, M. Schriver and M. Stojanac, *Inorg. Chem.*, **25**, 4451 (1986); (c) A. Apblett, T. Chivers and J. F. Fait, *Inorg. Chem.*, **29**, 1643 (1990).

67. J. Passmore and M. Schriver, *Inorg. Chem.*, **27**, 2751 (1988).

68. R. Mews, *Angew. Chem.*, **88**, 757 (1976).

69. S. Brownstein and B. Louie, *Can. J. Chem.*, **65**, 1361 (1987).

70. (a) R. Faggiani, R. J. Gillespie, C. J. L. Lock and J. D. Tyrer, *Inorg. Chem.*, **17**, 2975 (1978); (b) S. Parsons and J.Passmore, *Acc. Chem. Res.*, **27**, 101 (1994).

71. (a) E. G. Awere and J. Passmore, *J. Chem. Soc., Dalton Trans.*, 1343 (1992); (b) B. Ayres, A. J. Banister, P. D. Coates, M. I. Hansford, J. M. Rawson, C. E. F. Rickard, M. B. Hursthouse, K. M. A. Malik and M. Motevalli, *J. Chem. Soc., Dalton Trans.*, 3097 (1992).

72. (a) U. Thewalt, K. Berhalter and P. Müller, *Acta Crystallogr.*, **B38**, 1280 (1982); (b) J. P. Johnson, J. Passmore, P. S. White, A. J. Banister and A. G. Kendrick, *Acta Crystallogr.*, **C43**, 1651 (1987).

73. (a) S. Brownridge, J. Passmore and X. Sun, *Can. J. Chem.*, **76**, 1220 (1998); (b) S. Parsons, J. Passmore, M. J. Schriver and X. Sun, *Inorg. Chem.*, **30**, 3342 (1991).

74. S. Herler, P. Mayer, H. Nöth, A. Schulz, M. Suter and M. Vogt, *Angew. Chem., Int. Ed. Engl.*, **40**, 3173 (2001).

75. (a) S. A. Fairhurst, K. F. Preston and L. H. Sutcliffe, *Can. J. Chem.*, **62**, 1124 (1984): (b) K. M. Johnson, K. F. Preston and L. H. Sutcliffe, *Mag. Reson. Chem.*, **26**, 1015 (1988).

76. R. J. Gillespie, J. P. Kent and J. F. Sawyer, *Inorg. Chem.*, **20**, 3784 (1981).

77. E. Awere, J. Passmore, K. F. Preston and L. H. Sutcliffe, *Can. J. Chem.*, **66**, 1776 (1988).

78. R. Gleiter, R. Bartetzko and P. Hofmann, *Z. Naturforsch.*, **35B**, 1166 (1980).

79. F. Grein, *Can. J. Chem.*, **71**, 335 (1993).

80. W. V. F. Brooks, T. S. Cameron, S. Parsons, J. Passmore and M. J. Schriver, *Inorg. Chem.*, **33**, 6230 (1994).

81. J. Bojes, T. Chivers, W. G. Laidlaw and M. Trsic, *J. Am. Chem. Soc.*, **101**, 4517 (1979).

82. A. Apblett, T. Chivers, A. W. Cordes and R. Vollmerhaus, *Inorg. Chem.*, **30**, 1392 (1991).

83. W. L. Jolly and K. D. Maguire, *Inorg. Synth.*, **9**, 102 (1967).

84. J. W. Waluk and J. Michl, *Inorg. Chem.*, **21**, 556 (1982).

85. H. Johansen, *J. Am. Chem. Soc.*, **110**, 5322 (1988).

86. J-R. Galan-Mascaros, A. M. Z. Slawin, J. D. Williams and D. J. Williams, *Polyhedron*, **15**, 4603 (1996).

87. T. Chivers, R. T. Oakley, O. J. Scherer and G. Wolmerhäuser, *Inorg. Chem.*, **20**, 914 (1981).

88. R. J. Gillespie, J. P. Kent, J. F. Sawyer, D. R. Slim and J. D. Tyrer, *Inorg. Chem.*, **20**, 3799 (1981).

89. M. Trsic, W. G. Laidlaw and R. T. Oakley, *Can. J. Chem.*, **60**, 17 (1982).

90. T. Chivers, L. Fielding, W. G. Laidlaw and M. Trsic, *Inorg. Chem.*, **18**, 3379 (1979).

91. W. Isenberg, R. Mews, G. M. Sheldrick, R. Bartetzko and R. Gleiter, *Z. Naturforsch.*, **38B**, 1563 (1983).

92. H. W. Roesky, C. Graf and M. N. S. Rao, *Chem. Ber.*, **113**, 3815 (1980).

93. A. J. Banister, P. J. Dainty, A. C. Hazell, R. G. Hazell and J. G. Lomborg, *Chem. Commun.*, 1187 (1969).

94. A. J. Banister and H. G. Clarke, *Inorg. Synth.*, **17**, 188 (1977).

95. R. Bartetzko and R. Gleiter, *Inorg. Chem.*, **17**, 995 (1978).

96. (a) A. J. Banister, Z. V. Hauptman and A. G. Kendrick, *J. Chem. Soc., Dalton Trans.*, 915 (1987); (b) H. P. Fritz and R. Bruchhaus, *Z. Naturforsch.*, **38B**, 1375 (1983).

97. M. Herberhold and W. Ehrenreich, *Angew. Chem., Int. Ed. Engl.*, **21**, 633 (1982).

98. D. Hänssgen, H. Salz, S. Rheindorf and C. Schrage, *J. Organomet. Chem.*, **443**, 61 (1993).

99. T. Borrman, E. Lork, R. Mews, M. M. Shakirov and A. V. Zibarev, *Eur. J. Inorg. Chem.*, 2452 (2004).

100. (a) M. Herberhold, *Comments Inorg. Chem.* **7**, 53 (1988); (b) M. Herberhold, W. Jellen, W. Buhlemeyer, W. Ehrenreich and J. Reiner, *Z. Naturforsch.*, **40B**, 1229 (1985).

101. A. V. Zibarev, E. Lork and R. Mews, *Chem. Commun.*, 991 (1998).

102. T. Borrman, A. V. Zibarev, E. Lork, G. Knitter, S-J. Chen, P. G. Watson, E. Cutin, M. M. Shakirov, W-D. Stohrer and R. Mews, *Inorg. Chem.*, **39**, 3999 (2000).

103. J. Bojes, T. Chivers, W. G. Laidlaw and M. Trsic, *J. Am. Chem. Soc.*, **104**, 4837 (1982).

104. M. Iraqi, N. Goldberg and H. Schwarz, *Chem. Ber.*, **127**, 1171 (1994).

105. (a) T. S. Piper, *J. Am. Chem. Soc.*, **80**, 30 (1958); (b) J. Bojes, T. Chivers and P. W. Codding, *Chem. Commun.*, 1171 (1981).

106. T. Chivers and I. Drummond, *Inorg. Chem.*, **13**, 1222 (1974).

107. T. Chivers, W. G. Laidlaw, R. T. Oakley and M. Trsic, *J. Am. Chem. Soc.*, **102**, 5773 (1980).

108. P. Dubois, J. P. Lelieur and G. Lepoutre, *Inorg. Chem.*, **28**, 2489 (1989).

109. T. Chivers and K. J. Schmidt, *Can. J. Chem.*, **70**, 710 (1992).

110. S. Seelert and U. Schindewolf, *Phosphorus, Sulfur and Silicon*, **55**, 239 (1991).

111. P. Dubois, J. P. Lelieur and G. Lepoutre, *Inorg. Chem.*, **28**, 195 (1989).

112. J. Bojes and T. Chivers, *Inorg. Chem.*, **17**, 318 (1978).

113. J. Bojes, T. Chivers, W. G. Laidlaw and M. Trsic, *J. Am. Chem. Soc.*, **101**, 4517 (1979).

114. A. J. Banister, M. I. Hansford, Z. V. Hauptman, A. W. Luke, S. T. Wait, W. Clegg and K. A. Jorgensen, *J. Chem. Soc., Dalton Trans.*, 2793 (1990).

115. P. N. Jagg, P. F. Kelly, H. S. Rzepa, D. J. Williams, J. D. Woollins and W. Wylie, *J. Chem. Soc., Chem. Commun.*, 942 (1991).

116. (a) O. J. Scherer and G. Wolmershäuser, *Angew. Chem., Int. Ed. Engl.*, **14**, 485 (1975); (b) W. Flues, O. J. Scherer, J. Weiss and G. Wolmershäuser, *Angew. Chem., Int. Ed. Engl.*, **15**, 379 (1976).

117. O. J. Scherer and G. Wollmerhäuser, *Chem. Ber.*, **110**, 3241 (1977).

Chapter 6

Cyclic Chalcogen Imides

6.1 Introduction

The cyclic chalcogen imides are saturated ring systems involving two-coordinate chalcogen atoms and three-coordinate nitrogen atoms. Their formulae can be written without double bonds, but structural data indicate limited involvements of the nitrogen lone-pair electrons in π–bonding with the adjacent sulfur atoms. The best known examples of cyclic sulfur imides are eight-membered rings in which one or more of the sulfur atoms in *cyclo*-S_8 are replaced by an imido [NH or NR (R = alkyl)]. This class of compounds has an important place in the history of chalcogen–nitrogen chemistry and the developments in the field up to 1979 were covered comprehensively in the book by Heal.[1] Since that time a number of sulfur imides with ring sizes either smaller or larger than eight have been discovered (Section 6.2.2). In addition, a series of cyclic imides of the heavier chalcogens, notably selenium, have been prepared and structurally characterized (Section 6.3). This chapter will begin with a review of the salient features of the chemistry of cyclic sulfur imides. This summary will be followed by a discussion of the aforementioned recent aspects of the chemistry of cyclic chalcogen imides with an emphasis on the differences observed for the heavier chalcogen systems compared to their better known sulfur analogues.

6.2   Cyclic Sulfur Imides

6.2.1 Eight-membered rings

The formal replacement of a sulfur atom in $cyclo$-S_8 generates S_7NH (**6.1**), the first member of a series of cyclic sulfur imides that includes the three diimides 1,3-, 1,4-, and 1,5-$S_6(NH)_2$ (**6.2-6.4**), two triimides 1,3,5- and 1,3,6-$S_5(NH)_3$, and $S_4N_4H_4$ (**6.5**). Isomers of these ring systems containing adjacent NH groups are not known.

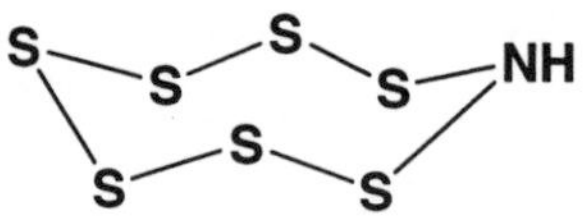

      6.1 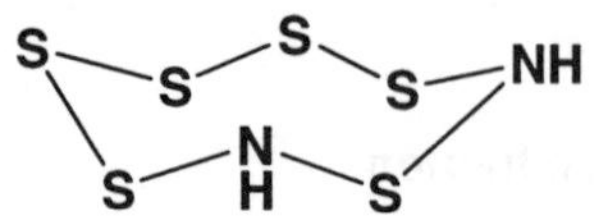   **6.2**

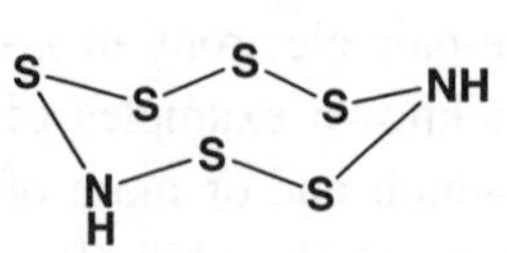

   6.3 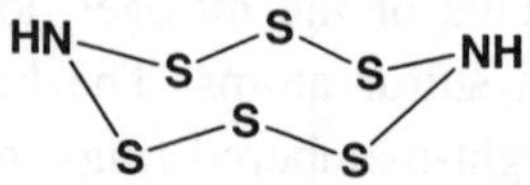   **6.4** 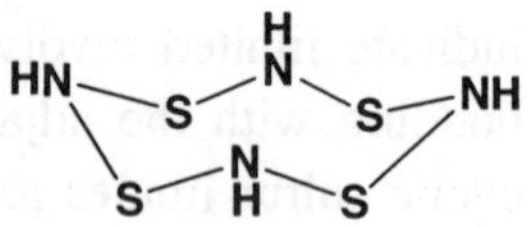   **6.5**

The standard preparation of these cyclic sulfur imides, with the exception of $S_4N_4H_4$, involves the reaction of S_2Cl_2 with gaseous ammonia in DMF at $ca.$ −10°C, followed by hydrolysis with cold dilute hydrochloric acid.[2] This method gives mainly S_7NH, but the three diimide isomers, **6.2-6.4**, and very small amounts of 1,3,6-$S_5(NH)_3$ can be separated by chromatography of CS_2 solutions on silica gel or by high-performance liquid chromatography.[3] The reaction of sodium azide with elemental sulfur in $(Me_2N)_3PO$ is an excellent source of S_7NH, and it has been employed for making the ^{15}N-enriched ring system.[4] Both of these methods for the synthesis of S_7NH proceed by the formation of the deep blue $[S_4N]^-$ anion (Section 5.4.3). The tetraimide $S_4N_4H_4$ is prepared in good yields from S_4N_4 by using methanolic $SnCl_2.2H_2O$ as the reducing agent.[5] The reduction of S_4N_4 by hydrazine is the only useful route to 1,3,5-$S_5(NH)_3$.[6]

The cyclic sulfur imides form colourless, air-stable crystals which undergo photochemical decomposition to give elemental sulfur. The pyrolysis of S_7NH in the temperature range 140-180°C produces S_8, S_4N_4, S_4N_2 and NH_3 gas.[7] The sulfur(IV) diimides RN=S=NR (R = H, Me) have been identified as one of the products of the flash vacuum pyrolysis of the tetraimides $S_4N_4R_4$ at 770-980°C by field ionization mass spectrometry.[8] The parent sulfur diimide HN=S=NH has also been generated in solution, but it is thermally unstable (Section 5.4.1).[9]

These ring systems all adopt the crown configuration of *cyclo*-S_8 with S–S bond lengths in the range 2.04-2.06 Å, typical of single bonds. The S–N bond distances (1.66-1.68 Å) are significantly shorter than a S–N single bond and the geometry around nitrogen is almost planar, indicating three-centre π-bonding in the S–N(H)–S units.[10] This conclusion is supported by experimental electron deformation density measurements of S_7NH[11] and $S_4N_4H_4$,[12] which reveal bent S–N bonds with endocyclic maxima.

Thin layer chromatography gives a quick indication of the composition of a mixture of the imides **6.1-6.4**. IR spectroscopy provides a better method for the analysis of such mixtures.[2] The IR spectra of CS_2 solutions exhibit bands in the S–N stretching region (775-840 cm^{-1}) that are characteristic of the individual imides. The vibrational assignments and force constants for the tetraimide **6.5** have been ascertained from the IR and Raman spectra of $S_4{}^{15}N_4H_4$ and $S_4{}^{15}N_4D_4$.[13] The ^{1}H and ^{15}N NMR chemical shifts of all the cyclic sulfur imides have been determined by inverse detection methods (Section 3.2).[14] The values of $^1J(^{15}N-^1H)$ fall within the narrow range 93-96 Hz, consistent with sp^2-hybridized nitrogen atoms.

The cyclic sulfur imides are weak Brønsted acids that are readily deprotonated by strong bases. In the case of S_7NH, deprotonation produces the thermally unstable, yellow $[S_7N]^-$ anion, which decomposes to the deep blue $[S_4N]^-$ anion (Eq. 6.1). The electrochemical reduction of S_7NH and $1,4$-$S_6(NH)_2$ generates the binary acyclic sulfur-nitrogen anions $[S_4N]^-$ and $[S_3N]^-$.[16] The deprotonation of $S_4N_4H_4$ with strong bases produces the acyclic $[S_2N_2H]^-$ anion and, subsequently, the six-membered ring $[S_3N_3]^-$,[17] while the exhaustive electrolysis of $S_4N_4H_4$ produces $[SN_2]^{2-}$.[16]

$$S_7NH + [Bu_4N]OH \rightarrow [Bu_4N][S_7N] \rightarrow [Bu_4N][S_4N] + \tfrac{3}{8} S_8 \quad (6.1)$$

The cyclic sulfur imides readily undergo condensation reactions in the presence of a base, *e.g.*, pyridine. For example, the reaction of S_7NH with sulfur halides, S_xCl_2 or $SOCl_2$, produces the series $(S_7N)_2S_x$ (x = 1, 2, 3, 5),[18] or $(S_7N)_2SO$,[19] respectively. The bicyclic compound $S_{11}N_2$ (**6.6**) is obtained by treatment of 1,3-$S_6(NH)_2$ with S_5Cl_2.[19] The reaction of S_7NH with phenylmercury acetate yields $PhHgNS_7$, which has been investigated as an S_7N transfer agent.[20] An interesting approach to covalent S_7N derivatives involves the synthesis of S_7NSNMe_2 by elimination of a secondary amine (Eq. 6.2).

$$S_7NH + S(NMe_2)_2 \rightarrow S_7NSNMe_2 + Me_2NH \quad (6.2)$$

The oxidation of S_7NH with trifluoroperoxyacetic acid yields $S_7NH(O)$. The S=O stretching frequency of 1110 cm^{-1} suggests that the oxidized sulfur atom is not attached to a nitrogen atom.[22] By contrast, oxidation with $SbCl_5$ in liquid sulfur dioxide results in rupture of the S_7N ring to generate the $[NS_2]^+$ cation.[23] This reaction provided the first synthesis of this important acyclic reagent (Section 5.3.2).

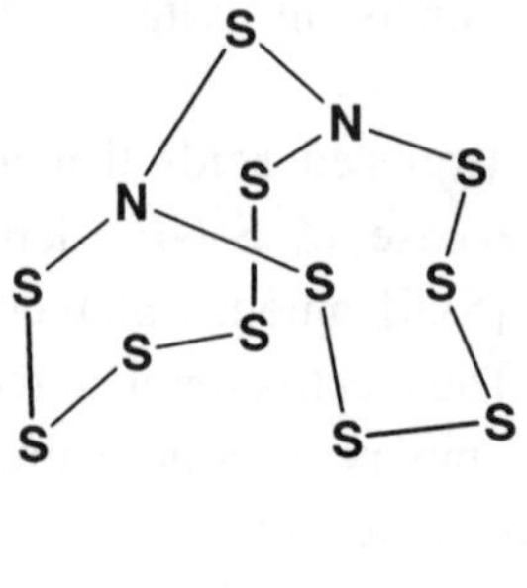

6.6 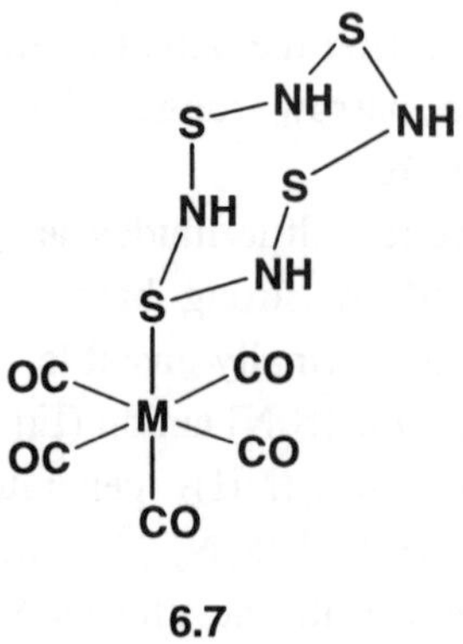   **6.7**

Two types of behaviour are observed in the interaction of cyclic sulfur imides with metal centres: (a) adduct formation and (b) oxidative addition. For example, a sandwich complex, $(S_4N_4H_4)_2 \cdot AgClO_4$, in which all four sulfur atoms of both ligands are bonded to silver has been structurally characterized.[24] A different type of ligand behaviour in which the tetraimide acts as an *S*-monodentate ligand is observed in complexes of type $(S_4N_4H_4)M(CO)_5$ (**6.7**, M = Cr, W), formed by the replacement of THF in $(THF)M(CO)_5$ by **6.5**.[25] By contrast, deprotonation and ring fragmentation occurs in the reaction of $S_4N_4H_4$ with the zerovalent platinum complex $Pt(PPh_3)_4$ to give $Pt(PPh_3)_2(S_2N_2)$ in which the $[S_2N_2]^{2-}$ ligand is coordinated to the platinum (II) centre (Section 7.3.2).[26]

Cyclic sulfur imides containing S–S bonds undergo oxidative addition reactions with $Cp_2Ti(CO)_2$. In the reaction with eight-membered rings S_7NR (R = H, Me) the metal inserts into a S–S bond to give the complexes $Cp_2Ti(S_7NR)$ (Eq. 6.3).[27] Surprisingly, the insertion into S_7NH involves a different S–S bond than the corresponding insertion for S_7NMe. Thus the parent compound (**6.8**, R = H) contains a bridging disulfido and pentasulfido units, whereas the methyl derivative (**6.9**, R=Me) involves bridging trisulfido and tetrasulfido fragments. A different transformation occurs in the reaction of $Cp_2Ti(CO)_2$ with the six-membered cyclic sulfur imides of the type $1,4-S_4(NR)_2$ (R = Me, Oct).[28] In this case the oxidative addition results in the replacement of one of the NR groups by a Cp_2Ti fragment to give the complexes Cp_2TiS_4NR (**6.10**), which have a chair conformation.

$$Cp_2Ti(CO)_2 \ + \ S_7NR \ \rightarrow \ Cp_2Ti(S_7NR) \ + \ 2CO \qquad (6.3)$$

6.8   **6.9**   **6.10**

6.2.2 Six-, seven-, nine- and ten-membered rings

Sulfur imides that are derived from the unstable cyclic sulfur allotropes S_6, S_7, S_9 and S_{10} have been structurally characterized. The best established examples are diimides of the type $S_4(NR)_2$ (**6.11**, R = Et, Bz, Cy, CH_2CH_2Ph), which are prepared in moderate yields (15-35%) from the reactions of the appropriate primary amine with S_2Cl_2 in diethyl ether under high dilution conditions. Other amines, *e.g.*, $^tBuNH_2$ and $PhNH_2$, give only polymeric materials. In contrast to the eight-membered rings **6.1-6.5**, the structural data for the six-membered ring systems **6.11** (R = Et, Bz) indicate no significant π-bonding in the S–N bonds.[29] These derivatives adopt chair conformations with S–N bond lengths that are *ca.* 0.05 Å longer than those in **6.1-6.5**. In addition, the bond angles at the nitrogen reveal a pyramidal (sp^3) geometry.

Sulfur imides with a single NR functionality, S_5NR (**6.12**),[28] S_6NR (**6.13**) (R = Oct),[28] S_8NH (**6.14**),[30] and S_9NH (**6.15**)[30] are obtained by a methodology similar to that which has been used for the preparation of unstable sulfur allotropes, *e.g.*, S_9 and S_{10}.[31] For example, the metathesis reaction between the bis(cyclopentadienyl)titanium complexes **6.8-6.10** and the appropriate dichlorosulfane yields **6.14** and **6.15** (Eq. 6.4).[30]

$$[Cp_2Ti(S_7NH)] \; + \; S_xCl_2 \; \rightarrow \; Cp_2TiCl_2 \; + \; S_{7+x}NH \quad\quad (6.4)$$

$$(x = 1, 2)$$

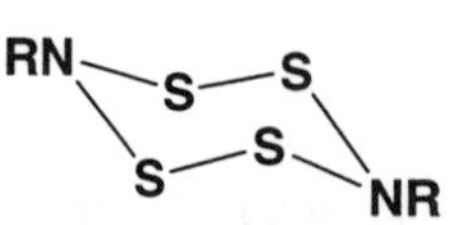

6.11 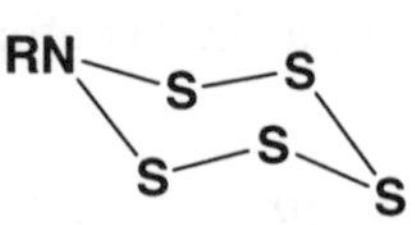   6.12 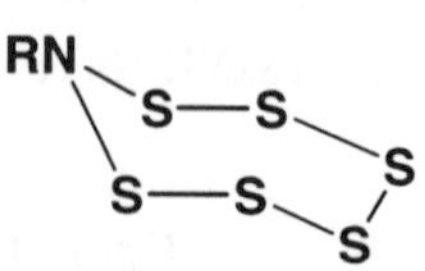   6.13

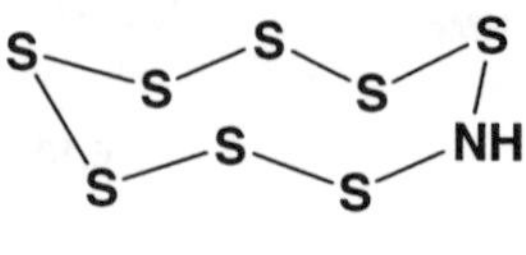

6.14 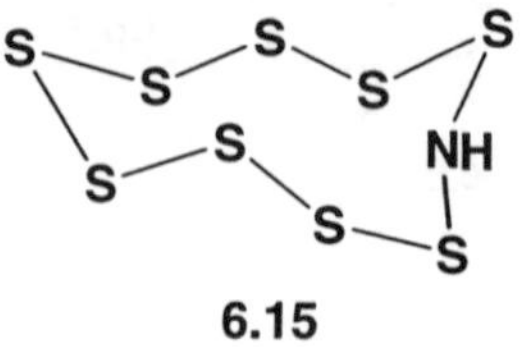   6.15

The six- and seven-membered rings **6.12** and **6.13** are obtained as yellow oils from the reactions of **6.10** with S_xCl_2 (x = 1, 2).[28] They exhibit molecular ions in their electron-impact mass spectra. The nine- and ten-membered rings **6.14** and **6.15** are obtained as pale yellow crystals that are soluble in CS_2.[30] They undergo photochemical decomposition in daylight after several days, but can be stored in the dark at 22°C for weeks. The nine-membered ring **6.14** shows a sequence of torsional angles similar to that found for *cyclo*-S_9. On the other hand, the motif of torsional angles for the ten-membered ring **6.15** resembles the crown structure of *cyclo*-S_8 with the insertion of an SN(H) unit rather than that of *cyclo*-S_{10}. The S–N bond distances of *ca.* 1.67 Å and the planarity at the nitrogen atoms in **6.14** and **6.15** implies the presence of π-bonding, as in **6.1-6.5**.

The twelve-membered ring $S_{11}NH$ is obtained via the reaction shown in Eq. 6.5. It was characterized by spectroscopic methods (IR, Raman and mass spectra). It is thermally more stable than **6.14** and **6.15**, but decomposes to S_7NH and S_8 after several weeks at 25°C.[30]

$$Cp_2Ti(S_7NH) + S_4(SCN)_2 \rightarrow S_{11}NH + Cp_2Ti(NCS)_2 \quad (6.5)$$

6.3 Cyclic Selenium and Tellurium Imides

The discovery of cyclic imides of the heavier chalcogens is a relatively recent development. The ring systems that have been structurally characterized include examples of five-, six-, eight- and fifteen-membered rings, all of which have bulky substituents attached to the nitrogen atoms: $Se_3(NR)_2$ (**6.16**, R = ^tBu, Ad),[32, 33] $Se_3(N^tBu)_3$ (**6.17**, E = Se),[34] $Se_6(N^tBu)_2$ (**6.18**)[35] and $Se_9(N^tBu)_6$ (**6.19**).[35] None of these ring systems have analogues among the cyclic sulfur imides. The only known cyclic tellurium imide is $Te_3(N^tBu)_3$ (**6.17**, E = Te), which is formed as a by-product in the synthesis of the tellurium(IV) diimide $^tBuNTe(\mu-N^tBu)_2TeN^tBu$ from the reaction of $TeCl_4$ with $LiNH^tBu$ when the reaction is carried out in toluene (Section 10.4).[36]

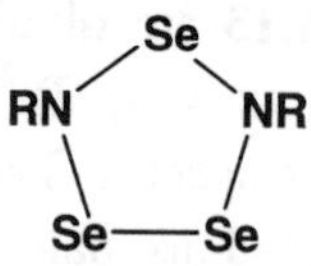

6.16

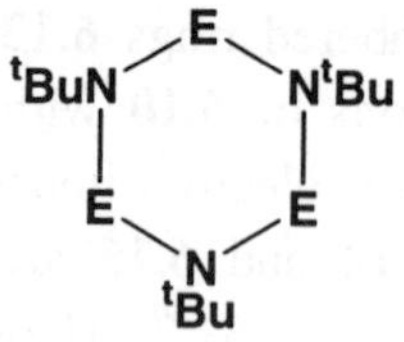

6.17

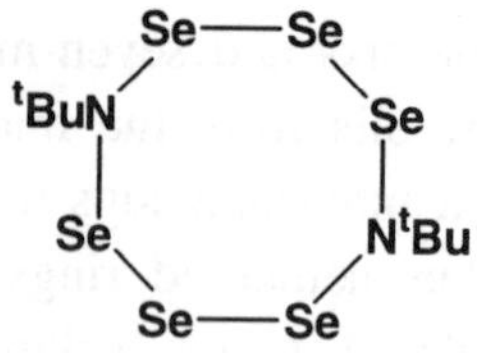

6.18

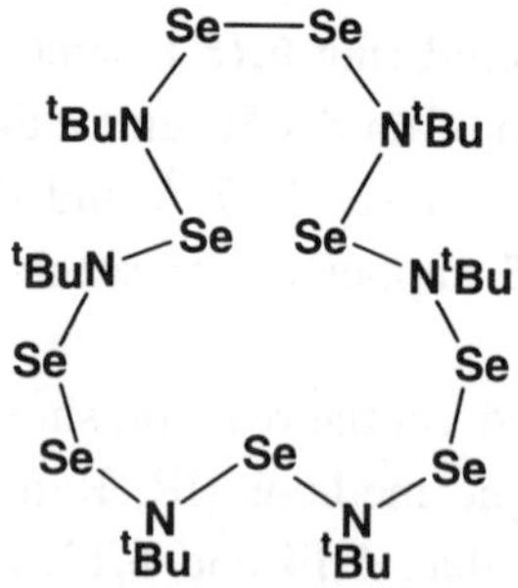

6.19

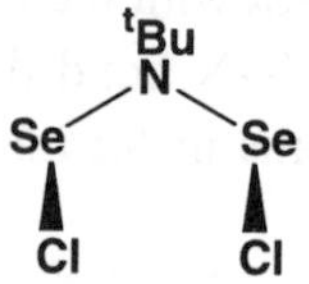

6.20

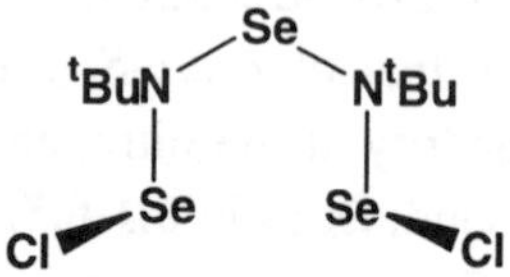

6.21

The first cyclic selenium imides, **6.18** and **6.19**, were obtained in low yields by the reaction of LiN(^{t}Bu)SiMe$_3$ with Se$_2$Cl$_2$ or SeOCl$_2$ in 1984.[35] The fifteen-membered ring **6.19** is the predominant product. The reaction of *tert*-butylamine with SeCl$_2$ in THF provides a source of the smaller ring systems **6.16** and **6.17**, in addition to **6.18** and **6.19**. The composition of the mixture of Se–N compounds obtained in this reaction is markedly dependent on the stoichiometry. When the reaction is carried out in a 2:3 stochiometry the acyclic imidoselenium(II) halides **6.20** and **6.21** are the major products.[34] The bifunctional reagent **6.21** can be viewed as a precursor to the six-membered ring **6.17** via reaction with *tert*-butylamine. When the SeCl$_2$/^{t}BuNH$_2$ reaction is carried out in a 1:3 molar ratio, the small ring systems **6.16** and **6.17** are formed in significant yields (in addition to **6.18** and **6.19**) and they can be isolated by fractional crystallization.[33] However, a better preparation of **6.17** involves the decomposition of the selenium(IV) diimide ^{t}BuN=Se=N^{t}Bu in toluene at 20°C; this decomposition also generates the five- and fifteen-membered rings, **6.15** and **6.19**, respectively.[33]

The ^{77}Se NMR spectra provide an informative analysis of the components of the reaction mixtures in these syntheses of cyclic selenium imides (Section 3.3). The five- and fifteen-membered rings **6.16** and **6.19**, show two resonances with relative intensities of 2:1 corresponding to the diselenido and monoselenido bridging units, respectively. The eight-membered ring **6.18** also exhibits two resonances attributable to the two different selenium environments. The characteristic chemical shifts for the different selenium environments in **6.16-6.19** show a marked upfield shift as the electronegativity of the neighbouring atoms decreases: δ 1620-1725 (NSeCl), 1400-1625 (NSeN), 1100-1200(NSeSe), and 500-600 (SeSeSe).^{33}

The cyclic chalcogen imides **6.16** (R = Ad),^{32} **6.17** (E = Se,^{34} Te^{36}), **6.18**^{35} and **6.19**^{35} have been structurally characterized by X-ray crystallography. The metrical parameters for the puckered five-membered ring **6.16** are significantly different from those of the larger rings as a result of ring strain. Thus the Se–Se distance in **6.16** is *ca.* 0.07 Å longer than the mean Se–Se bond distance in **6.18**, which has typical Se–Se single bond values of 2.33 Å (*cf.* 2.34 Å in *cyclo*-Se$_8$).^{37} The mean Se–N distance in the N–Se–N unit of **6.16** is *ca.* 0.06 Å longer than the corresponding bonds in the six-membered ring **6.17** (E = Se), which adopts a chair conformation with Se–N bond distances of 1.83 Å indicative of single bonds. The tellurium analogue **6.17** (E = Te) has a similar chair conformation with Te–N distances of 2.03 Å, also consistent with single bonds. Another structural difference between **6.16** and the other cyclic selenium imides involves the geometry at the nitrogen atoms, which is distinctly pyramidal (Σ<N = 343-344°) in **6.16** compared to the almost planar configurations in **6.17** and **6.18**. The eight-membered ring **6.18** crystallizes in a crown conformation similar to that of *cyclo*-Se$_8$.

The chemistry of this interesting series of cyclic selenium imides, *e.g.*, ligand behaviour, has not been investigated.

References

1. H. G. Heal, *The Inorganic Heterocyclic Chemistry of Sulfur, Nitrogen and Phosphorus*, Academic Press, London (1980).

2. H. G. Heal and J. Kane, *Inorg. Synth.*, **11**, 184 (1968).

3. R. Steudel and F. Rose, *Journal of Chromatography*, **216**, 399 (1981).

4. J. Bojes, T. Chivers and I. Drummond, *Inorg. Synth.*, **18**, 203 (1978).

5. G. Brauer, *Handbook of Preparative Inorganic Chemistry*, 2nd Ed., Academic Press, New York, Vol. 1, p.411 (1963).

6. H. Garcia-Fernandez, *Bull. Soc. Chim. Fr.*, 1210 (1973).

7. Y. Kudo and S. Hamada, *Bull. Chem. Soc. Jpn.*, **60**, 2391 (1987).

8. L. Carlsen, H. Egsgaard and S. Elbel, *Sulfur Letters*, **3**, 87 (1985).

9. M. Herberhold, W. Jellen, W. Bühlmeyer, W. Ehrenreich and J. Reiner, *Z. Naturforsch.*, **40B**, 1229 (1985).

10. H-J. Hecht, R. Reinhardt, R. Steudel and H. Bradacek, *Z. Anorg. Allg. Chem.*, **426**, 43 (1976).

11. C-C. Wang, Y-Y. Hong, C-H. Ueng and Y. Wang, *J. Chem. Soc., Dalton Trans.*, 3331 (1992).

12. D. Gregson, G. Klebe and H. Fuess, *J. Am. Chem. Soc.*, **110**, 8488 (1988).

13. R. Steudel, *J. Mol. Struct.*, **87**, 97 (1982).

14. T. Chivers, M. Edwards, D. D. McIntyre, K. J. Schmidt and H. J. Vogel, *Mag. Reson. Chem.*, **30**, 177 (1992).

15. T. Chivers and I. Drummond, *Inorg. Chem.*, **13**, 1222 (1974).

16. T. Chivers and M. Hojo, *Inorg. Chem.*, **23**, 2738 (1984).

17. T. Chivers and K. J. Schmidt, *J. Chem. Soc., Chem. Comm.*, 1342 (1990).

18. H. Garcia-Fernandez, H. G. Heal and G. Teste de Sagey, *C. R. Acad. Sc. Paris*, **C282**, 241 (1976).

19. R. Steudel and F. Rose, *Z. Naturforsch.*, **30B**, 810 (1975).

20. R. J. Ramsay, H. G. Heal and H. Garcia-Fernandez, *J. Chem. Soc., Dalton Trans.*, 237 (1976).

21. R. B. Bruce, R. J. Gillespie and D. R. Slim, *Can. J. Chem.*, **56**, 2927 (1978).

22. R. Steudel and F. Rose, *Z. Naturforsch.*, **33B**, 122 (1978).

23. R. Faggiani, R. J. Gillespie, C. J. L. Lock and J. D. Tyrer, *Inorg. Chem.*, **17**, 2975 (1978).

24. M. B. Hursthouse, K. M. A. Malik and S. N. Nabi, *J. Chem. Soc., Dalton Trans.*, 355 (1980).

25. G. Schmid, R. Greese and R. Boese, *Z. Naturforsch.*, **37B**, 620 (1982).

26. T. Chivers, F. Edelmann, U. Behrens and R. Drews, *Inorg. Chim. Acta*, **116**, 145 (1986).

27. K. Bergemann, M. Kustos. P. Krüger and R. Steudel, *Angew. Chem., Int. Ed. Engl.*, **34**, 1330 (1995).

28. R. Steudel, O. Schumann, J. Buschmann and P. Luger, *Angew. Chem., Int. Ed. Engl.*, **37**, 492 (1998).

29. R. Jones, D. J. Williams and J. D. Woollins, *Angew. Chem., Int. Ed. Engl.*, **24**, 760 (1985).

30. R. Steudel, K. Bergemann, J. Buschmann and P. Luger, *Angew. Chem., Int. Ed. Engl.*, **35**, 2537 (1996).

31. R. Steudel, *Topics Current Chem.*, **102**, 149 (1982).

32. T. Maaninen, H. M. Tuononen, G. Schatte, R. Suontamo, J. Valkonen, R. Laitinen and T. Chivers, *Inorg. Chem.*, **43**, 2097 (2004).

33. T. Maaninen, T. Chivers, R. Laitinen, G. Schatte and M. Nissinen, *Inorg. Chem.*, **39**, 5341 (2000).

34. T. Maaninen, T. Chivers, R. Laitinen and E. Wegelius, *Chem. Commun.*, 759 (2000).

35. H. W. Roesky, K-L. Weber and J. W. Bats, *Chem. Ber.*, **117**, 2686 (1984).

36. T. Chivers, X. Gao and M. Parvez, *J. Am. Chem. Soc.*, **117**, 2359 (1995).

37. A. Maaninen, J. Konu, R. S. Laitinen, T. Chivers, G. Schatte, J. Pietikäinen and M. Ahlgrén, *Inorg. Chem.*, **40**, 3529 (2001).

Chapter 7

Metal Complexes

7.1 Introduction

Investigations of the coordination chemistry of simple sulfur–nitrogen ligands have produced a wide variety of complexes.[1,2] The first metal complexes, $MoCl_5.S_4N_4$ and $TiCl_4.S_4N_4$ were described more than 100 years ago. Although the topic enjoyed some popularity in the 1950s, primarily due to the pioneering work of Becke-Goehring and co-workers, progress was slow. The advent of NMR spectroscopy in the 1960s had little impact on the subject. However, in recent years X-ray crystallography has played a important role in providing key information about the bonding modes and structural parameters of metal complexes. Several aspects of the coordination chemistry of chalcogen–nitrogen ligands warrant attention: (a) the ability of metals to stabilize unstable neutral and anionic binary S–N ligands, (b) the applications of metal complexes as reagents for the preparation of other S–N compounds, (c) comparison of the behaviour of of S–N ligands with valence isoelectronic ligands, $e.g.$, NO, NO_3^-, and (d) the possible incorporation of metals into sulfur-nitrogen chains to produce conducting materials.

This chapter will begin with a discussion of metal complexes of binary sulfur nitrides. The ring systems S_2N_2 and S_4N_4 form typical Lewis acid–Lewis base adducts in which the metal is bonded to the ligand through a nitrogen atom and these have been described in Sections 5.2.4 and 5.2.6, respectively. Consequently, the first section will focus on complexes of the thionitrosyl (NS) ligand. This will be followed by a discussion of complexes formed with binary anionic S–N ligands in which the metal is part of an S–N ring system. These complexes, known as cyclometallathiazenes, are represented by the general formula MS_xN_y. The next, short section recounts metal complexes of the monomeric sulfur imide (SNH) ligand, which bonds in a π–fashion to metal centres. The final sections compare the behaviour of monomeric thiazyl halides, NSX (X = F, Cl, Br) and the valence isoelectronic thionyl imide [NSO]⁻ and sulfur diimide anions, [NSNR]⁻ and [NSN]²⁻, as ligands in metal complexes. Although the vast majority of metal–chalcogen–nitrogen complexes involve sulfur as the chalcogen site in the ligand, there is some information about analogous selenium- and tellurium-containing complexes that will be included in the appropriate sections.[3,4] A discussion of metal complexes of C-N-E (E = S, Se) heterocycles is included in Sections 11.3.1 and 12.4.1; P-N-E (E = S, Se) ligands are considered in Section 13.2.1. The coordination chemistry of the chalcogen diimides RN=S=NR and $^tBuNTe(\mu-N^tBu)_2TeN^tBu$, is covered in Section 10.4.3.

7.2 Thionitrosyl and Selenonitrosyl Complexes

Unlike nitric oxide, NO, the monomeric radical sulfur nitride, NS, is only known as a short-lived intermediate in the gas phase. Nevertheless the properties of this important diatomic molecule have been thoroughly investigated by a variety of spectroscopic and other physical techniques (Section 5.2.1). The NS molecule is stabilized by coordination to a transition metal and a large number of complexes, primarily with metals from Groups 6, 7, 8 and 9, are known. Several detailed reviews of the topic have been published.[5-7]

A variety of routes is available for the preparation of metal–thionitrosyl complexes. The most common of these are (a) reaction of nitride complexes with a sulfur source, *e.g.*, elemental sulfur, propylene sulfide or sulfur halides, (b) reaction of $(NSCl)_3$ with transition-metal complexes, and (c) reaction of [SN]⁺ salts with transition-metal complexes. An example of each of these approaches is given in Eq. 7.1, 7.2 and 7.3, respectively. The second method was employed to generate the first thionitrosyl complex.[8] Halide abstraction from complexes of thiazyl halides has also been used in a few cases. Surprisingly, the use of

S_4N_4 as a source thionitrosyl-metal complexes is limited to a single example. The first selenonitrosyl-metal complex $TpOs(NSe)Cl_2$ (Tp = hydrotris(1-pyrazolyl)borate) was prepared recently by method (a).[9]

$$MoN(S_2CNR_2)_3 + \tfrac{1}{8}S_8 \rightarrow Mo(NS)(S_2CNR_2)_3 \qquad (7.1)$$

$$(R = Me, Et)$$

$$Na[Cr(CO)_3Cp] + \tfrac{1}{3}(NSCl)_3 \rightarrow Cr(CO)_2Cp(NS) + NaCl + CO \quad (7.2)$$

$$[Re(CO)_5SO_2][AsF_6] + [SN][AsF_6] \rightarrow$$
$$[Re(CO)_5NS][AsF_6]_2 + SO_2 \qquad (7.3)$$

The thionitrosyl ligand is generally coordinated as a terminal, linear unit in which the bond angle <MNS is in the range 169-180°.[6] The only known selenonitrosyl–metal complex has a bond angle <MNSe of 165°.[9] In contrast to transition-metal nitrosyl complexes, for which both bent and bridging NO ligands are well known, these bonding modes are uncommon for the NS ligand. Complexes containing more than one thionitrosyl ligand, *e.g.*, *cis*-$MCl_4(NS)_2$ (M = Ru, Os), are also rare.[10] The characteristic NS stretching vibration in the infrared spectrum is lowered from 1437 cm^{-1} in the cation [SN]$^+$ to somewhere in the range 1050-1400 cm^{-1} in metal complexes (*cf.* 1205 cm^{-1} in neutral NS). The charge of the complex has a marked influence on the metal–NS back donation as indicated by the trend observed for the isoelectronic half-sandwich complexes of the type [CpM(CO)$_2$MS]$^{n+}$ (Table 7.1).[11]

Table 7.1 IR stretching frequencies for [CpM(CO)$_2$]$^{n+}$

	$\nu(NS)[cm^{-1}]$
CpCr(CO)$_2$(NS)	1154
[CpMn(CO)$_2$(NS)][AsF$_6$]	1284
[CpFe(CO)$_2$(NS)][AsF$_6$]$_2$	1388

The nitrogen atom in the (almost) linear metal–thionitrosyl complexes is *sp*-hybridized and the NS ligand behaves as a three-electron donor. The N–S bond distances in metal complexes vary between 1.45

and 1.59 Å (*cf.* 1.44 and 1.495 Å for $[SN]^+$ and NS, respectively). The metal–nitrogen distances are short indicating that, in addition to **A**, the resonance form **B** is an important contributor to the overall bonding in these complexes. The N–Se bond length of 1.63 Å in $TpOs(NSe)Cl_2$ is slightly shorter that that in gas phase NSe (1.65 Å).[9]

$$M–N^+\equiv S^- \quad \leftrightarrow \quad M^-=N^+=S$$
$$\textbf{A} \qquad\qquad \textbf{B}$$

Comparison of the photoelectron spectra and electronic structures of M-NS and M-NO complexes, *e.g.*, $[CpCr(CO)_2(NX)]$ (X = S, O), indicates that NS is a better σ-donor and a stronger π-acceptor ligand than NO.[12] This conclusion is supported by ^{14}N and ^{95}Mo NMR data, and by the UV-visible spectra of molybdenum complexes.[13]

A few reactions of metal complexes that involve the NS ligand rather than the metal centre have been reported. For example, the reaction of the dication $[Re(CO)_5(NS)]^{2+}$ with cesium halides CsX (X = Cl, Br) converts the NS ligand to a thiazyl halide NSX.[14] Oxygen transfer from an NO_2 to an NS ligand on the same metal centre to give a thiazate (NSO) complex occurs in ruthenium porphyrin complexes.[15] Treatment of an osmium thionitrosyl complex with triphenylphosphine converts the NS ligand to an $NPPh_3$ group with the formation of Ph_3PS (Eq. 7.4).[16] Mechanistic studies of this transformation indicate that it occurs via the intermediate formation of an $NSPPh_3$ ligand, which reacts rapidly with a second equivalent of PPh_3. The NSe ligand is deselenized almost instantly at room temperature by PPh_3.[9]

$$[Os(tpm)Cl_2(NS)]^+ + 2PPh_3 \rightarrow [Os(tpm)Cl_2(NPPh_3)]^+ + SPPh_3 \quad (7.4)$$

(tpm = tris(pyrazol-1-yl)methane)

7.3 Cyclometallathiazenes

7.3.1 *Introduction*

The term cyclometallathiazene refers to ring systems that contain only sulfur, nitrogen and one (or more) metal atoms. As indicated in Table 7.2,

Table 7.2   Cyclometallathiazenes

Corresponding S-N anion	Metal	Year of Discovery (first X-ray structure)
 $S_2N_2^{2-}$	Co, Ir, Ni, Pd, Pt (Pb)	1982 (1982)
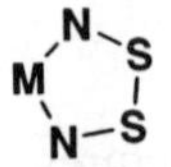 $S_2N_2^{4-}$	Re	1984 (1984)
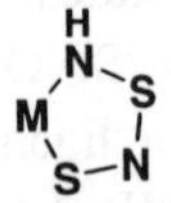 $[S_2N_2H]^-$	Co, Ni, Pd, Pt	1953 (1958)
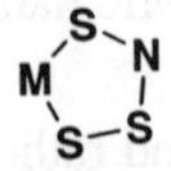 S_3N^-	Fe, Ru, Co, Rh, Ni, Pd, Pt, Cu, Ag, Au	1958 (1966)
 $S_2N_3^-$	Pd	1992 (1992)
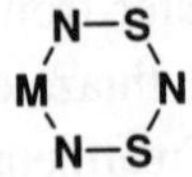 $S_2N_3^{3-}$	V, Mo, W	1983 (1983)
 $S_3N_4^{2-}$	Ti	1983 (1983)
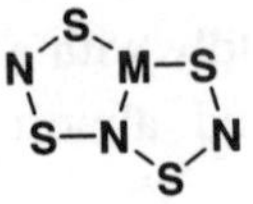 $S_4N_3^-$	Pt	1981 (1981)
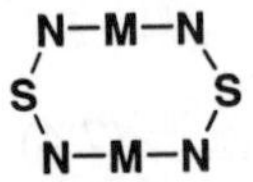 SN_2^{2-}	Hg, Zr, Hf	1904 (1984)
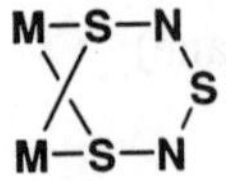 $S_3N_2^{2-}$	Pd	1966 (1982)
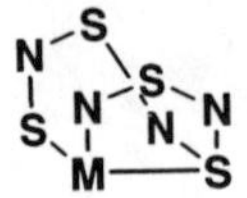 $S_4N_4^{2-}$	Ir, Pt	1970 (1986)

these heterocycles can formally be considered as complexes between cationic metal fragments and binary S–N anions in which, with the exception of the $[S_2N_2H]^-$ and $[S_4N_4]^{2-}$ ions, both the sulfur and nitrogen atoms are two-coordinate. Table 1 indicates which metals are known to form complexes with individual S–N anions, as well as the year of discovery and first X-ray structure for each type of cyclometallathiazene. The unknown selenium–nitrogen anions $[Se_3N]^-$, $[Se_2N_2]^{2-}$ and $[Se_2N_2H]^-$ have also been stabilized by coordination to platinum.[17-19] A few examples of complexes involving anionic ligands with two different chalcogens, *e.g.*, $[ESN_2]^{2-}$ and $[ESN_2H]^-$ (E = Se, Te), are known.[20, 21]

Since only two of these chalcogen-nitrogen anions, $[S_3N]^-$ (Section 5.4.2) and $[SN_2]^{2-}$ (Section 5.4.1), have been isolated as ionic salts, the synthesis of cyclometallathiazenes usually employs S_4N_4 (or other S–N ring systems) as a source of the S–N fragment. Some representative examples of these redox transformations are shown in Scheme 7.1. The explosive regent Se_4N_4 has also been used to generate complexes of the $[Se_3N]^-$ and $[Se_2N_2H]^-$ anions.[17, 19] A safer route for the *in situ* generation of the anions $[Se_2N_2]^{2-}$ and $[Se_2N_2H]^-$ involves the use of $SeCl_4$ or, preferably, $SeOCl_2$, in liquid ammonia.[18]

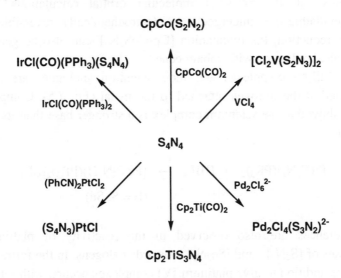

Scheme 7.1 Preparation of cyclometallathiazenes from S_4N_4

7.3.2 Complexes of the $[E_2N_2H]^-$ and $[E_2N_2]^{2-}$ anions (E = S, Se)

Complexes containing the $[S_2N_2H]^-$ ion comprise the best studied group of cyclometallathiazenes. The dark green nickel(II) derivative is readily obtained in good yields from the reaction of S_4N_4 with anhydrous $NiCl_2$ in methanol.[22] The individual MS_2N_2H rings in this spirocyclic complex are planar. The platinum(II) complexes $[Pt(S_2N_2H)(PR_3)]X$ (X = halide) (and their selenium analogues) are especially interesting since they form continuous stacking structures.[23] The NH groups in $M(S_2N_2H)_2$ complexes are readily deprotonated to give mono- or di-anions,[24a] which can be converted to organic derivatives by treatment with alkyl halides.[2] An interesting application of these dianionic metal complexes involves the preparation of the trinuclear complex $[Ni_3(N_2S_2)_4]^{2-}$ (**7.1**) by the reaction of two equivalents of $[Ni(N_2S_2)_2]^{2-}$ with $NiCl_2$.[24b]

In addition to S_4N_4, a variety of S-N reagents, including $[S_3N_3]^-$, $S_4N_4H_4$/diazobicycloundecene, $[R_2Sn(S_2N_2)]_2$ (R = Me, ^{n}Bu), and solutions of $[S_4N_3]Cl$ in liquid ammonia, provide a source of $[S_2N_2]^{2-}$ for metal complexes. Although X-ray structural data for MS_2N_2 complexes indicate a partial localization of π-bonding within the –N=S=N– unit, delocalization over the entire ring system has been inferred for $Cp_2CoS_2N_2$ on the basis of molecular orbital calculations.[25] This cyclometallathiazene undergoes an electrochemically reversible one-electron reduction; the monoanion $[Cp_2CoS_2N_2]^-$ can also be generated by chemical reduction with cobaltocene.[26]

The MS_2N_2 complexes and their selenium analogues are readily protonated at the nitrogen attached to the metal (Eq. 7.5). Competitive studies show that the selenium complex is a stronger base than its sulfur analogue.

$$Pt(E_2N_2)(PR_3)_2 \; + \; HBF_4 \; \rightarrow \; [Pt(E_2N_2H)(PR_3)_2]BF_4 \qquad (7.5)$$

$$(E = S, Se)$$

Differences are also observed in the reactivity of platinum(II) complexes of $[S_2N_2]^{2-}$ and $[Se_2N_2]^{2-}$ towards halogens. In the former case oxidative addition to give platinum(IV) complexes occurs, with retention of the S–N ligand. By contrast, the Se–N ligand is removed in the form of Se_4N_4.

7.3.3 Complexes of the [E₃N]⁻ anion (E = S, Se)

Complexes of the $[S_3N]^-$ anion with a wide variety of transition metals are known (see Table 7.2).[6] Although S_4N_4 is not an efficient reagent for the production of these complexes, the first representative, $Ni(S_3N)_2$ was obtained in a minuscule yield (9 mg, 0.04%) from the reaction of $NiCl_2$ and S_4N_4 in methanol and correctly identified on the basis of analytical data.[27] Such complexes are now obtained in good yields via deprotonation of *cyclo*-S_7NH in the presence of metal halides or by using mercury(II) complexes $Hg(S_7N)_2$ or $PhHg(S_7N)_2$ as a source of $[S_3N]^-$. The $[SSNS]^-$ ligand is invariably chelated to the metal via the two terminal sulfur atoms.[28] The formation of the trigonal bipyramidal anion $\{[Ni(S_3N)]_3S_2\}^-$ (**7.2**) by the former process is of particular interest.[29] In the anionic copper(I) complex $[Cu(S_3N)(S_7N)]^-$ (**7.3**) an S_7N ring is preserved and coordinated to the metal through the nitrogen atom.[30]

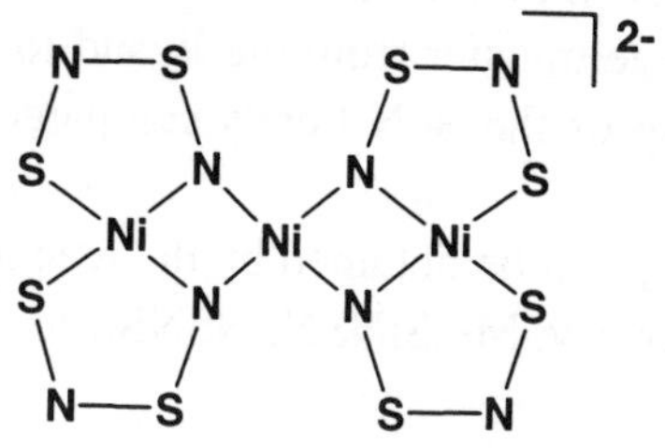

7.1

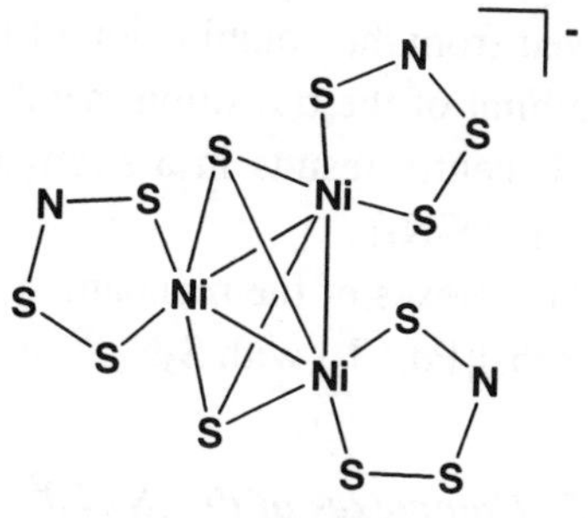

7.2

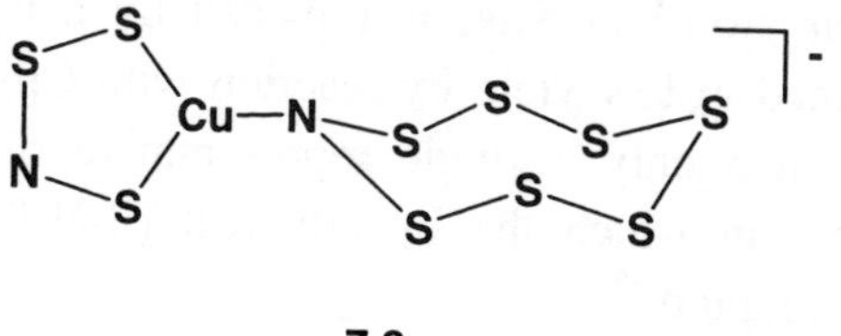

7.3

Cyclic selenium imides, *e.g.*, Se_7NH, are not available as a source of Se–N anions (Section 6.3). The only example of a metal complex of the $[Se_3N]^-$ anion, $Pt(Se_3N)Cl(PMe_2Ph)$, was obtained by the reaction of Se_4N_4 with $[PtCl_2(PMe_2Ph)]_2$.[17]

7.3.4 Complexes of the $[S_2N_3]^{3-}$ anion

The trianionic chelating ligand $[S_2N_3]^{3-}$ stabilizes early transition metals in high oxidation states. The known complexes are limited to V(V), Mo(VI) and W(VI). The MS_2N_3 ring, *e.g.*, $[VCl_2(S_2N_3)]_\infty$,[31] is generated in reactions of high oxidation state early transition-metal halides with S_4N_4, $(NSCl)_3$ or $[S_3N_2Cl]Cl$. The thiazene segment of the ring is usually planar but, in some cases, the metal is tilted out of this plane. The electronic structures of six-membered cyclometallathiazenes (MS_2N_3) have been investigated by density functional theory calculations.[32] For anions of the type $[Cl_4MS_2N_3]^-$ (M = Mo, W), in which a sulfur atom in $[S_3N_3]^-$ is formally replaced by the MCl_4 group, the HOMO-LUMO gap is predicted to be small. The closed shell (singlet) state is preferred over the open shell (triplet) state by only *ca.* 7 kcal mol^{-1}. The HOMO is derived from the combination of an S–N antibonding (π^*) orbital and the d_{xy} orbital of the transition metal. Charge transfer from the ligand to the M(VI) centre results in a strengthening of the S–N bonds compared to those in $[S_3N_3]^-$.

Complexes of the monoanion $[S_2N_3]^-$ can be obtained by the reactions of $[PPh_4][Pd_2Cl_6]$ with S_5N_6[33a] or, preferably, $Me_3SiNSNSNSNSiMe_3$.[33b]

7.3.5 Complexes of the $[S_3N_4]^{2-}$, $[S_4N_3]^-$ and $[S_4N_4]^{2-}$ anions

The only example of the eight-membered MS_3N_4 ring system, in which one of the sulfur atoms in S_4N_4 is replaced by a transition metal, is $Cp_2TiS_3N_4$ obtained in low yield by reaction with $Cp_2Ti(CO)_2$ (Scheme 7.1).[34] There is also only a single representative of the MS_4N_3 ring system, $ClPtS_4N_3$, in which the hypothetical $[S_4N_3]^-$ ligand acts as a tridentate (S,N,S') ligand.[35]

Coordination of the dianion $[S_4N_4]^{2-}$ as a tridentate ligand has been established in a few cases. The complexes $IrCl(CO)(S_4N_4)PPh_3$[36] and

$[PtCl_3(S_4N_4)]^-$ [37] contain this ligand *facially* bonded to the metal through two sulfur and one nitrogen atoms (Table 1). By contrast, the same ligand adopts a *meridional* (*S,S,N*) geometry in the platinum(IV) complex $PtCl_2(S_4N_4)(PMe_2Ph)$ which, according to ^{15}N NMR investigations is converted to a *facial S,N,N* isomer in boiling chloroform.[38]

7.4 Complexes of SNH and the [SN]⁻ Anion

Salts of the [SN]⁻ anion, isolectronic with sulfur monoxide SO, are unknown and the monomeric sulfur imide SNH is unstable. These simple sulfur–nitrogen ligands can be stabilized, however, by coordination in a π-fashion to transition metals The complex $Fe_2(CO)_6(HNS)$ (**7.4**) is obtained from the reaction of $Fe_3(CO)_{12}$ with $Me_3SiNSNSiMe_3$, followed by chromatography on silica, which converts a $N-SiMe_3$ linkage to an NH group.[39] Complex **7.4** is readily converted to the corresponding anion **7.5** with ^nBuLi. This anion reacts with trialkyloxonium salts to form alkyl derivatives[40a] and with a variety of main group element halides to form E–N bonds (E = P, As, Si, Ge, Sn and B).[40b]

7.4　　　　　　　**7.5**

The HNS molecule behaves as six-electron ligand. It can be methylated with diazomethane to give $Fe_2(CO)_6(MeNS)$. Similar tetrahedrane complexes with alkyl or aryl substituents attached to nitrogen are known for ruthenium as well as iron, *i.e.*, $M_2(CO)_6(RNS)$ (M = Ru, Fe; R = alkyl, aryl). The N–S bond length in $Ru_2(CO)_6(^tBuNS)$ is 1.72 Å suggesting an essentially single bond that is arranged perpendicular to the Ru–Ru axis.[41] Sulfur imide, HNS, is the parent compound of organic thionitroso compounds which are, in general, unstable (Section 10.2). However, *N*-thionitrosamines such as Me_2NNS

have been synthesized, and several complexes of this ligand with metals, *e.g.*, Cr(CO)$_5$(SNNMe$_2$), have been structurally characterized.[42] In such complexes the Me$_2$NNS ligand is always end-on coordinated through the sulfur atom.

7.5   Complexes of Thiazyl Halides NSX (X = F, Cl, Br), NSF$_3$ and the [NSF$_2$]$^-$ Anion

Monomeric thiazyl halides NSX (X = F, Cl Br) have been characterized in the gas phase, but oligomerization to cyclic species, *e.g.*, (NSX)$_3$ (X = F, Cl) and (NSF)$_4$, occurs in the condensed phase (Section 8.7). These ligands can be stabilized, however, by coordination to a transition metal. The NSF complexes are conveniently prepared in SO$_2$ (Eq. 7.6).[43] The monomeric fluoride NSF is conveniently generated *in situ* by thermal decomposition of FC(O)NSF$_2$ or Hg(NSF$_2$)$_2$ (Section 8.2).

$$[M(SO_2)][[AsF_6]_2 + 6NSF \rightarrow [M(NSF)_6][AsF_6]_2 + SO_2 \quad (7.6)$$
$$(M = Co, Ni)$$

The most general route to M-NSCl complexes involves the reactions of the cyclic trimer (NSCl)$_3$ with anhydrous metal chorides in solvents such as CH$_2$Cl$_2$ or CCl$_4$.[44] For example, 1:1 complexes, MCl$_5$(NSCl) (**7.6**, M = Nb, Ta) are formed from MCl$_5$ and (NSCl)$_3$, whereas VCl$_4$ yields a dimeric structure [VCl$_3$(NSCl)$_2$] (**7.7**).[45] However, these reactions are quite complicated and give rise to other S–N complexes, *e.g.* cyclometallathiazenes. The nature of the products from these reactions is also markedly dependent on changes in the reaction conditions. The dimeric complexes [MCl$_4$(NSCl)]$_2$ (M = Mo, W) are obtained from (NSCl)$_3$ and Mo$_2$Cl$_{10}$ or WCl$_6$, respectively, in CH$_2$Cl$_2$.[46] When the MoCl$_5$:NSCl ratio is 1:2, a complex containing a bridging S$_2$N$_2$ ligand (**7.8**, M = Mo) is obtained.[47] In CCl$_4$ solution the reaction of WCl$_6$ with (NSCl)$_3$ produces the tungsten analogue **7.8** (M = W), in addition to [WCl$_4$(NSCl)]$_2$ and the cyclometallathiazene [Cl$_3$WS$_2$N$_3$]$_2$.[46b]

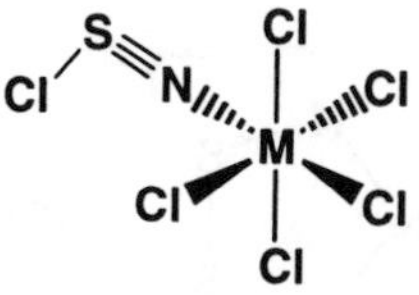

7.6 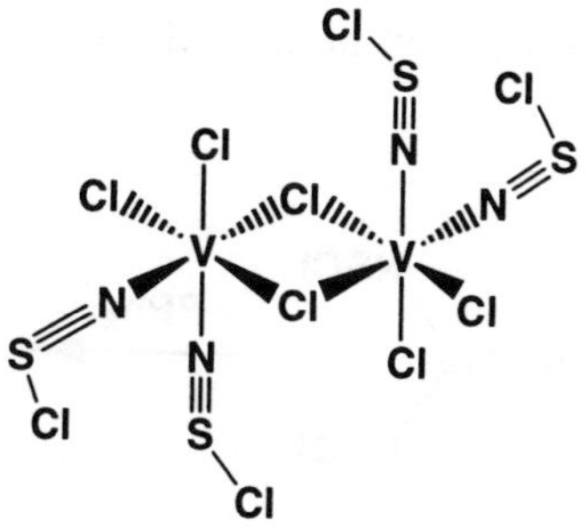   **7.7**

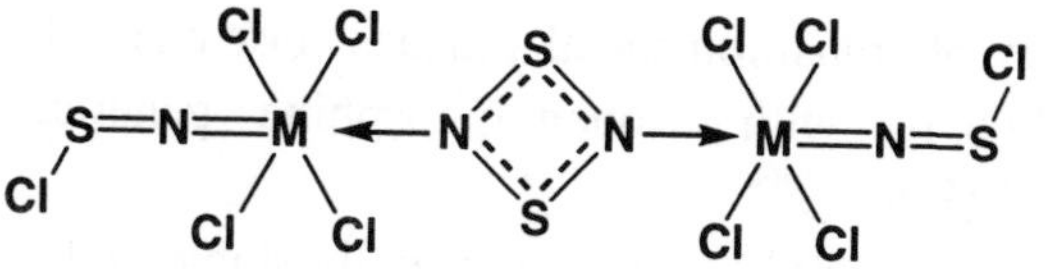

7.8

Halogen exchange at coordinated NSF, using silicon tetrahalides as the reagent and liquid SO_2 as solvent, has also been used to prepare both NSCl and NSBr complexes (Eq. 7.7).[48a]   Reaction with Me_3SiNMe_2 generates the thiazyl amide ligand $NSNMe_2$.[48b]

$$[Re(CO)_5NSF][AsF_6] + \tfrac{1}{4}SiX_4 \rightarrow [Re(CO)_5NSX][AsF_6] + \tfrac{1}{4}SiF_4 \quad (7.7)$$
$$(X = Cl, Br)$$

The IR stretching frequencies $v(SN)$ and $v(SF)$ occur at higher wave numbers and the S–N and S–F distances are significantly shorter in the octahedral cations $[M(NSF)_6]^{2+}$ than those in the free ligand, indicating an increase in S–N and S–F bond strength upon coordination.

Except for the halide exchange process (Eq. 7.7), reactions of coordinated NSX ligands have received limited attention. The most interesting transformation involves the reduction of two juxtaposed NSCl ligands in $ReCl_3(NSCl)_2(POCl_3)$ with triphenylphosphine to form a novel cyclometallathiazene $[Ph_3PCl][Cl_4ReN_2S_2]$ (Scheme 7.2).[49] Extended Hückel molecular orbital calculations indicate that coupling of two NS ligands by formation of an S–S bond is more favourable than N–N bond formation.[50]

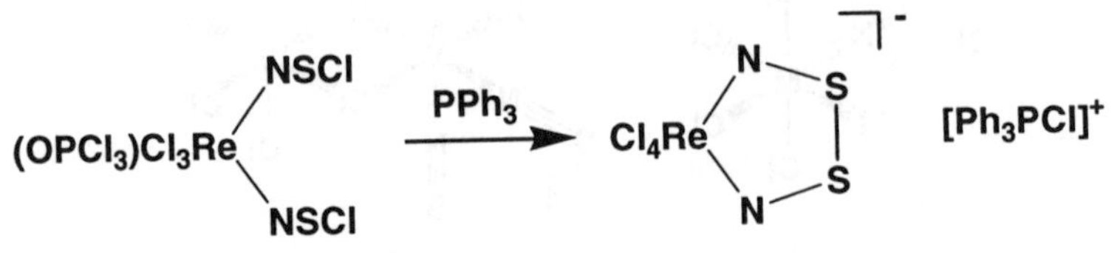

Scheme 7.2　Formation of a ReNSSN ring

The conversion of coordinated NSCl into a nitrido ligand provides a useful synthesis of transition-metal nitrides. For example, treatment of $ReCl_4(NSCl)(POCl_3)$ with triphenylphosphine generates the nitrido complex $ReNCl_2(PPh_3)_2$.[46a]

Thiazyl trifluoride NSF_3 is more thermally stable and easier to handle than NSF (Section 8.3). It can be introduced into transition-metal complexes by SO_2 displacement reactions (Eq. 7.8).[51] An alternative route to metal–NSF_3 complexes is shown in Eq. 7.9.[51]

$$[M(CO)_5SO_2][AsF_6] \ + \ NSF_3 \ \rightarrow \ [M(CO)_5(NSF_3)][AsF_6] \quad (7.8)$$
$$(M = Mn, Re)$$

$$Ag[AsF_6] \ + \ 2NSF_3 \ \rightarrow \ [Ag(NSF_3)_2][AsF_6] \quad (7.9)$$

In these complexes the S–N and S–F bonds are both shortened by *ca.* 0.05 Å compared to their values in free NSF_3 (1.416 and 1.552 Å, respectively). Consistently, the IR stretching frequencies $v(NS)$ and $v(SF)$ occur at higher wave numbers, 1575-1650 cm^{-1} and *ca.* 860 and 825 cm^{-1}, compared to the corresponding values of 1515 cm^{-1} and 811 and 775 cm^{-1} for NSF_3.

The reagent NSF_3 may also react via oxidative addition of an S–F bond to a metal centre to give a complex of the thiazyl difluoride anion $[NSF_2]^-$, which is readily hydrolyzed to an NSO complex (Scheme 7.3).[53]

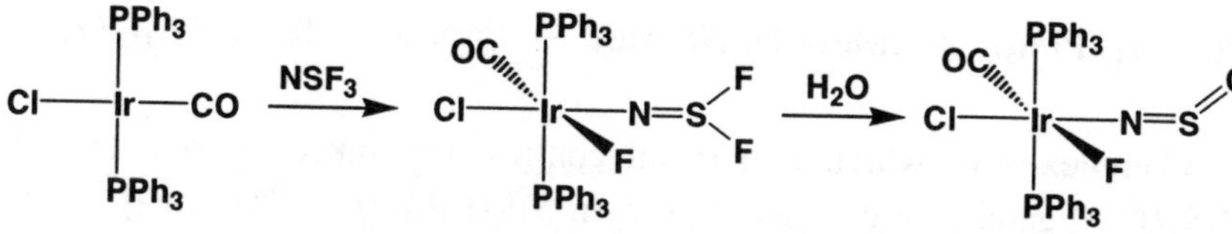

Scheme 7.3 Formation of an [NSF₂]⁻ complex

7.6 Complexes of the [NSO]⁻, [NSNR]⁻ and [NSN]²⁻ Anions

A variety of complexes of the thionyl imide anion [NSO]⁻ with both early and late transition-metal complexes have been prepared and structurally characterized. Since both ionic and covalent derivatives of this anion are readily prepared, *e.g.*, K[NSO], Me_3MNSO (M = Si, Sn) or $Hg(NSO)_2$, metathetical reactions of these reagents with transition-metal halide complexes represent the most general synthetic method for the preparation of these complexes (Eq. 7.10 and 7.11).[54, 55]

$$Cp_2ZrCl_2 + 2K[NSO] \rightarrow Cp_2Zr(NSO)_2 + 2KCl \quad (7.10)$$

$$cis\text{-}[PtCl_2(PR_3)_2] + Hg(NSO)_2 \rightarrow cis\text{-}[Pt(NSO)_2(PR_3)_2] + HgCl_2 (7.11)$$

Transition-metal complexes of the thionylimide anion exhibit characteristic vibrations in the regions 1260-1120, 1090-1010 and 630-515 cm⁻¹, which are assigned to $v_{as}(NSO)$, $v_s(NSO)$ and $\delta(NSO)$, respectively. X-ray structural data for several M-NSO complexes reveal N–S and S–O bond lengths of *ca.* 1.46 ± 0.04 Å indicative of double bond character in both of these bonds.

The coordinated [NSO]⁻ ligand reacts with $Li[N(SiMe_3)_2]$ to give the [NSNSiMe₃]⁻ ligand (Eq. 7.12).[54] This ligand system may also be generated by oxidative addition of $Me_3SiNSNSiMe_3$ to a zerovalent platinum complex (Eq. 7.13).[56]

$$Cp_2Ti(NSO)_2 + 2Li[N(SiMe_3)_2] \rightarrow Cp_2Ti(NSNSiMe_3)_2 + (Me_3Si)_2O$$
$$+ Li_2O \quad (7.12)$$

$$Pt(C_2H_4)(PPh_3)_2 + 2Me_3SiNSNSiMe_3 \rightarrow Pt(NSNSiMe_3)_2(PPh_3)_2 \ (7.13)$$

Complexes in which two metal centres are linked by one or two $[NSN]^{2-}$ ligands, *e.g.*, $[Na(15\text{-crown-5})]_2[F_5Mo(\mu\text{-NSN})MoF_5]^{57}$ and $Cp_2Zr(\mu\text{-NSN})_2ZrCp_2,^{58}$ are known. The cyclic zirconium system is prepared by a metathetical reaction (Eq. 7.14). However, the formation of polymers in which metal centres are linked by NSN units has not been achieved.

$$2Cp_2ZrCl_2 + 2Me_3SnNSNSnMe_3 \rightarrow$$
$$Cp_2Zr(\mu\text{-NSN})_2ZrCp_2 + 4Me_3SnCl \qquad (7.14)$$

References

1. T. Chivers and F. Edelmann, *Polyhedron*, **5**, 1661 (1986).

2. P. F. Kelly and J. D. Woollins, *Polyhedron*, **5**, 607 (1986).

3. P. F. Kelly, A. M. Z. Slawin, D. J. Williams and J. D. Woollins, *Chem. Soc. Rev.*, 245 (1992).

4. J. D. Woollins, The Preparation and Structure of Metalla-sulfur/selenium Nitrogen Complexes and Cages, in R. Steudel (ed.) *The Chemistry of Inorganic Heterocycles*, Elsevier, pp. 349-372 (1992).

5. H. W. Roesky and K. K. Pandey, *Adv. Inorg. Chem. Radiochem.*, **26**, 337 (1983).

6. K. K. Pandey, *Prog. Inorg. Chem.*, **40**, 445 (1992).

7. M. Herberhold, *Comments Inorg. Chem.*, **7**, 53 (1988).

8. B. W. S. Kolthammer and P. Legzdins, *J. Am. Chem. Soc.*, **100**, 2247 (1978).

9. T. J. Crevier, S. Lovell, J. M. Mayer, A. L. Rheingold and I. A. Guzei, *J. Am. Chem. Soc.*, **120**, 6607 (1998).

10. U. Demont, W. Willing, U. Müller and K. Dehnicke, *Z. Anorg. Allg. Chem.*, **532**, 175 (1986).

11. G. Hartmann and R. Mews, *Angew. Chem., Int. Ed. Engl.*, **24**, 202 (1985).

12. D. L. Lichtenberger and J. l. Hubbard, *Inorg. Chem.*, **24**, 3835 (1985).

13. M. Minelli, J. L. Hubbard, D. L. Lichtenberger and J. H. Enemark, *Inorg. Chem.*, **23**, 2721 (1984).

14. G. Hartmann and R. Mews, *Z. Naturforsch.*, **40B**, 343 (1985).

15. D. S. Bohle, C-H. Hung, A. K. Powell, B. D. Smith and S. Wocadlo, *Inorg. Chem.*, **36**, 1992 (1997).

16. M. H. V. Huynh, P. S. White and T. J. Meyer, *Inorg. Chem.*, **39**, 2825 (2000).

17. R. Jones, P. F. Kelly, D. J. Williams and J. D. Woollins, *J. Chem. Soc., Chem. Commun.*, 408 (1989).

18. (a) I. P. Parkin and J. D. Woollins, *J. Chem. Soc., Dalton Trans.*, 925 (1990); (b) V. C. Ginn, P. F. Kelly and J. D. Woollins, *J. Chem. Soc., Dalton Trans.*, 2129 (1992).

19. P. F. Kelly, I. P. Parkin, A. M. Z. Slawin, D. J. Williams and J. D. Woollins, *Angew. Chem., Int. Ed. Engl.*, **28**, 1047 (1989).

20. C. A. O'Mahoney, I. P. Parkin, D. J. Williams and J. D. Woollins, *Polyhedron*, **8**, 2215 (1989).

21. P. F. Kelly, I. P. Parkin, R. N. Sheppard and J. D. Woollins, *Heteroatom Chem.*, **2**, 301 (1991).

22. D. T. Haworth, J. D. Brown and Y. Chen, *Inorg. Synth.*, **18**, 124 (1978).

23. R. Jones, P. F. Kelly, D. J. Williams and J. D. Woollins, *J. Chem. Soc., Dalton Trans.*, 803 (1988).

24. (a) J. Weiss, *Z. Anorg. Allg. Chem.*, **502**, 165 (1985); (b) J. Weiss, *Angew. Chem., Int. Ed. Engl.*, **23**, 225 (1984).

25. G. Granozzi, M. Casarin, F. Edelmann, T. A. Albright and, E. D. Jemmis, *Organometallics*, **6**, 2223 (1987).

26. R. T. Boeré, B. Klassen and K. H. Moock, *J. Organomet. Chem.*, **467**, 127 (1994).

27. T. S. Piper, *J. Am. Chem. Soc.*, **80**, 30 (1958).

28. J. Bojes, T. Chivers and P. W. Codding, *J. Chem. Soc., Chem. Commun.*, 1171 (1981).

29. J. Weiss, *Angew. Chem., Int. Ed. Engl.*, **21**, 705 (1982).

30. J.Weiss, Z. Anorg. Allg. Chem., **532**, 184 (1986).

31. J. Hanich, M. Krestel, U. Müller, K. Dehnicke and D. Rehder, Z. Naturforsch., **39B**, 1686 (1984).

32. A. Sundermann and W. W. Schoeller, Inorg. Chem., **38**, 6261 (1999).

33. (a) P. F. Kelly, A. M. Z. Slawin, D. J. Williams and J. D. Woollins, Angew. Chem., Int. Ed. Engl., **31**, 616 (1992); (b) P. F. Kelly, A. M. Z. Slawin and A. Soriano-Rama, J. Chem. Soc., Dalton Trans., 53 (1996).

34. C. G. Marcellus, R. T. Oakley, W. T. Pennington and A. W. Cordes, Organometallics, **5**, 1395 (1986).

35. H. Endres and E. Galantai, Angew. Chem., Int. Ed. Engl., **19**, 653 (1980).

36. F. Edelmann, H. W. Roesky, C. Spang, M. Noltemeyer and. G. M. Sheldrick, Angew. Chem., Int. Ed. Engl., **25**, 931 (1986).

37. P. S. Belton, V. C. Ginn, P. F. Kelly and J. D. Woollins, J. Chem. Soc., Dalton Trans., 1135 (1992).

38. P. F. Kelly, R. N. Sheppard and J. D. Woollins, Polyhedron, **11**, 2605 (1992).

39. (a) M. Herberhold and W. Bühlmeyer, Angew. Chem., Int. Ed. Engl., **23**, 80 (1984); (b) M. Herberhold, Comments Inorg. Chem., **7**, 53 (1988).

40. (a) M. Herberhold, U. Bertholdt and W. Milius, Z. Naturforsch., **50B**, 1252 (1995); (b) M. Herberhold, U. Bertholdt, W. Milius and B. Wrackmeyer, Z. Naturforsch., **51B**, 1283 (1996)..

41. M. Herberhold, W. Bühlmeyer, A. Gieren, T. Hubner and J. Wu, Z. Naturforsch., **42B**, 65 (1987)

42. H. W. Roesky, R. Emmert, W. Clegg, W. Isenberg and G. M. Sheldrick, Angew. Chem., Int. Ed. Engl., **20**, 591 (1981).

43. B. Buss, P.G. Jones, R. Mews, M. Noltemeyer and G. M. Sheldrick, Angew. Chem., Int. Ed. Engl., **18**, 231 (1979).

44. K. Dehnicke and U. Müller, Comments Inorg. Chem., **4**, 213 (1985).

45. (a) J. Hanich and K. Dehnicke, Z. Naturforsch., **39B**, 1467 (1984); (b) G. Beber, J. Hanich and K. Dehnicke, Z. Naturforsch, **40B**, 9 (1985).

46. (a) U. Kynast and K. Dehnicke, *Z. Anorg. Allg. Chem.*, **502**, 29 (1983); (b) U. Kynast, U. Müller and K. Dehnicke, *Z. Anorg. Allg. Chem.*, **508**, 26 (1984).

47. U. Kynast, P. Klingelhofer, U. Müller and K. Dehnicke, *Z. Anorg. Allg. Chem.*, **515**, 61 (1984).

48. (a) G. Hartmann and R. Mews, *Z. Naturforsch.*, **40B**, 343 (1985); (b) G. Hartmann, R. Mews and G. M. Sheldrick, *Angew. Chem., Int. Ed. Engl.*, **22**, 723 (1983).

49. W. Hiller, J. Mohlya, J. Strähle, H. G. Hauck and K. Dehnicke, *Z. Anorg. Allg. Chem.*, **514**, 72 (1984).

50. M. Kersting and R. Hoffman, *Inorg. Chem.*, **29**, 279 (1990).

51. R. Mews, *J. Chem. Soc., Chem. Commun.*, 278 (1979).

52. R. Mews and O. Glemser, *Angew. Chem., Int. Ed. Engl.*, **14**, 186 (1975).

53. P. G. Watson, E. Lork and R. Mews, *J. Chem. Soc., Chem. Commun.*, 1069 (1994).

54. H. Plenio, H. W. Roesky, F. T. Edelmann and M. Noltemeyer, *J. Chem. Soc., Dalton Trans.*, 1815 (1989).

55. R. Short, M. B. Hursthouse, T. G. Purcell and J. D. Woollins, *J. Chem. Soc., Chem., Commun.*, 407 (1987).

56. N. P. C. Walker, M. B. Hursthouse, C. P. Warrens and J. D. Woollins, *J. Chem. Soc., Chem. Commun.*, 227 (1985).

57. A. El-Kholi, K. Völp, U. Müller and K. Dehnicke, *Z. Anorg. Allg. Chem.*, **572**, 18 (1989).

58. H. Plenio and H. W. Roesky, *Z. Naturforsch.*, **43B**, 1575 (1988).

Chapter 8

Chalcogen–Nitrogen Halides

8.1 Introduction

This chapter will deal with both acyclic and cyclic chalcogen–nitrogen compounds that contain chalcogen–halogen bonds. The simplest examples of these ternary systems are the monomeric thiazyl halides NSX (X = F, Cl, Br), in which the atomic arrangement differs from that of nitrosyl halides ONX. In these triatomic molecules the least electronegative atom assumes the central position. The thiazyl halides with a formal N≡S triple bond, readily oligomerize to cyclic trimers (X = F, Cl). Other important acyclic derivatives include the anion $[NSCl_2]^-$, the cations $[N(ECl)_2]^+$ (E = S, Se) and $[N(EX_2)_2]^+$ (E = S, Se; X = F, Cl), and the neutral compound NSF_3. Tellurium analogues of these simple chalcogen–nitrogen halides are unknown, but the complex halides $[Te_4N_2Cl_8]^{2+}$ and $Te_{11}N_6Cl_{26}$ have been structurally characterized. The cyclotrithiazyl halides $(NSX)_3$ (X = F, Cl) are six-membered rings which, in the case of the chloro derivative, have found applications as a source of a thiazyl unit in the synthesis of organic heterocycles. The sulfanuric halides $[NS(O)X]_3$ (X = F, Cl) constitute a related class of ring systems in which the sulfur atoms are in the +6 oxidation state.

8.2 Thiazyl Halides NSX (X = F, Cl, Br) and the $[NSX_2]^-$ Anions (X = F, Cl)

Thiazyl fluoride is a moisture-sensitive, thermally unstable gas.[1] It is conveniently generated by decomposition of compounds which already contain the NSF group, *e.g.*, $FC(O)NSF_2$ (Eq. 8.1) or $Hg(NSF_2)_2$ (Eq. 8.2).[2] It forms the cyclic trimer $(NSF)_3$ at room temperature. A high yield synthesis of NSF from $(NSCl)_3$ and KF in tetramethylsulfone at 80°C has been reported.[3]

$$FC(O)NSF_2 \rightarrow NSF + COF_2 \qquad (8.1)$$

$$Hg(NSF_2)_2 \rightarrow 2NSF + HgF_2 \qquad (8.2)$$

Monomeric NSCl is formed as a greenish-yellow gas by heating the cyclic trimer $(NSCl)_3$ under vacuum or in an inert gas stream.[4] NSCl monomer may also be generated in solutions of $(NSCl)_3$ in liquid SO_2 at room temperature or in CCl_4 at 70°C.[5] It trimerizes in the condensed state. Self-consistent field calculations predict that (a) the dimers $(NSX)_2$ (X= F, Cl) are thermodynamically unstable with respect to 2NSX and (b) the trimer $(NSF)_3$ is stable compared to 3NSF.[6] Monomeric thiazyl halides NSX (X = Cl, Br) have also been generated by the pyrolysis of $[S_4N_3]X$ at 120°C (X = Cl) or 90°C (X = Br).[7] This method has been used to produce NSX in an argon matrix at 15 K for the determination of infra-red spectra. The gas-phase structures of NSF[8a] and NSCl[8b] have been determined by microwave spectroscopy. They are bent molecules (<NSF = 117°, <NSCl = 118°)) with sulfur–nitrogen bond lengths of 1.448 and 1.450 Å, respectively, consistent with substantial triple bond character. The sulfur-halogen distances are 1.643 Å (X= F) and 2.161 Å (X = Cl). *Ab initio* molecular orbital calculations show that NSCl is 18.4 kcal mol^{-1} more stable than the hypothetical thionitrosyl chloride SNCl.[9] The energy difference is reduced to 8.6 kcal mol^{-1} for the corresponding cation radicals; $[ClNS]^{+\cdot}$ has been tentatively identified in the gas phase by mass spectrometric methods.[9]

Monomeric thiazyl halides can be stabilized by coordination to transition metals and a large number of such complexes are known (Section 7.5). In addition, NSX monomers undergo several types of reactions that can be classified as follows: (a) reactions involving the π–system of the N≡S bond (b) reactions at the nitrogen centre (c) nucleophilic substitution reactions (d) halide abstraction, and (e) halide addition. Examples of each type of behaviour are illustrated below.

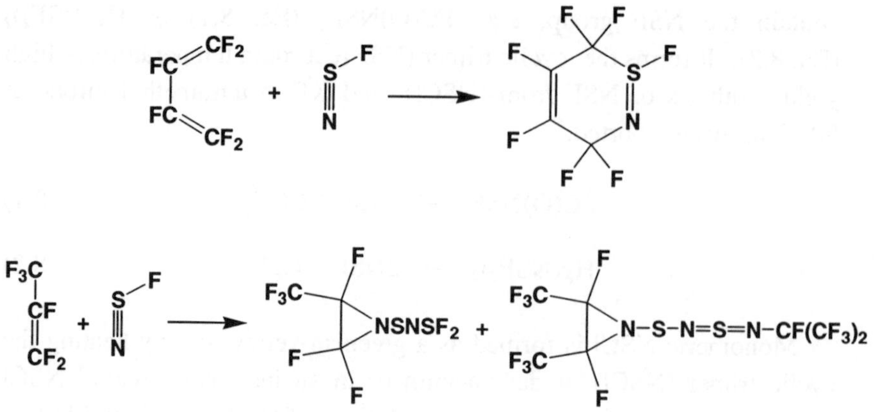

Scheme 8.1 Cycloaddition reactions of NSF

(a) NSF undergoes cycloaddition with hexafluorobuta-1,3-butadiene to form a six-membered ring (Scheme 8.1).[10]

(b) There is no clear evidence for nitrene reactions of NSF. However, irradiation of excess thiazyl fluoride in the presence of hexafluoropropene generates sulfenylaziridines (Scheme 8.1).

(c) Nucleophilic replacement of the fluoride substituent in NSF usually results in a rearrangement with loss of the formal N≡S triple bond. For example, hydrolysis of moisture-sensitive NSF results in replacement of F by OH$^-$ and subsequent isomerization of thiazyl hydroxide to thionyl imide HNSO (Eq. 8.3) which, in turn, is hydrolyzed to give sulfur oxoanions.[11]

$$\text{NSF} + \text{OH}^- \ \rightarrow \ [\text{N}{\equiv}\text{SOH}] \ \rightarrow \ \text{HNSO} + \text{F}^- \qquad (8.3)$$

(d) The reaction of NSF with strong fluoride ion acceptors, *e.g.*, MF$_5$ (M = As, Sb) in liquid SO$_2$ was the first synthesis of [SN]$^+$ salts (Eq. 8.4).[12] Although other preparative routes to these important reagents have subsequently been developed (Section 5.3.1), the original method gives the salt in highest purity.

$$\text{NSF} + \text{MF}_5 \ \rightarrow \ [\text{SN}][\text{MF}_6] \qquad (8.4)$$

(e)   The thiazyl halide monomers NSX also undergo nucleophilic addition with halide ions to give ternary anions of the type [NSX$_2$]$^-$. The [NSF$_2$]$^-$ ion in Cs[NSF$_2$] and [(Me$_2$N)$_3$S][NSF$_2$] has been characterized by vibrational spectra.[13] The [NSCl$_2$]$^-$ anion, obtained by chloride addition to NSCl (generated from the cyclic trimer), has been isolated in salts with large counter-anions, *e.g.*, [Ph$_4$P]$^+$ and [Me$_4$N]$^+$.[14] The [NSCl$_2$]$^-$ anion in the [(Ph$_3$PN)$_2$SCl]$^+$ salt has a slightly distorted C_s structure with a very short S–N bond (1.44 Å) and two loosely bound chlorine atoms [d(S–Cl) = 2.42 Å]. The structure is best described by the resonance forms depicted in Scheme 8.2.

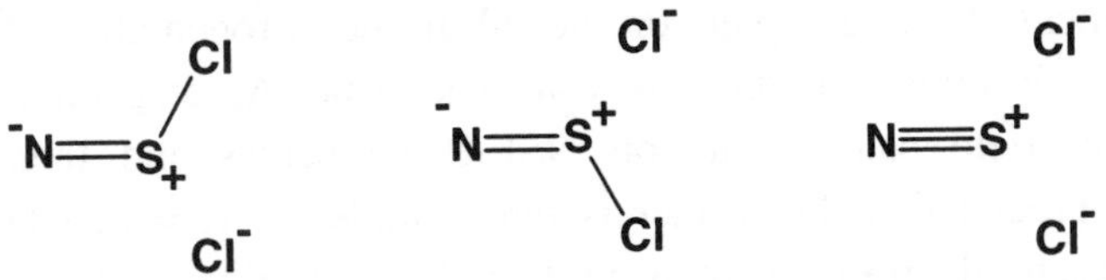

Scheme 8.2   Resonance structures of [NSCl$_2$]$^-$

Selenazyl halides NSeX have not been characterized either as monomers or cyclic oligomers. However, the monomeric ligand is stabilized in metal complexes of the type [Cl$_4$M(NSeCl)]$_2$ (M = Mo, W), which are obtained from the reactions of MoCl$_5$ or WCl$_6$ with Se$_4$N$_4$ in dichloromethane.[15] The bonding features in the anion [Cl$_5$W(NSeCl)]$^-$ are similar to those in NSCl complexes (Section 7.5) with a short W=N bond, a Se=N bond length of 1.77 Å, and a bond angle <NSeCl of *ca.* 92°.

8.3   Thiazyl Trifluoride NSF$_3$ and Haloiminosulfur Difluorides XNSF$_2$ (X = F, Cl)

Thiazyl trifluoride, a colourless gas with a pungent odour, is prepared by the oxidative decomposition of FC(O)NSF$_2$ with AgF$_2$ (Eq. 8.5).[16a] NSF$_3$ is kinetically very stable even in the liquid form. The chemical inertness of NSF$_3$ resembles that of SF$_6$. For example, it does not react with sodium metal below 200°C.[1]

$$FC(O)NSF_2 + 2AgF_2 \rightarrow NSF_3 + 2AgF + COF_2 \quad (8.5)$$

The structure of NSF_3 (**8.1**) in the gas phase has been determined by microwave spectroscopy to be a distorted tetrahedron with C_{3v} geometry and a $S\equiv N$ bond length of 1.416 Å, consistent with triple bond character.[16b] The bond angle <FSF is ca. 94° and the S–F distance is 1.55 Å indicating a much stronger bond than that in NSF [d(S–F) = 1.64 Å. The mixed halide derivatives $XNSF_2$ (X = Cl, Br, I) are obtained by the reaction of $Hg(NSF_2)_2$ with halogens (Eq. 8.6).[2] This reaction also gives rise to $FNSF_2$, a structural isomer of NSF_3.[17] The chloro derivative $ClNSF_2$ (**8.2**) has been shown by electron diffraction to have a cis arrangement of the lone pairs on the sulfur and nitrogen atoms.[18] The S–N and S–F bond lengths are 1.48 and 1.60 Å, respectively. Self-consistent field calculations on NSF_2R molecules and their $RNSF_2$ isomers predict that the former is more stable for R = F, whereas the latter structural arrangement is preferred for R = Me, CF_3, consistent with experimental observations. The isomer $FNSF_2$ is less stable than NSF_3 by 3.8 kcal mol^{-1}.[19]

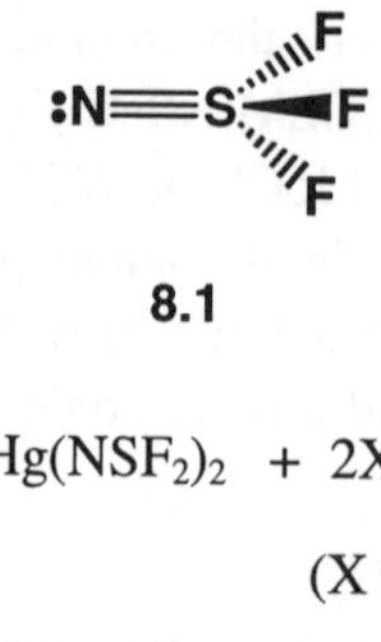

8.1 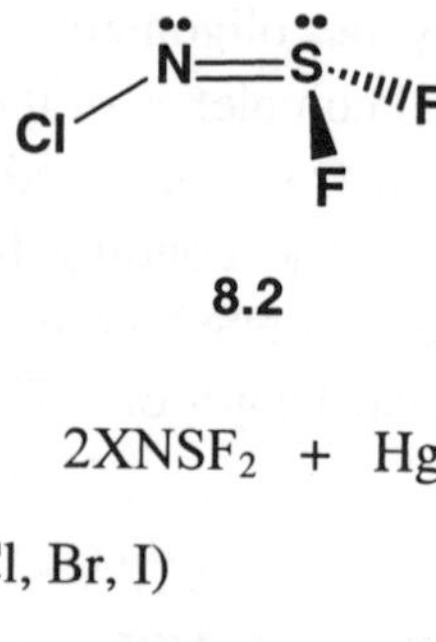   8.2

$$Hg(NSF_2)_2 + 2X_2 \rightarrow 2XNSF_2 + HgX_2 \quad (8.6)$$

$$(X = F, Cl, Br, I)$$

The reactions of NSF_3 have been investigated in considerable detail. They can be classified under the following categories: (a) reactions with electrophiles (b) addition to the SN triple bond and (c) reactions with nucleophiles. Some examples of these different types of behaviour are discussed below.

(a) A variety of metal complexes with up to four NSF_3 ligands coordinated to the M^{2+} centre have been characterized (Section 7.5). In a similar fashion NSF_3 forms *N*-bonded adducts with fluoro Lewis acids (Eq. 8.7).[1] The $S{\equiv}N$ and, especially, the S–F bond lengths are significantly shortened upon adduct formation as revealed by the X-ray structure of $F_5As{\cdot}NSF_3$.[1] These structural changes have been attributed to rehybridization of the lone pair on nitrogen from primarily *s* to *sp* character upon complex formation and a cocomitant increase in π-bonding in the S–F bonds.

$$A + NSF_3 \rightarrow A{\cdot}NSF_3 \qquad (8.7)$$

$$(A = BF_3, AsF_5, SbF_5)$$

A different type of behaviour is observed with the chloro Lewis acid BCl_3. With this reagent halogen exchange occurs to produce the acyclic cation $[N(SCl)_2]^+$, as the $[BCl_4]^-$ salt, rather than $NSCl_3$.[20] Thiazyl trichloride $NSCl_3$ is predicted to be unstable with respect to $NSCl + Cl_2$.[19]

A singular example of the oxidative addition of an S–F bond in NSF_3 to an electrophilic metal centre has been reported in the formation of the thiadiazyl difluoride complex $[Ir(CO)ClF(NSF_2)(PPh_3)_2]$ (Scheme 7.3).[21]

Protonation of NSF_3 by the superacid HSO_3F/SbF_5 produces the thermally unstable $[HNSF_3]^+$ cation.[1] Similarly, alkylation of NSF_3 with $[RSO_2][AsF_6]$ (R = Me, Et) gives $[RNSF_3]^+$ cations, which can be isolated as stable salts in the presence of weakly nucleophilic anions, *e.g.*, $[MF_6]^-$ (M = As, Sb).[22] The reaction of $[MeNSF_3]^+$ with sodium fluoride at elevated temperatures yields $MeN{=}SF_4$ (**8.3**), which was shown by electron diffraction to have a distorted trigonal pyramidal structure analogous to that of the isoelectronic molecule OSF_4.[1]

(b) Polar species XF (X = H, Cl) add to the $S{\equiv}N$ triple bond. The reaction of NSF_3 with an excess of HF results in a double addition to give the stable pentafluorosulfur amine H_2NSF_5 (Eq. 8.8),[23a] which has an extensive derivative chemistry involving the NH_2 group.[1] For example, the reactions with OCF_2 and SCl_2 produce F_5SNCO an $F_5SN=S=NSF_5$, respectively.

$$NSF_3 + 2HF \rightarrow H_2NSF_5 \qquad (8.8)$$

(c) The reaction of NSF_3 with nucleophilic reagents such as secondary amines, alcohols or organolithium reagents results in the formation of monosubstituted products with retention of the $N\equiv S$ bond (Eq. 8.9).[1, 23b] Hydrolysis of the kinetically inert NSF_3 molecule requires hot alkali and generates $HNSOF_2$ as the initial product. The structure of $HNSOF_2$ (**8.4**), as determined by microwave spectroscopy,[24] shows the NH and SF_2 groups to be in a *cis* arrangement with respect to the S=N bond, which has a bond length of 1.47 Å.

$$NSF_3 + ArLi \rightarrow NSF_2Ar + LiF \qquad (8.9)$$

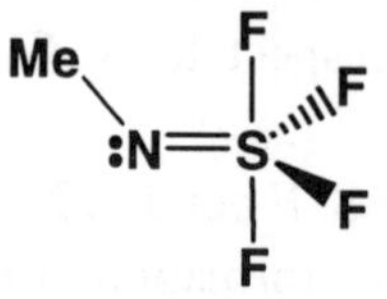

8.3 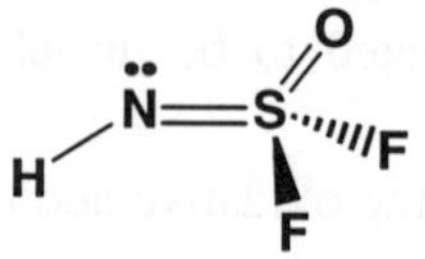 **8.4** 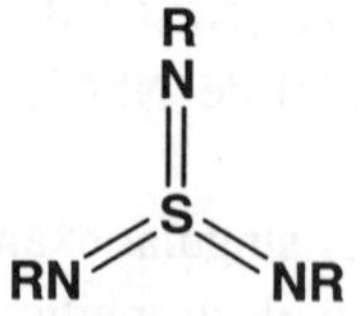 **8.5a** (R = SiMe_3)
 8.5b (R = ^{t}Bu)

The reactions of NSF_3 with the amido-lithium reagents $LiN(SiMe_3)R$ (R = ^{t}Bu, SiMe_3) results in an interesting rearrangement to produce the sulfur triimides $S(NR)_3$, isoelectronic with SO_3.[25] The first sulfur triimide **8.5a** (R = SiMe_3) was obtained in this manner (Eq. 8.10). Bis(trimethylsilylimino)sulfur difluoride $(Me_3SiN)_2SF_2$ is a by-product of this reaction. The X-ray structure of the *tert*-butyl derivative **8.5b** reveals the anticipated trigonal planar arrangement of N^tBu groups around the sulfur(VI) atom with S=N bond lengths of 1.51 Å.[25a] The oxidation of the $[S(N^tBu)_3]^{2-}$ dianion with halogens is a better synthesis of **8.5b** (Section 10.5).

$$NSF_3 + 2LiN(SiMe_3)_2 \rightarrow S(NSiMe_3)_3 + 2LiF + Me_3SiF \qquad (8.10)$$

8.4 Acyclic Chalcogen-Nitrogen–Halogen Cations $[N(ECl)_2]^+$ (E = S, Se) and $[N(SeCl_2)_2]^+$

The most straightforward route to the acyclic cation $[N(SCl)_2]^+$ (**8.6a**) is the reaction of $[NS]^+$ with SCl_2 (Eq. 8.11).[26] Other preparative methods include the reactions of (a) $(NSCl)_3$ with SCl_2 in the presence of a metal chloride (*e.g.*, $AlCl_3$ or $SbCl_5$) or $AgAsF_6$[27] or (b) an $[SCl_3]^+$ salt with $N(SiMe_3)_3$ in CCl_4.[28] Dechlorination of **8.6a** with $SnCl_2$ produces the $[NS_2]^+$ cation (Section 5.3.2).

$$[NS][SbCl_6] + SCl_2 \rightarrow [N(SCl)_2][SbCl_6] \qquad (8.11)$$

The reaction of $N(SiMe_3)_3$ with $SeCl_4$ in boiling CH_2Cl_2 yields Se_2NCl_3 (**8.7a**), which reacts with $GaCl_3$ to produce $[N(SeCl)_2][GaCl_4]$ containing the selenium analogue **8.6b**.[29] The bromo derivative **8.7b** is prepared in a similar manner from $SeBr_4$ and $N(SiMe_3)_3$.[30] The attempted preparation of $[SeN]^+$ from the reaction of equimolar quantities of $[SeCl_3][AsF_6]$ with $N(SiMe_3)_3$ in $CFCl_3$ produced instead the novel cation $[N(SeCl_2)_2]^+$ (**8.8**).[31] Surprisingly, $[SeCl_3][SbCl_6]$ reacts with $N(SiMe_3)_3$ in a different way to give the cation **8.6b** which is isolated as the *cis, cis* isomer.[32] The reagent $[SeCl_3][FeCl_4]$ produces $[N(SeCl)_2]_2[FeCl_4]$ in which the cation **8.6b'** has a *cis, tran* geometry.[33]

8.6a (E = S) **8.6b'** **8.7a** (X = Cl) **8.8**
8.6b (E = Se) **8.7b** (X = Br)

The structures of various salts of **8.6a** have been determined by X-ray diffraction. The cation adopts a U-shaped (C_{2v}) geometry with an <NSN bond angle of 150 ± 1° in the absence of strong cation-anion interactions. The S–N bond lengths are *ca.* 1.53 Å and the S–Cl distances are relatively short at 1.91-1.99 Å. The structures of **8.6a**, **8.7a,b** and **8.8** exhibit Se–N bond lengths that are substantially shorter than the single

bond value of 1.86 Å. Negative hyperconjugation [lone pair (N) → σ^* (Se–Cl)] accounts for the short S–N and Se–N bond lengths in these cations[32, 34] and, in the case of **8.6b′**, explains the inequality of the Se–N distances.[32] The Se–Cl bond distances of 2.14-2.17 Å found for the two cations **8.6b** and **8.8** are normal for terminal Se–Cl bonds, while the long distances (2.52 and 2.68 Å) for the bridging Se–Cl bonds in the neutral molecule **8.7a** suggest ionic character for this chlorine atom. This conclusion is supported by calculations of the Mulliken charge distributions in **8.6b** and **8.7a**.[32]

8.5   Tellurium–Nitrogen–Chlorides [Te$_4$N$_2$Cl$_8$]$^{2+}$ and Te$_{11}$N$_6$Cl$_{26}$

There are no tellurium analogues of the chalcogen–nitrogen halides described in Sections 8.2 and 8.3. However, the dication [Te$_4$N$_2$Cl$_8$]$^{2+}$ (**8.9**) is obtained, as the [AsF$_6$]$^-$ salt, from the reaction of TeCl$_4$ with N(SiMe$_3$)$_3$ in a 2:1 molar ratio in acetonitrile.[35] The formation of the four-membered Te$_2$N$_2$ ring in **8.9** provides an illustration of the facile self-association of multiply bonded TeN species (*cf.*, **8.21a** and Section 10.4.2). This dication is a dimer of the hypothetical tellurium(IV) imide [Cl$_3$Te–N=TeCl]$^+$, which is a structural isomer of [N(TeCl$_2$)$_2$]$^+$ (the tellurium analogue of **8.8**). The compound [Te$_{11}$N$_6$Cl$_{26}$]·C$_7$H$_8$ is isolated from the reaction of TeCl$_4$ with N(SiMe$_3$)$_3$ in boiling toluene.[36] Each half of this centrosymmetric dimer contains a [Te$_5$N$_3$Cl$_{10}$]$^+$ cation (**8.10**) and a [Te$_5$N$_3$Cl$_{12}$]$^-$ anion linked to a TeCl$_4$ molecule. A Te$_5$N$_3$ structural core is common to both the anion and cation in this complex structure. The structure of cation **8.10** is comprised of two [Cl$_3$Te–N=TeCl]$^+$ cations (*cf.*, **8.9**) bridged by a monomeric NTeCl unit and a chloride ion.

        8.9   **8.10**

8.6 Thiodithiazyl and Selenadiselenazyl Dichloride [E_3N_2Cl]Cl (E = S, Se)

The [S_3N_2Cl]$^+$ cation (**8.11a**) is an important intermediate in the synthesis of other S–N compounds, *e.g.*, (NSCl)$_3$, [S_4N_3]$^+$, S_4N_4 and S_3N_2O.[37] It is conveniently prepared by refluxing S_2Cl_2 with dry, finely ground ammonium chloride (Eq. 8.12).[38] [S_3N_2Cl]Cl may also be prepared from urea and S_2Cl_2.[39] The other halo derivatives, [S_3N_2Br]$^+$ and [S_3N_2F]$^+$, are obtained by treatment of [S_3N_2][AsF_6] with Br_2 and by cycloaddition of NSF to [S_2N]$^+$, respectively.[40] The selenium analogue [Se_3N_2Cl]$^+$ (**8.11b**) is prepared by the reduction of the acyclic cation **8.6b** with Ph_3Sb.[41] The explosive and insoluble compound $Se_3N_2Cl_2$, which also contains the cyclic cation **8.11b**, is formed in the reaction of Se_2Cl_2 with trimethylsilyl azide in CH_2Cl_2 (Eq. 8.13).[42] X-ray diffraction studies show that **8.11a**, in the [$FeCl_4$]$^-$ salt,[43] and **8.11b**, in the [$SbCl_6$]$^-$ salt,[44] consist of slightly puckered five-membered rings.

$$4S_2Cl_2 + 2NH_4Cl \rightarrow [S_3N_2Cl]Cl + 8HCl + \tfrac{5}{8}S_8 \quad (8.12)$$

$$3Se_2Cl_2 + 2Me_3SiN_3 \rightarrow 2Se_3N_2Cl_2 + 2Me_3SiCl + N_2 \quad (8.13)$$

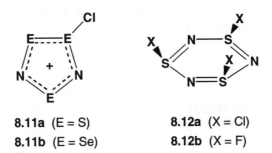

8.11a (E = S)   **8.12a** (X = Cl)
8.11b (E = Se)   **8.12b** (X = F)

8.7 Cyclotrithiazyl Halides (NSX)$_3$ (X = Cl, F)

A safe and convenient procedure for the preparation of (NSCl)$_3$ (**8.12a**) is the chlorination of [S_3N_2Cl]Cl with either Cl_2 or SO_2Cl_2 (Eq. 8.14).[38,45] The moisture-sensitive, pale-yellow product may be recrystallized from CCl_4 without decomposition provided that the temperature is kept below

50°C. The fluoride $(NSF)_3$ (**8.12b**) can be made in high yield by stirring **8.12a** with AgF_2 in CCl_4 at room temperature.[46] Halogen exchange between **8.12a** and Me_3SiBr produces the polymer $(NSBr_{0.4})_x$ rather than $(NSBr)_3$.[47] The cyclotrithiazyl halides $(NSX)_3$ (X = Br, I) are unknown. There are no selenium analogues of **8.12a** or **8.12b**.

$$3[S_3N_2Cl]Cl + 3SO_2Cl_2 \rightarrow 2(NSCl)_3 + 3SCl_2 + 3SO_2 \qquad (8.14)$$

The six-membered rings **8.12a** and **8.12b** adopt chair conformations with all three halogen atoms in axial positions. This arrangement is stabilized by the delocalization of the nitrogen lone pair into an S-X σ^* bond (the anomeric effect).[48] All the S–N distances are equal within experimental error [ld(S–N)l = 1.60 (**8.12a**),[49] 1.59 Å (**8.12b**)[50]].

The cyclic trimer **8.12a** is an important reagent in S–N chemistry as a source of both cyclic and acyclic S-N compounds (Scheme 8.3).[37] In part, this synthetic utility stems from the ease with which **8.12a** dissociates into monomeric NSCl in solution. Thus the trimer provides a facile source of the $[SN]^+$ cation (Section 5.3.1). Monomeric NSCl, generated from $(NSCl)_3$ undergoes a [2 + 4] cycloaddition reaction with hexafluorobutadiene to give a six-membered ring (*cf.* Scheme 8.1), but it reacts as a nitrene with fluorinated alkenes to give *N*-(chlorosulfenyl)aziridines (Eq. 8.15).[51]

$$\underset{F}{\overset{R}{\diagdown}} C = C \underset{F}{\overset{F}{\diagup}} \; + \; 1/3 \; (NSCl)_3 \; \xrightarrow{50\ °C} \qquad (8.15)$$

R = SF_5, CF_3, F, Cl

The reactions of $(NSCl)_3$ with sodium alkoxides to give $(NSOR)_3$[52] and with AgF_2 to produce **8.12b**[46] are two examples of transformations that occur with retention of the six-membered ring. The S_3N_3 ring in **8.12b** is more robust than that in **8.12a**. For example, the salts

$[N_3S_3F_2][MF_6]$ can be isolated from the reaction of **8.12b** with MF_5 (M = As, Sb), but they decompose to give the bicyclic cation $[S_4N_5]^+$ (Section 5.3.8).[53] Treatment of **8.12a** with trimethylsilyl azide generates the polymer $(SN)_x$,[54a] presumably via the intermediate formation of thiazyl azide, NSN_3. *Ab initio* molecular orbital calculations indicate that this azide is a possible candidate for experimental observation.[54b]

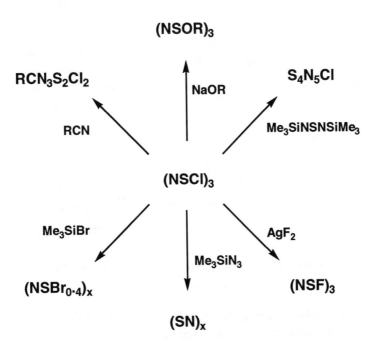

Scheme 8.3   Some reactions of $(NSCl)_3$

Cyclotrithiazyl chloride is also a useful reagent in organic chemistry in the fusion of 1,2,5-thiadiazoles to quinones[55] as well as the synthesis of (a) isothiazoles from 2,5-disubstituted furans[56a] and (b) bis-1,2,5-thiadiazoles from *N*-alkylpyrroles[56b] (Scheme 8.4). Alkenes and alkynes react readily with $(NSCl)_3$ to give 1,2,5-thiadiazoles,[57a] while 1,4-diphenyl-1,3-butadiene gives a variety of heterocyclic products including a bis(1,2,5-thiadiazole).[57b]

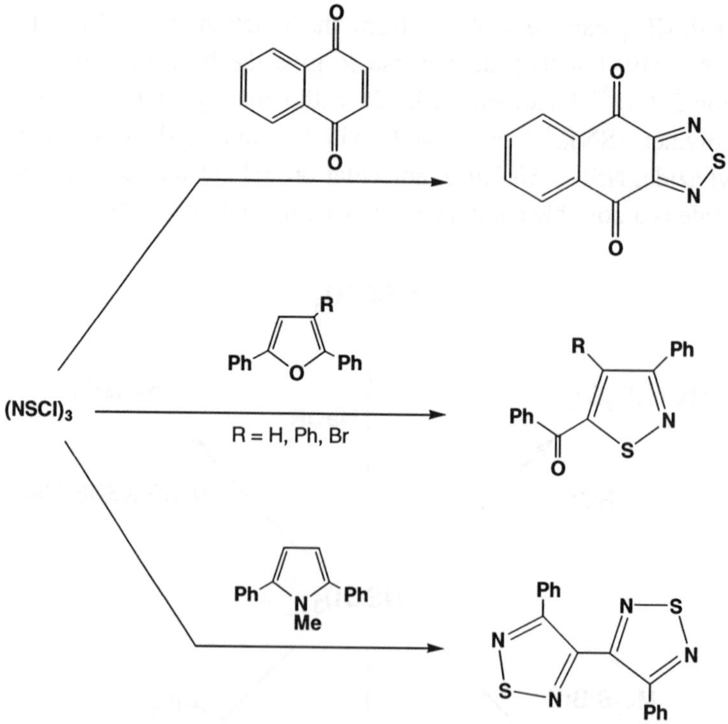

Scheme 8.4   Synthesis of S–N heterocycles from (NSCl)$_3$

8.8   Dihalocyclotetrathiazenes S$_4$N$_4$X$_2$ (X = Cl, F) and Cyclotetrathiazyl Fluoride (NSF)$_4$

The oxidative addition of one equivalent of X$_2$ (X = Cl, F) to S$_4$N$_4$ under mild conditions produces 1,5-S$_4$N$_4$X$_2$ as moisture-sensitive, thermally unstable compounds.[58,59] The structure of 1,5-S$_4$N$_4$Cl$_2$ (**8.13**) consists of a folded eight-membered ring [d(S•••S) = 2.45 Å] with the exocyclic substituents in *exo,endo* positions.[60] The halogen atoms in **8.13** can be readily replaced by NR$_2$ groups (R = alkyl) with retention of the eight-membered ring by using trimethylsilylated reagents.[61] The reaction of **8.13** with Me$_3$SiN=S=NSiMe$_3$ is the best route to the binary sulfur nitride S$_5$N$_6$ (Section 5.2.7).

8.13   **8.14**

The fluorination of S_4N_4 with an excess of AgF_2 in CCl_4 under reflux gives $(NSF)_4$ (**8.14**).[62] The heterocyclic ring of **8.14** is boat-shaped with juxtaposed long (1.66 Å) and short (1.54 Å) S–N bonds as a result of Jahn-Teller distortion.[63] The fluorine substituents occupy alternate axial and equatorial positions. The tetramer **8.14** dissociates into NSF monomer when heated under vacuum at 300°C. Treatment of **8.14** with AsF_5 results in ring contraction to give $[S_3N_3F_2][AsF_6]$ via elimination of NSF from the thermally unstable salt $[S_4N_4F_3][AsF_6]$.[1] The chloro derivative $(NSCl)_4$ is unstable. Halogen exchange between $(NSF)_4$ and $SiCl_4$ yields the six-membered ring $(NSCl)_3$ through loss of NSCl.

8.9 Sulfanuric Halides [NS(O)X]$_3$ (X = Cl, F)

Sulfanuric halides contain the characteristic group –N=S(O)X– (X = Cl, F). Unlike the isoelectronic cyclophosphazenes $(NPCl_2)_x$ (x = 3-17),[64] only six-membered rings have been well characterized. The sulfanuric halides are colourless solids (X = Cl) or liquids (X = F), which are stable in dry air. Sulfanuric chloride [NS(O)Cl]$_3$ is best prepared by treatment of $SOCl_2$ with sodium azide in acetonitrile at -35°C (Eq. 8.16).[65] It may also be obtained as a mixture of α– and β-isomers in a two-stage reaction from H_2NSO_3H and PCl_5.[66] The fluoride [NS(O)F]$_3$ is formed as a mixture of isomers by the fluorination of [NS(O)Cl]$_3$ with SbF_3.[67]

$$SOCl_2 + NaN_3 \rightarrow \frac{1}{3}[NS(O)Cl]_3 + NaCl + N_2 \qquad (8.16)$$

The isomer α-[NS(O)Cl]$_3$ (**8.15a**) is a six-membered ring in the chair form with equal S–N bond lengths (1.57 Å). The three chlorine atoms are in axial positions on the same side of the ring.[49] The β-isomer **8.15b** has two axial and one equatorial chlorine atoms.[68]

8.15a   **8.15b**   **8.16**

The fluoride [NS(O)F]$_3$ is more stable thermally and towards nucleophilic reagents than the corresponding chloride. For example, **8.15a** is hydrolyzed by water to NH(SO$_2$NH$_2$)$_2$, whereas sulfanuric fluoride is unaffected by cold water. In warm water, however, hydrolysis occurs to give the [N$_3$S$_3$O$_4$F$_2$]$^-$ anion.[69] All three fluorine atoms in [NS(O)F]$_3$ can be replaced by primary or secondary amines at 80–90°C in the absence of a solvent.[70] Mono- or diphenyl derivatives can be prepared by treatment of [NS(O)F]$_3$ with PhLi in diethyl ether at –70°C, while the Friedel–Crafts reaction with benzene at reflux in the presence of AlCl$_3$ gives two isomers of [NS(O)Ph]$_3$.[71]

Fluoride addition to [NS(O)F]$_3$ to give the anion [N$_3$S$_3$O$_3$F$_4$]$^-$ (**8.16**) is accomplished quantitatively by reaction with [(Me$_2$N)$_3$S][Me$_3$SiF$_2$].[72] The geometry at the five-coordinate sulfur(VI) centre in **8.16** is distorted trigonal bipyramidal with the two fluorine atoms in the axial positions. The S–F bonds involving this sulfur atom are 1.66 and 1.72 Å, *cf.* 1.54 Å in [NS(O)F]$_3$, as expected for a three-centre four-electron system. The S–N bonds at the five-coordinate sulfur atom are elongated by *ca.* 0.06-0.07 Å.

8.10   Chalcogen–Nitrogen Halides Containing Two Chalcogens

The reaction of Me$_3$SiNSNSiMe$_3$ with TeCl$_4$ is an especially fruitful source of chalcogen-nitrogen halides that contain both sulfur and tellurium.[73] The initial product of this reaction is the bicyclic compound **8.17**, which is obtained when the reaction is carried out in a 1:2 molar ratio in CH_2Cl_2 at –50°C. The subsequent reaction of **8.17** with one equivalent of $Me_3SiNSNSiMe_3$ at 20°C generates **8.18**, which is assumed to be an eight-membered ring.[74] The decomposition of **8.18** in THF yields the polycyclic compound **8.19a** via elimination of NSCl and $TeCl_4$.[73a] Compound **8.19a** was originally obtained from the reaction of $TeCl_4$ with $Me_3SiNSNSiMe_3$ in a 1:1 molar ratio in toluene.[75a] The fluoro analogue of **8.19b** is obtained in a similar manner by using TeF_4.[75b] X-ray structural determinations for **8.17** and **8.19a,b** reveal Te–N single bonds and S=N double bonds. The μ_3-nitrido function is an interesting feature of the structures of **8.19a,b** that was subsequently observed in the binary cluster Te_6N_8 (Section 5.2.8).

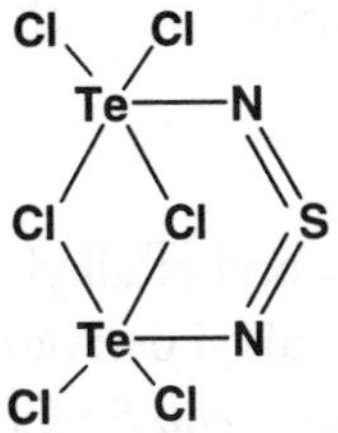

8.17 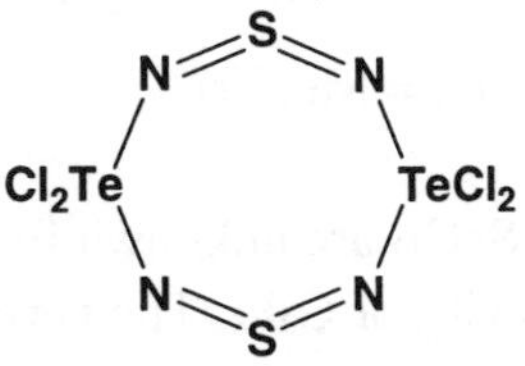   **8.18** 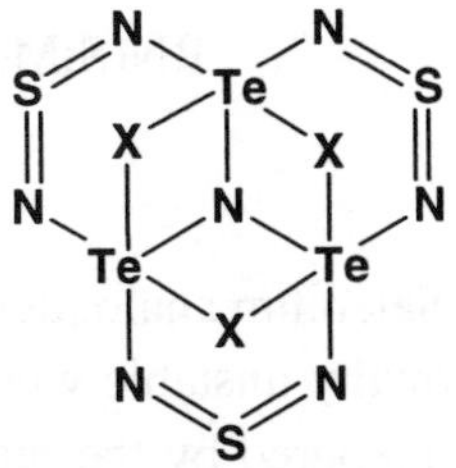   **8.19a** X = Cl
                                    8.19b X = F

The bicyclic compound **8.17** also serves as a source of the five-membered ring **8.20** upon reduction with $SbPh_3$.[76] In contrast to the related S or Se systems, **8.11a** and **8.11b**, both Cl substituents are attached covalently to Te in **8.20**. Reaction of **8.20** with an excess of AsF_5 in SO_2 produces the eight-membered cyclic $[Te_2S_2N_4]^{2+}$ dication, which exhibits a Te–Te bond length of 2.88 Å (*cf.* 2.70 Å for a Te–Te single bond) and no transannular S•••S bonding .[76]

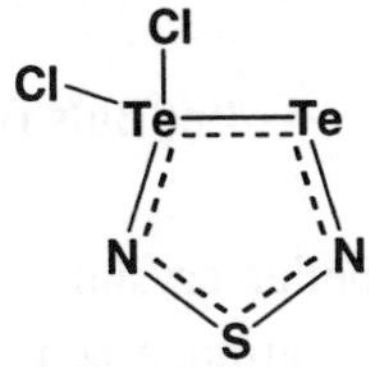

8.20 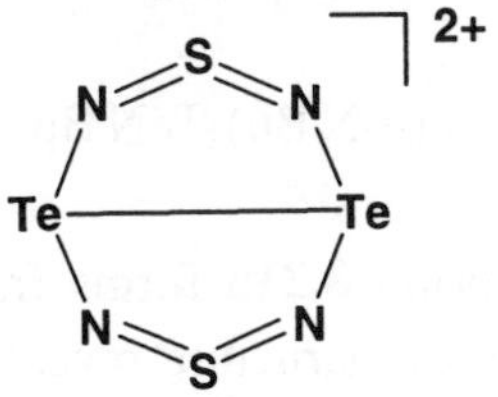   **8.21**

8.11   Chalcogen–Nitrogen Halides Containing Three-Coordinate Nitrogen

Imido chalcogen halides of the type $RNECl_2$ (E = S, Se, Te) provide an interesting illustration of the reluctance of the heavier chalcogens to form –N=E< double bonds. The sulfur derivatives $RNSX_2$ (X = F, Cl) are stable, monomeric compounds.[77,78] The difluorides are organic derivatives of **8.2**, which may be obtained in good yields from SF_4 and N-silylated primary amines (Eq. 8.17). Halogen exchange, *e.g.*, with $AlCl_3$ in nitromethane, yields the corresponding dichlorides $RN=SCl_2$;[77] $^tBuNSCl_2$ is a yellow oil.[79]

$$RN(SiMe_3)_2 + SF_4 \rightarrow RN=SF_2 + 2Me_3SiF \qquad (8.17)$$

$$(R = Me, Et)$$

Selenium analogues $RN=SeCl_2$ are unknown for R = aryl or alkyl and thermally unstable when R = CF_3 or C_2F_5. The perfluoroalkyl derivatives are prepared by the reaction of dichloroamino compounds and Se_2Cl_2 in CCl_3F (Eq. 8.18), but they decompose at room temperature to the corresponding diazene and a mixture of selenium chlorides.[80]

$$3RNCl_2 + 2Se_2Cl_2 \rightarrow 3RN=SeCl_2 + SeCl_4 \qquad (8.18)$$

$$(R = CF_3, C_2F_5)$$

By contrast to the selenium systems, *tert*-butylimidotellurium dihalides ($^tBuNTeX_2)_n$ (X = Cl, Br) are thermally stable in the solid state. They are obtained in good yields in THF solution by the redistribution reaction depicted in Eq. 8.19.[81]

$$^tBuNTe(\mu-N^tBu)_2TeN^tBu + 2TeX_4 \rightarrow 4/n(^tBuNTeX_2)_n \quad (8.19)$$

The dichloride **8.21a** forms fragile, lamellar crystals with a golden colour. The X-ray structure reveals a layered arrangement of hexameric units formed by linking three ($^tBuNTeCl_2)_2$ dimers by chloride bridges. The reaction of **8.21a** with potassium *tert*-butoxide yields

^{t}BuN=Te(O^{t}Bu)$_2$, which forms the dimer **8.22** with significantly different Te–N bond lengths (1.94 and 2.22 Å) in the solid state.[81]

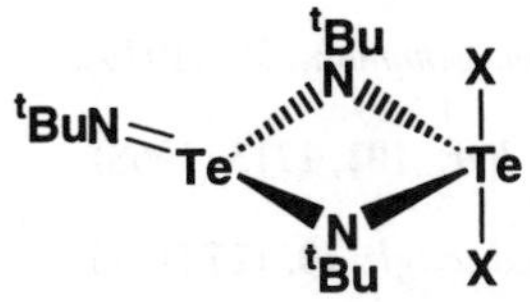

8.21a X = Cl
8.21b X = Br

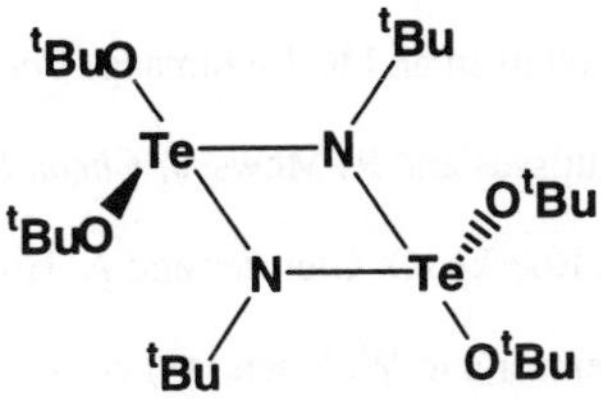

8.22

The acyclic imidoselenium(II) dihalides ClSe[N(^{t}Bu)Se]$_n$Cl (**8.23**, n =1; **8.24**, n = 2) are obtained from the reaction SeCl$_2$ with *tert*-butylamine in a 2:3 molar ratio in THF.[82] There are no sulfur or tellurium analogues of this class of chalcogen–nitrogen halide.

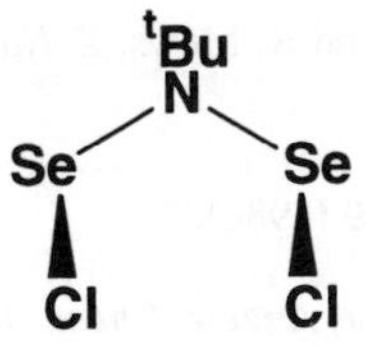

8.23 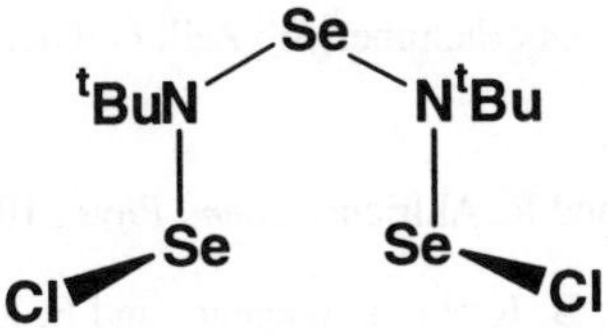   **8.24**

References

1. O. Glemser and R. Mews, *Angew. Chem., Int. Ed. Engl.*, **19**, 883 (1980).

2. O. Glemser, R. Mews and H. W. Roesky, *Chem. Ber.*, **102**, 1523 (1969).

3. A. Haas and M. Rieland, *Chimia*, **42**, 67 (1988).

4. R. C. Patton and W. L. Jolly, *Inorg. Chem.*, **9**, 1079 (1970).

5. J. Passmore and M. J. Schriver, *Inorg. Chem.*, **27**, 2749 (1988).

6. R. Ahlrichs and C. Ehrhardt, *Chem. Phys.*, **107**, 1 (1986).

7. S.C. Peake and A. J. Downs, *J.Chem.. Soc. Dalton Trans.*, 859 (1974).

8. (a) W. H. Kirchoff and E. B. Wilson, Jr., *J. Am. Chem. Soc.*, **85**, 1726 (1963);
   (b) T. Beppu, E. Hirota and Y. Morino, *J. Mol. Spectroscopy*, **36**, 386 (1970).

9. M. T. Nguyen and R. Flammang, *Chem. Ber.*, **129**, 1379 (1996).

10. W. Bludssus and R. Mews, *J. Chem. Soc., Chem. Commun.*, 35 (1979).

11. H. W. Roesky, O. Glemser and A. Hoff, *Chem. Ber.*, **101**, 1215 (1968).

12. O. Glemser and W. Koch, *Angew. Chem., Int. Ed. Engl.*, **10**, 127 (1971).

13. W. Heilemann and R. Mews, *Chem. Ber.*, **121**, 461 (1988).

14. E. Kessenich, F. Kopp, P. Mayer and A. Schulz, *Angew. Chem., Int. Ed. Engl.*, **40**, 1904 (2001).

15. J. Adel and K. Dehnicke, *Chimia*, **42**, 413 (1988).

16. (a) R. Mews, K. Keller and O. Glemser, *Inorg. Synth.*, **24**, 12 (1986); (b) W. H. Kirchoff and E. B. Wilson, Jr., *J. Am. Chem. Soc.*, **84**, 334 (1962).

17. O. Glemser, R. Mews and H. W. Roesky, *Chem. Commun.*, 914 (1969).

18. J. Haase, H. Oberhammer, W. Zeil, O. Glemser and R. Mews, *Z. Naturforsch.*, **25A**, 153 (1970).

19. C. Erhardt and R. Ahlrichs, *Chem. Phys.*, **108**, 429 (1986).

20. O. Glemser, B. Krebs, J. Wegener and E. Kindler, *Angew. Chem., Int. Ed. Engl.*, **8**, 598 (1969).

21. P.G. Watson, E. Lork and R. Mews, *J. Chem. Soc., Chem. Commun.*, 1069 (1994).

22. (a) R. Mews, *Angew. Chem., Int. Ed. Engl.*, **17**, 530 (1978); (b) R. Bartsch, H. Henle, T. Meier and R. Mews, *Chem. Ber.*, **121**, 451 (1988).

23. (a) A. F. Clifford and G. R. Zeilinga, *Inorg. Chem.*, **8**, 979 (1969); (b) A. F. Clifford, J. L. Howell and D. L. Wooton, *J. Fluorine Chem.*, **11**, 433 (1978).

24. R. Cassoux, R. L. Kuczowski and R. A. Creswell, *Inorg. Chem.*, **16**, 2959 (1977).

25. (a) O. Glemser and J. Wegener, *Angew. Chem., Int. Ed. Engl.*, **9**, 309 (1970);
    (b) R. Mews, P. G. Watson and E. Lork, *Coord. Chem. Rev.*, **158**, 233 (1997);
    (c) S. Pohl, B. Krebs, U. Seyer, and G. Henkel, *Chem. Ber.*, **112**, 1751 (1979).

26. R. Mews, *Angew. Chem., Int. Ed. Engl.*, **15**, 691 (1976).

27. B. Ayres, A. J. Banister, P. D. Coates, M. I. Hansford, J. M. Rawson, C. E. F. Rickard. M. B. Hursthouse, K. M. Abdul Malik and M. Motevalli, *J. Chem.. Soc. Dalton Trans.*, 3097 (1992).

28. M. Borschag, A. Schulz and T. M. Klapötke, *Chem. Ber.*, 127, 2187 (1994).

29. R. Wollert, A. H. Ilworth, G. Frenking, D. Fenske, H. Goesman and K. Dehnicke, *Angew. Chem., Int. Ed. Engl.*, 31 1251 (1992).

30. C. Lau, B. Neumüller, W. Hiller, M. Herker, S. F. Vyboishchikov, G. Frenking and K. Dehnicke, *Chem. Eur. J.*, 2, 1373 (1996).

31. M. Broschag, T. M. Klapötke, I. C. Tornieporth-Oetting and P. S. White, *J. Chem. Soc., Chem. Commun.*, 1390 (1992).

32. M. Broschag, T. M. Klapötke, A. Schulz and P. S. White, *Inorg. Chem.*, 32, 5734 (1993).

33. M. Broschag, T. M. Klapötke, A. Schulz and P. S. White, *Chem. Ber.*, 127, 2177 (1994).

34. A. Schulz, P. Buzek, P. von R. Schleyer, M. Broschag, I. C. Tornieporth-Oetting, T. M. Klapötke and P. S. White, *Chem. Ber.*, 128, 35 (1995).

35. J. Passmore, G. Schatte and T. S. Cameron, *J. Chem. Soc., Chem. Commun.*, 2311 (1995).

36. C. Lau, B. Neumüller and K. Dehnicke, *Z. Anorg. Allg. Chem.*, 622, 739 (1996).

37. T. Chivers, Sulfur-Nitrogen Heterocycles, in I. Haiduc and D. B. Sowerby (ed.) *The Chemistry of Inorganic Homo- and Heterocycles*, Vol. 2, Academic Press, London, 822 (1987).

38. W. L. Jolly and K. D. Maguire, *Inorg. Synth.*, 9, 102 (1967).

39. H. W. Roesky, W. Schaper, O. Petersen and T. Müller, *Chem. Ber.*, 110, 2695 (1977).

40. D. K. Padma and R. Mews, *Z. Naturforsch.*, 42B, 699 (1987).

41. R. Wollert, B. Neumüler and K. Dehnicke, *Z. Anorg. Allg. Chem.*, 616, 191 (1992).

42. T. Chivers, J. Siivari and R. S. Laitinen, *Inorg. Chem.*, 32, 4391 (1993).

43. (a) H. M. M. Shearer, A. J. Banister, J. Halfpenny and G. Whitehead, *Polyhedron*, **2**, 149 (1983); (b) W. Isenberg, N. K. Homsy, J. Anhaus, H. W. Roesky and G. M. Sheldrick, *Z. Naturforsch.*, **38B**, 808 (1983).

44. C. Lau, B. Neumüller and K. Dehnicke, *Z. Naturforsch.*, **52B**, 543 (1997).

45. G. G. Alange, A. J. Banister and B. Bell, *J. Chem. Soc., Dalton Trans.*, 2399 (1972).

46. H. Schröder and O. Glemser, *Z. Anorg. Allg. Chem.*, **298**, 78 (1959).

47. U. Demant and K. Dehnicke, *Z. Naturforsch.*, **41B**, 929 (1986).

48. E. Jaudas-Prezel, R. Maggiulli, R. Mews, H. Oberhammer and W-D. Stohrer, *Chem. Ber.*, **123**, 2117 (1990).

49. A. C. Hazell, G. Wiegers and A. Vos, *Acta Crystallogr.*, **20**, 186 (1966).

50. B. Krebs and S. Pohl, *Chem. Ber.*, **106**, 1069 (1973).

51. A. Lork, G. Gard, M. Hare, R. Mews, W-D. Stohrer and R. Winter, *J. Chem. Soc., Chem. Commun.*, 898 (1992).

52. R. Jones, I. P. Parkin, D. J. Williams and J. D. Woollins, *Polyhedron*, **6**, 2161 (1987).

53. W. Isenberg and R. Mews, *Z. Naturforsch.*, **37B**, 1388 (1982).

54. (a) F. A. Kennett, G. K. MacLean, J. Passmore and M. N. S. Rao, *J. Chem. Soc., Dalton Trans.*, 851 (1982); (b) M. T. Nguyen and R. Flammang, *Chem. Ber.*, **129**, 1373 (1996).

55. S. Shi, T. J. Katz, B. V. Yang and L. Liu, *J. Org. Chem.*, **60**, 1285 (1995).

56. (a) X. L. Duan, C. W. Rees and T. Y. Yue, *Chem. Commun.*, 367 (1997); (b) X. L. Duan and C. W. Rees, *Chem. Commun.*, 1493 (1997).

57. (a) X. G. Duan, X. L. Duan, C. W. Rees and T. Y. Yue, *J. Chem. Soc., Perkin Trans.*, *1*, 2597 (1997); (b) C. W. Rees and T. Y. Yue, *Chem. Commun.*, 1202 (1998).

58. L. Zborilova and P. Gebauer, *Z. Anorg. Allg. Chem.*, **448**, 5 (1979).

59. I. Ruppert, *J. Fluorine Chem.*, **20**, 241 (1982).

60. Z. Zak, *Acta Crystallogr.*, **B37**, 23 (1981).

61. (a) H. W. Roesky, M. N. S. Rao, C. Graf, A. Gieren and E. Hädicke, *Angew. Chem., Int. Ed. Engl.*, **20**, 592 (1981); (b) H. W. Roesky, C. Pelz, B. Krebs and G. Henkel, *Chem. Ber.*, **115**, 1448 (1982).

62. O. Glemser, H. Schröder and H. Haeseler, *Z. Anorg. Allg. Chem.*, **279**, 28 (1955).

63. (a) D. Gregson, G. Klebe and H. Fuess, *Acta Crystallogr.*, **C47**, 1784 (1991); (b) M. H. Palmer, R. T. Oakley and N. P. C. Westwood, *Chem. Phys.*, **131**, 255 (1989).

64. H. R. Allcock, *Phosphorus-Nitrogen Compounds*, Academic Press, New York (1972).

65. H. Kluver and O. Glemser, *Z. Naturforsch.*, **32B**, 1209 (1977).

66. (a) A. V. Kirsanov, *Zh. Obshch. Chim.*, **22**, 81 (1952); (b) T. J. Maricich and M. H. Khalil, *Inorg. Chem.*, **18**, 912 (1979).

67. (a) F. Seel and G. Simon, *Z. Naturforsch.*, **19B**, 354 (1964); (b) T-P. Lin, U. Klingebiel and O. Glemser, *Angew. Chem., Int. Ed. Engl.*, **11**, 1095 (1972).

68. E. Lork, U. Behrens, G. Steinke and R. Mews, *Z. Naturforsch.*, **49B**, 437 (1994).

69. D. L. Wagner, H. Wagner and O. Glemser, *Chem. Ber.*, **108**, 2469 (1975).

70. H. Wagner, D. L. Wagner, and O. Glemser, *Chem. Ber.*, **110**, 683 (1977).

71. T. Moeller and A. Ouchi, *J. Inorg. Nucl. Chem.*, **28**, 2147 (1966).

72. E. Lork and R. Mews, *J. Chem. Soc., Chem. Commun.*, 1113 (1995).

73. (a) A. Haas, *J. Organomet. Chem.*, **646**, 80 (2002); (b) A. Haas, *Adv. Heterocyclic Chem.*, **71**, 115 (1998).

74. R. Boese, J. Dworak, A. Haas and M. Pryka, *Chem. Ber.*, **128**, 477 (1995).

75. (a) H. W. Roesky, J. Münzenberg and M. Noltemeyer, *Angew. Chem., Int. Ed. Engl.*, **29**, 61 (1990); (b) J. Münzenberg, H. W. Roesky, S. Besser, R. Herbst-Irmer and G. M. Sheldrick, *Inorg. Chem.*, **31**, 2986 (1992).

76. (a) A. Haas and M. Pryka, *J. Chem. Soc., Chem. Commun.*, 391 (1994); (b) A. Haas and M. Pryka, *Chem. Ber.*, **128**, 11 (1995).

77. O. Glemser and R. Mews, *Adv. Inorg. Chem. Radiochem.*, **14**, 333 (1970).

78. H. W. Roesky, The Sulfur-Nitrogen Bond, in A. Senning (ed.) *Sulfur in Organic and Inorganic Chemistry*, Vol. 4, Marcel Dekker, Inc., New York, pp. 15-45, (1982).

79. O. Scherer and G. Wolmershäuser, *Z. Anorg. Allg. Chem.* **432**, 173 (1977).

80. J. S. Thrasher, C. W. Bauknight, Jr. and D. S. Desmarteau, *Inorg. Chem.*, **24**, 1598 (1985).

81. T. Chivers, G. D. Enright, N. Sandblom, G. Schatte and M. Parvez, *Inorg. Chem.*, **38**, 5431 (1999).

82. T. Maaninen, T. Chivers, R. S. Laitinen and E. Wegelius, *Chem. Commun.*, 759 (2000).

Chapter 9

Chalcogen–Nitrogen Oxides

9.1 Introduction

In this chapter the chemistry of both acyclic and cyclic species involving a chalcogen, nitrogen and oxygen linked together will be discussed. The simplest examples are ternary anionic species, which exist as the structural isomers [NSO]⁻ (thionylimide) and [SNO]⁻ (monothionitrite). The former anion is isoelectronic with SO_2 while the latter is isovalent with [NO₂]⁻. The related thionitrate ion [SNO₂]⁻ is especially interesting because of its role as an ignition agent in the gunpowder reaction.[1]

Although the Se and Te analogues of these ternary anions are unknown, a few examples of seleninylamines RNSeO have been reported and their behaviour is compared with the more common thionylamines RNSO. Organic derivatives of the type RSNO (*S*-nitrosothiols) are attracting significant attention in view of their involvement in the biological storage and transportation of NO.[2] This property leads to the possible application of these simple organic sulfur-nitrogen compounds as NO donors for the treatment of blood circulation problems.

The chapter will begin with a discussion of these acyclic species including the coordination chemistry of neutral RNSO ligands[3] and the applications of E(NSO)₂ (E = S, Se, Te) in the synthesis of chalcogen-nitrogen heterocycles.[4] The final sections will be concerned with five-, six- and eight-membered S-N rings and bicyclic systems that have one (or more) oxygen atoms attached to one (or more) of the sulfur centres.

9.2   The Thionyl Imide Anion, [NSO]⁻

Alkali-metal thionylimides are prepared by the reaction of Me_3SiNSO with the appropriate alkali-metal *tert*-butoxide in THF (Eq. 9.1).[5,6] The more soluble $[(Me_2N)_3S]^+$ salt has also been reported (Eq. 9.2).[7]

$$MO^tBu + Me_3SiNSO \rightarrow M[NSO] + Me_3SiO^tBu \qquad (9.1)$$

$$(M = Na, K, Rb, Cs)$$

$$[(Me_2N)_3S][Me_3SiF_2] + Me_3SiNSO \rightarrow$$

$$[(Me_2N)_3S][NSO] + 2Me_3SiF \qquad (9.2)$$

The tetramethylammonium salt $[Me_4N][NSO]$ is obtained by cation exchange between $M[NSO]$ (M = Rb, Cs) and tetramethylammonium chloride in liquid ammonia.[8] An X-ray structural determination reveals approximately equal bond lengths of 1.43 and 1.44 Å for the S–N and S–O bonds, respectively, and a bond angle <NSO of *ca.* 127°. The [NSO]⁻ anion (**9.1**) exhibits three characteristic bands in the IR spectrum at *ca.* 1270-1280, 985-1000 and 505-530 cm^{-1}, corresponding to v(S-N), v(S-O) and δ(NSO), respectively.[6] *Ab initio* molecular orbital calculations, including a correlation energy correction, indicate that the [NSO]⁻ anion is more stable than the isomer [SNO]⁻ by at least 9.1 kcal mol^{-1}.[9]

Salts of the [NSO]⁻ anion can be used for the synthesis of both transition-metal and main group element thionyl imides by metathetical reactions, *e.g.*, $Cp_2Ti(NSO)_2$ and $Ph_{3-x}As(NSO)_x$ (x = 1,2), respectively (Section 7.6).

9.3   The Thionitrite and Perthionitrite Anions, [SNO]⁻ and [SSNO]⁻

The red [SSNO]⁻ anion (**9.2**) (λ_{max} 448 nm) is produced by the reaction of an ionic nitrite with elemental sulfur or a polysulfide in acetone, DMF or DMSO.[10] The formation of **9.2** probably proceeds via an intermediate such as the $[S_2NO_2]^-$ anion.[11] This process is thought to occur in the gunpowder reaction, which also entails the reaction of potassium nitrite (produced by reduction of potassium nitrate with charcoal) and sulfur.

Potassium thionitrate, $K[SNO_2]$, causes a rapid ignition through its strongly exothermic decomposition into thiosulfate and dinitrogen oxide (Eq. 9.3), which is consumed by its explosive reaction with carbon monoxide to give dinitrogen and carbon dioxide.[1] In solution the [SSNO]⁻ anion decomposes to give the blue radical anion $S_3^{-\bullet}$.

$$2K[SNO_2] \rightarrow K_2[S_2O_3] + N_2O \qquad (9.3)$$

The [SSNO]⁻ anion has been isolated as the $[N(PPh_3)_2]^+$ salt and shown to have a planar *cis* structure with a short S–S distance (1.99 Å) and S–N and N–O distances of 1.67 Å and 1.22 Å, respectively.[10] The treatment of [SSNO]⁻ with triphenylphosphine produces the thionitrite anion [SNO]⁻ (**9.3**), which has a bent structure with a bond angle of *ca.* 120° at nitrogen and S–N and N–O bond lengths of 1.69 Å and 1.21 Å, respectively.[10]

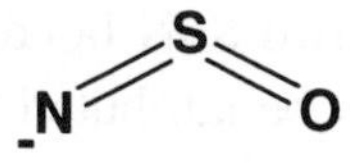

9.1 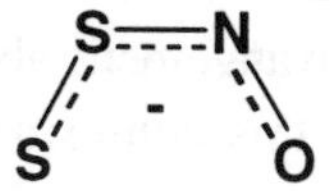   **9.2** 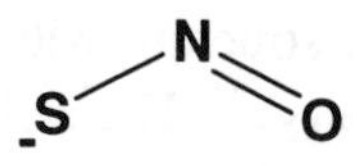   **9.3**

9.4   The $[SO_xN_3]^-$ (x = 2, 3) Anions

Yellow $[(SO_2)_2N_3]^-$ salts, obtained from SO_2 solutions of CsN_3 or $[Me_4N]N_3$, lose SO_2 to give the corresponding $[SO_2N_3]^-$ salts.[12a] The $[SO_3N_3]^-$ anion is prepared by the reaction of $Cs[SO_3Cl]$ with trimethylsilyl azide (Eq. 9.4).[12a]

$$Cs[SO_3Cl] + Me_3SiN_3 \rightarrow Cs[SO_3N_3] + Me_3SiCl \qquad (9.4)$$

The S–N bond in $[SO_2N_3]^-$ is much weaker than that in $[SO_3N_3]^-$ as reflected in an increase in the S–N bond distance by 0.23 Å in the Cs^+ salts.[12a] The S-N bond length in $[Me_4N][SO_2N_3]$ is 2.00 Å.[12b] This marked difference can be attributed to the much weaker Lewis acidity of SO_2 compared to that of SO_3 which also results in lower thermal stability for the $[SO_2N_3]^-$ anion.

9.5   Bis(sulfinylamino)chalcogenanes E(NSO)₂ (E = S, Se, Te)

Bis(sulfinylamino)chalcogenanes $E(NSO)_2$ (E = S, Se. Te) are useful reagents for the synthesis of chalcogen-nitrogen ring systems.[4] The best method to prepare $S(NSO)_2$ is by treatment of Me_3SiNSO with SCl_2[13] at reflux in the absence of solvent (Eq. 9.5). This method can be adapted to the synthesis of $Se(NSO)_2$ (Eq. 9.6).[14] The recommended procedure for preparing the tellurium analogue involves the metathesis reaction of bis(trifluoromethylthio)tellurium with $Hg(NSO)_2$ at 50°C (Eq. 9.7).[4,15]

$$2Me_3SiNSO + SCl_2 \rightarrow S(NSO)_2 + 2Me_3SiCl \qquad (9.5)$$

$$2Me_3SiNSO + Se_2Cl_2 \rightarrow Se(NSO)_2 + Se + 2Me_3SiCl \qquad (9.6)$$

$$Hg(NSO)_2 + Te(SCF_3)_2 \rightarrow Te(NSO)_2 + Hg(SCF_3)_2 \qquad (9.7)$$

In 1961 $S(NSO)_2$ (**9.4a**, E = S) was determined to have a planar, acyclic structure with a *cis* arrangement about the two S=N bonds (C_{2v} symmetry).[16] The stability of this conformation can be attributed to the electrostatic interaction between the negatively charged oxygen centres and the positively charged sulfur atom.[17] Accurate structural parameters for **9.4a** have subsequently been determined[18] and the selenium and tellurium analogues, **9.4b** and **9.4c**, have been shown to adopt similar structures.[14,19] The central E–N bonds are essentially single bonds. The S=N bond distances fall in the double bond range with a significant shortening of the bond along the series **9.4a-c**, suggesting increased ionic character in this bond as a result of the decreasing electronegativity of the central chalcogen E.

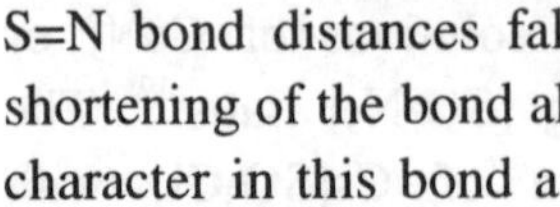

9.4a (E = S)
9.4b (E = Se)
9.4c (E = Te)

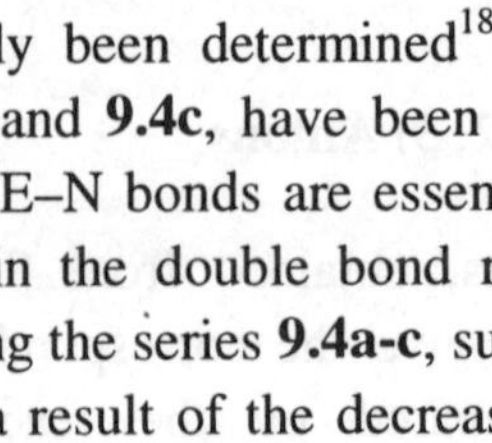

9.5

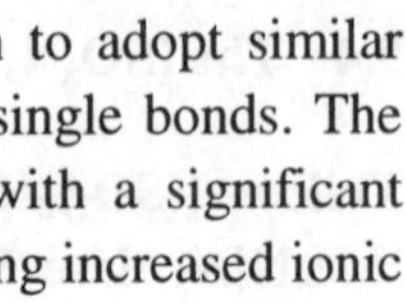

9.6

Complexes of **9.4a** with silver(I) or zinc(II), in which the ligand bonds to the metal through either oxygen or nitrogen atoms, have been structurally characterized.[20] A common reaction of chalcogenylamines RNSO is the thermal or base-promoted elimination of SO_2 to give the corresponding sulfur diimide (Section 9.6). In the case of $E(NSO)_2$ this process gives rise to chalcogen-nitrogen ring systems. For example, the reactions of **9.4a** or **9.4b** with $TiCl_4$ produce the adducts $ESN_2 \cdot TiCl_4$ (E = S, Se), which are assumed to have polymeric structures with bridging ESN_2 ligands.[4,14,21] The reaction of **9.4a** with Cl_2 produces the six-membered ring **9.5** with one S(VI) and two S(IV) centres.[22] The reagent **9.4a** may also be used to incorporate metal centres into S-N rings. For example, the reaction of $(Ph_3P)_2Pt(CH_2=CH_2)$ with **9.4a** yields the cyclometallathiazene $(Ph_3P)_2PtS_2N_2$ (**9.6**) (Section 7.3.2).[23]

The selenium-containing reagent **9.4b** is also a useful precursor for the preparation of chalcogen-nitrogen heterocycles containing two chalcogens. For example, the reactions of **9.4b** with Lewis acids MF_5 (M = As, Sb) in liquid SO_2 produce the dimeric dication $[Se_2SN_2]_2^{2+}$ as $[MF_6]^-$ salts.[14] A different type of behaviour is observed with PCl_5 or $SeCl_4$, which yield the cation $[ClSe_2SN_2]^+$ (**9.7**) as the chloride salt.[14] The corresponding reactions with tellurium tetrahalides TeX_4 (X = Cl, Br) generate $X_2TeSeSN_2$ (**9.8**), a five-membered ring that contains all three chalcogens.[24]

9.7

9.8a (X= Cl)
9.8b (X = Br)

In contrast to the behaviour of **9.4a**, the acyclic Te(IV) compound $Cl_2Te(NSO)_2$ can be isolated from the reaction of **9.4c** with Cl_2 under mild conditions.[25] Subsequent reaction of $Cl_2Te(NSO)_2$ with Cl_2 generates the bicyclic compound $Cl_6Te_2N_2S$ (Section 8.10).

9.6 Organic Chalcogenylamines RNEO (E = S, Se, Te)

Thionyl imide, HNSO, is a thermally unstable gas, which polymerizes readily. It can be prepared by the reaction of thionyl chloride with ammonia in the gas phase.[26] Organic derivatives RNSO have higher thermal stability, especially when R = Ar. The typical synthesis involves the reaction of a primary amine or, preferably, a silylated amine with thionyl chloride. A recent example is the preparation of FcNSO (Fc = ferrocenyl) shown in Eq. 9.8.[27] In common with other thionylimines, FcNSO readily undergoes SO_2 elimination in the presence of a base, *e.g.*, KO^{t}Bu, to give the corresponding sulfur diimide FcNSNFc.

$$FcNH(SiMe_3) + SOCl_2 + NEt_3 \rightarrow FcNSO + Me_3SiCl + [Et_3NH]Cl \quad (9.8)$$

The important reagent Me_3SiNSO is obtained by the reaction of thionyl chloride with tris(trimethylsilyl)amine at 70°C in the presence of $AlCl_3$ (Eq. 9.9).[28a] It may also be prepared by the interaction of sulfur dioxide with $HN(SiMe_3)_2$.[28b]

$$(Me_3Si)_3N + SOCl_2 \rightarrow Me_3SiNSO + 2Me_3SiCl \quad (9.9)$$

The reaction of Me_3SiNSO with HgF_2 produces the covalent mercury(II) derivative $Hg(NSO)_2$.[29] Carefully controlled reactions of halogens with Me_3SiNSO or $Hg(NSO)_2$ generate *N*-halosulfinylamines XNSO (X = F, Cl, Br) (Eq. 9.10).[29]

$$Me_3SiNSO + X_2 \rightarrow XNSO + Me_3SiX \quad (9.10)$$

$$(X = F, Cl, Br)$$

Organic thionylamines have planar, *cis* structures (**9.9**) in the solid state and in solution, as determined by X-ray crystallography and ^{15}N NMR spectroscopy, respectively.[30] The gas-phase structures of the parent compound HNSO and MeNSO have been determined by microwave spectroscopy.[31,32] The S=N and S=O double bond lengths are 1.51-1.52 and 1.45-1.47 Å, respectively. The bond angle <NSO is *ca.* 117° in

MeNSO and 120° in HNSO, *cf.* 119° in SO₂. The trityl derivative Ph₃CNSO is the only alkyl derivative to have been structurally characterized in the solid state.[33]

The seleninylamine (ᵗBuNSeO)ₙ can be prepared by the reaction of *tert*-butylamine with SeOCl₂.[34] In contrast to ᵗBuNSO, however, the selenium analogue has a dimeric structure (**9.10**) in which the two exocyclic oxo substituents are in a *cis* configuration with respect to the Se₂N₂ ring.[35] The unsymmetrical imido-oxo system *cis*-ᵗBuNSe(μ–NᵗBu)₂SeO has also been structurally characterized.[36] The Se–N bond lengths in these dimers are in the range 1.86-1.94 Å, slightly longer than the single-bond value of 1.86 Å; the Se=O bond lengths of 1.62-1.63 Å are indicative of double bonds. The tellurium analogue (ᵗBuNTeO)ₙ has not been isolated, but the tetrameric unit **9.11** has been obtained as a complex in which the strong Lewis acid B(C₆F₅)₃ is coordinated to the terminal oxygen atoms.[37] The aggregation via Te=O→Te interactions observed in **9.11** suggest that the imidotelluroxane (ᵗBuNTeO)ₙ will have a polymeric structure with alternating Te₂N₂ and Te₂O₂ rings. The increasing reluctance for the heavier chalcogens to form double bonds with NR or O in chalcogenylamines parallels the well-established trend for chalcogen dioxides, *viz.*, SO₂ is a monomeric gas, SeO₂ is a two-dimensional polymer with both single and double SeO bonds, and (TeO₂)∞ is a three-dimensional polymer with only Te–O single bonds.

9.9 **9.10** **9.11**

A number of transition-metal complexes of RNSO ligands have been structurally characterized.[3,30] Three bonding modes, π(*N,S*), σ(*S*)-trigonal and σ(*S*)-pyramidal, have been observed (Scheme 9.1). Side-on (*N,S*) coordination is favoured by electron-rich (d^8 or d^{10}) metal centers, while the σ(*S*)-trigonal mode is preferred for less electron-rich metal centres (or those with competitive strong π-acid co-ligands). As expected π(*N,S*) coordination results in a significant (*ca.* 6%) lengthening of the S–N bond,[38] whereas there are no substantial changes in the structural parameters of the RNSO ligand in the $\sigma(S)$-trigonal bonding mode.[39]

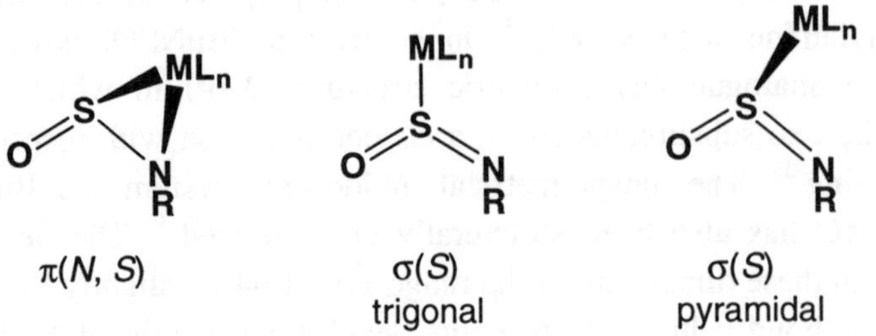

$$\pi(N, S) \qquad \sigma(S) \text{ trigonal} \qquad \sigma(S) \text{ pyramidal}$$

Scheme 9.1 Common bonding modes for RNSO ligands

The parent ligand forms complexes of the type $[M(CO)_5(HNSO)][AsF_6]$ (M = Mn, Re) (Eq. 9.11).[40] The rhenium complex can also be prepared by nucleophilic displacement of F^- from coordinated NSF using Me_3SnOH as the source of OH^-.[40]

$$[M(CO)_5SO_2][AsF_6] + HNSO \rightarrow [M(CO)_5HNSO][AsF_6] + SO_2 \qquad (9.11)$$

$$(M = Mn, Re)$$

The hydrolytic sensitivity of thionylimines is also displayed by L_nM-RNSO complexes, which produce the corresponding L_nM-SO_2 complexes. The ease of hydrolysis of these metal complexes follows the order: $\sigma(S)$-pyramidal > $\sigma(S)$-trigonal > $\pi(N,S)$.[3]

A characteristic reaction of sulfinylimines RNSO is the quantitative addition of R'Li reagents to form adducts of the type $Li[RNS(R')NR]$.[30] The structures of these sulfinimidinates are discussed in Section 10.4.4.[41] The reactions of RNSO derivatives with two equivalents of lithium *tert*-butylamide result in the formation of diazasulfite anions $[OSNR(N^tBu)]^{2-}$ (**9.12**) (Eq. 9.12).[42] The dilithium derivatives of these dianions form hexameric thirty-six atom ($Li_{12}N_{12}O_6S_6$) clusters with structures that are dependent on the nature of the R group.

$$6RNSO + 12LiNH^tBu \rightarrow \{Li_2[OS(N^tBu)(NR)]\}_6 + 6H_2N^tBu \quad (9.12)$$

$$(R = {}^tBu, SiMe_3)$$

The corresponding reaction of SO_2 with lithium derivatives of primary amines yields dilithium derivatives of the azasulfite anions $[O_2SNR]^{2-}$ (**9.13**) that have not been structurally characterized.[43] The reaction of hexameric magnesium anilide $[(thf)MgNPh]_6$ with SO_2 results in twelvefold insertion of SO_2 into the Mg–N bonds to form a forty-eight atom $(Mg_6S_{12}N_6O_{24})$ cluster that contains the azadisulfite dianion $[O_2S(\mu\text{-}NPh)SO_2]^{2-}$ (**9.14**), which acts as a bis-chelating ligand towards the Mg^{2+} centres.[44]

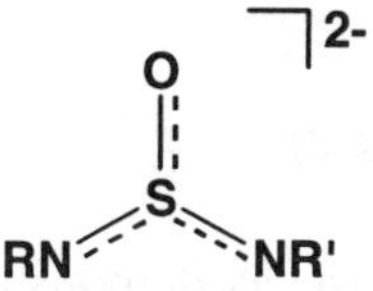

9.12 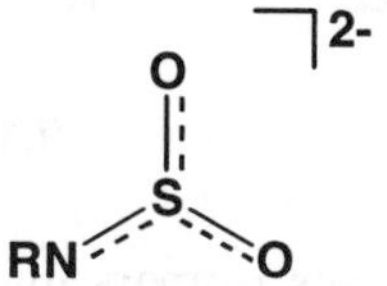   9.13 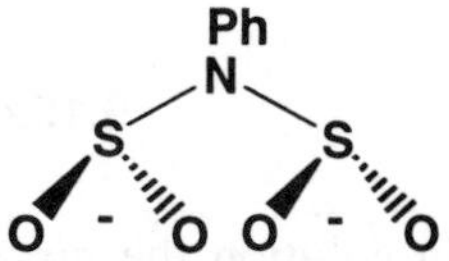   9.14

9.7 *S*-Nitrosothiols RSNO

S-Nitrosothiols RSNO are structural isomers of thionyl imines RNSO. The formal reversal of the order of the S and N atoms in these organic derivatives of ternary S/N/O anions results in completely different properties. S-Nitrosothiols, several of which occur naturally, *e.g.*, *S*-nitrosocysteine and S-nitrosoglutathione, have been the subject of intense investigation because of their important role in NO transport and regulation in biological systems. Potential applications of RSNO compounds under scrutiny include their use as vasodilators in the treatment of angina, and in the search for a cure for male impotence.[2]

The most convenient route to *S*-nitrosothiol formation is the nitrosation of thiols by nitrous acid (Eq. 9.13).

$$RSH + HNO_2 \leftrightarrow RSNO + H_2O \quad (9.13)$$

S-nitrosothiols may adopt either *cis* (**9.15a**) or *trans* (**9.15b**) conformations with respect to the S–N bond. *S*-Nitrosocaptopril[45] adopts a *cis* arrangement, whilst the *trans* isomer pertains in the solid state for *S*-nitroso-*N*-acetylpenicillamine[46] and Ph$_3$CSNO[47] as determined by X-ray crystallography. The S–N bond lengths are in the range 1.76-1.79 Å typical for a single bond. Nevertheless, there is a significant barrier to rotation about the S–N bond of 11-12 kcal mol^{-1}. The S–N bond dissociation energies are estimated to be about 20-30 kcal mol^{-1}.[45, 47-49]

9.15a (*cis*)   **9.15b** (*trans*)

In solution the *cis* and *trans* isomers may co-exist, as demonstrated by ^{15}N NMR and UV-visible spectra. The ^{15}N NMR chemical shift of the *trans* isomer is shifted *ca.* 60 ppm downfield relative to the *cis* isomer.[45] The visible absorption band of *S*-nitrosothiols corresponds to a weak $n \rightarrow \pi^*$ transition in the 520-590 nm region. The absorption maxima of *trans* conformers are red-shifted by *ca.* 30 nm relative to those of the *cis* isomer. Two absorptions are observed in the 520-590 nm region in the experimental spectra of RSNO derivatives.[45]

Stable *S*-nitrosothiols have also been generated by attaching bowl-shaped aryl substituents to the nitrogen centre.[50a] This approach has also been successful for the isolation of the first *S*-nitrososelenol as stable purple crystals which, like the sulfur analogue, adopts a *cis* conformation.[50b] The related alkyl derivative (Me$_3$Si)$_3$CSeNO is prepared by nitrosation of (Me$_3$Si)$_3$CSeH with tert-butylnitrite.[50c] This nitrosated selenium species is much less stable that the corresponding *S*-nitrosothiol. It is a red compound, which decomposes above −78°C. EPR spectra indicate that the formation of the radical [(Me$_3$Si)$_3$CSe(NO)$_2$]$^{\bullet}$ occurs during this decomposition, which produces the corresponding di- and tri-selenides.

The ability of *S*-nitrosothiols to mimic many of the biological properties of NO itself may emanate from *in vivo* decomposition to generate NO. This decomposition is catalyzed by Cu^{2+},[2] and it has been demonstrated that PVC or polyurethane films doped with a lipophilic copper(II) complex are capable of catalytically decomposing endogenous RSNO species to NO. This observation may be important in the development of thrombo-resistant devices used in kidney dialysis or coronary by-pass surgery.[51] It is also possible that direct transfer of NO from RSNO may occur in biological systems, since this process has been established to take place at an iron centre with the use of model non-haem iron systems.[52]

9.8 Cyclic Chalcogen-Nitrogen Oxides

Five-, six- and eight-membered chalcogen-nitrogen ring systems in which oxygen is attached to one or more of the chalcogen atoms, as well as bicyclic species, have been structurally characterized. The discussion of these heterocycles is arranged into two sections: (a) neutral compounds and (b) anions. The neutral systems are limited to those that contain sulfur while the anions include the selenium-nitrogen oxo-anion $[Se_3N_3O_6]^{3-}$.

9.8.1 *Neutral compounds*

The sulfur-nitrogen oxide S_3N_2O (thiodithiazyl oxide) is a red liquid. It was first prepared in 1975 by the reaction of $(Me_2SnS_2N_2)_2$ with SOF_2,[53a] but it is more conveniently prepared from $[S_3N_2Cl]Cl$ (Eq. 9.14).[53b]

$$[S_3N_2Cl]Cl + HCOOH \rightarrow S_3N_2O + 2HCl + CO \qquad (9.14)$$

The ^{15}N NMR spectrum of S_3N_2O exhibits two resonances consistent with the five-membered ring structure (**9.16**).[54] Lewis acids such as $SnCl_4$ form 2:1 adducts with **9.16**, in which the ligands are coordinated to tin in a *cis* configuration through their oxygen atoms.[54] The neutral compound $S_3N_2O_5$ is obtained when S_4N_4 or $S(NSO)_2$ is treated with an excess of SO_2.[55] It has a cyclic structure (**9.17**) in which the ring oxygen atom is displaced by *ca.* 0.6 Å out of the mean plane of the S_3N_2 unit.

Cyclotetrathiazene dioxide $S_4N_4O_2$ (**9.18**) is obtained as orange-yellow needles by the condensation of $[S_3N_2Cl]Cl$ with sulfamide $SO_2(NH_2)_2$.[56] The S–N bond lengths in **9.18** indicate a structure in which two –N=S=N– units bridge a sulfur atom and an SO_2 group to form a boat-shaped eight-membered ring. Sodium methoxide attacks **9.18** at the sulfur atom opposite to the SO_2 group to give $[S_4N_4O_2(OMe)]^-$, an eight-membered ring with a transannular S•••S interaction.[57] Other nucleophiles promote ring contraction. For example, the reaction of **9.18** with $AsPh_3$ produces the five-membered ring $S_3N_2=NSO_2N=AsPh_3$, which undergoes a self-condensation to give the eleven-atom sulfur-nitrogen chain $Ph_3As=N–SO_2–N=S=N–S–N=S=N–SO_2–N=AsPh_3$.[58]

9.16	**9.17**	**9.18**

9.8.2 *The anions $[S_3N_3O_x]^-$ (x = 1, 2, 4), $[E_3N_3O_6]^{3-}$ (E = S, Se) and $[S_4N_5O_x]^-$ (x = 1,2)*

An early application of ^{15}N NMR spectroscopy for monitoring the reactions of sulfur-nitrogen compounds involved the oxidation of the anion $[S_3N_3]^-$.[59] It was shown that controlled air oxidation allows the introduction of one or two oxygens on one of the sulfur atoms to give red $[S_3N_3O]^-$ (**9.19**) and purple $[S_3N_3O_2]^-$ (**9.20**), respectively. The $[S_3N_3O_2]^-$ anion is also formed by ring contraction of **9.18** under the influence of azide anion. Another member of the series of cyclic N-S-O anions, the yellow anion $[S_3N_3O_4]^-$ (**9.21**), is produced as the $[S_6N_4]^{2+}$ salt from the treatment of **9.18** with SO_3.[60] The six-membered rings in the anions **9.19** and **9.20** exhibit significant differences in chemically equivalent bond lengths and bond angles. The structural features of the anion **9.21** are similar to those of the isoelectronic neutral molecule **9.17**.

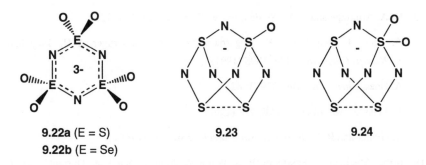

9.19　　　　**9.20**　　　　**9.21**

The [NSO$_2$]$^-$ anion, isoelectronic with SO$_3$, may be obtained as the ammonium salt by the reaction of SO$_2$(NH$_2$)$_2$ with sulfuryl chloride followed by treatment with ammonia. It forms a trimer in the form of a chair-shaped, six-membered ring [N$_3$S$_3$O$_6$]$^{3-}$ (**9.22a**), which exhibits mean S–N bond lengths of 1.60 Å and approximately equal S–O bond lengths.[61] The selenium analogue [N$_3$Se$_3$O$_6$]$^{3-}$ (**9.22b**) adopts a similar conformation in the tripotassium salt with mean Se–N bond lengths of 1.77 Å.[62]

The yellow [S$_4$N$_5$O]$^-$ anion (**9.23**) is obtained as the water-soluble ammonium salt from the reaction of SOCl$_2$ with liquid ammonia.[63] The structure of **9.23** is related to that of the binary anion [S$_4$N$_5$]$^-$ (Section 5.4.7); the oxygen atom in **9.23** is attached to a sulfur that is connected to three nitrogen atoms and the unbridged S•••S distance is 2.63 Å.[63] The [S$_4$N$_5$O$_2$]$^-$ ion (**9.24**) has been prepared from **9.20** by oxidative-addition with PhICl$_2$, followed by treatment with Me$_3$SiNSNSiMe$_3$.[64] Mild heating converts **9.24** back to **9.20** via extrusion of an –NSN– bridging unit.

9.22a (E = S)　　　　**9.23**　　　　**9.24**
9.22b (E = Se)

References

1. F. Seel, Sulfur in History: The Role of Sulfur in "Black Powder", in A. Müller and B. Krebs (ed.) *Sulfur: Its Significance for Chemistry, for the Geo-, Bio- and Cosmophere and Technology"*, Elsevier, Amsterdam, pp. 55-66 (1984).

2. D. L. H. Williams, *Acc. Chem. Res.*, **32**, 869 (1999).

3. A. F. Hill, *Adv. Organomet. Chem.*, **36**, 159 (1994).

4. A. Haas, *Adv. Heterocycl. Chem.*, **71**, 115 (1998).

5. D. A. Armitage and J. C. Brand, *J. Chem. Soc., Chem. Commun.*, 1078 (1979).

6. S. Mann and M. Jansen, *Z. Anorg. Allg. Chem.*, **621**, 153 (1995).

7. W. Heilemann and R. Mews, *Chem. Ber.*, **121**, 461 (1988).

8. S. Mann and M. Jansen, *Z. Naturforsch.*, **49B**, 1503 (1994).

9. S. P. So, *Inorg. Chem.*, **28**, 2888 (1989).

10. F. Seel, R. Kuhn, G. Simon, M. Wagner, B. Krebs and M. Dartmann, *Z. Naturforsch.*, **40B**, 1607 (1985).

11. F. Seel, G. Simon. J. Schuh, M. Wagner, B. Wolf, I. Ruppert and A. B. Wieckowski, *Z. Anorg. Allg. Chem.*, **538**, 177 (1986).

12. (a) K. O. Christe, J. A. Boatz, M. Gerken, R. Haiges, S. Schneider. T. Schroer, F. S. Tham. A. Vij, V. Vij, R. I. Wagner and W. W. Wilson, *Inorg. Chem.*, **41**, 4275 (2002); (b) A. Kornath, O. Blecher and R, Ludwig, *Z. Anorg. Allg. Chem.*, **628**, 183 (2002).

13. D. A. Armitage and A. W. Sinden, *Inorg. Chem.*, **11**, 1151 (1972).

14. A. Haas, J. Kasprowski, K. Angermund, P. Betz, C. Kruger, Y-H. Tsay and S. Werner, *Chem. Ber.*, **124**, 1895 (1991).

15. A. Haas, *Chem.-Ztg.*, **106**, 239 (1982).

16. J. Weiss, *Z. Naturforsch.*, **16B**, 477 (1961).

17. R. Gleiter and R. Bartetzko, *Z. Naturforsch.*, **36B**, 492 (1981).

18. (a) G. MacLean, J. Passmore, P. S. White, A. Banister and J. A. Durrant, *Can. J. Chem.*, **59**, 187 (1981); (b) R. Steudel, J. Steidel and N. Rautenberg, *Z. Naturforsch.*, **35B**, 792 (1980).

19. A. Haas and R. Pohl, *Chimia*, **43**, 261 (1989).

20. (a) H. W. Roesky, M. Thomas, P. G. Jones, W. Pinkert and G. M. Sheldrick, *J. Chem. Soc., Dalton Trans.*, 1211 (1983): (b) H. W. Roesky, M. Thomas, J. W. Bats and H. Fuess, *Inorg. Chem.*, **22**, 2342 (1983).

21. H. W. Roesky, J. Anhaus and W. S. Sheldrick, *Inorg. Chem.*, **23**, 75 (1984).

22. J. Weiss, R. Mews and O. Glemser, *J. Inorg. Nucl. Chem., Suppl.*, 213 (1976).

23. T. Chivers, F. Edelmann, U. Behrens and R. Drews, *Inorg. Chim. Acta*, **116**, 145 (1986).

24. A. Haas, J. Kasprowski and M. Pryka, *Chem. Ber.*, **125**, 789 (1992).

25. R. Boese, J. Dworak, A. Haas and M. Pryka, *Chem. Ber.*, **128**, 477 (1995).

26. P. W. Schenk, *Ber. Dtsch. Chem. Ges.*, **75**, 94 (1942).

27. M. Herberhold, B. Distler, H. Maisel, W. Milius, B. Wrackmeyer and P. Zanello, *Z. Anorg. Allg. Chem.*, **622**, 1515 (1996).

28. (a) E. Parkes and J. D. Woollins, *Inorg. Synth.*, **25**, 48 (1989); (b) J. F. Davis and L. D. Spicer, *Inorg. Chem.*, **19**, 2191 (1980).

29. W. Verbeek and W. Sundermeyer, *Angew. Chem., Int. Ed. Engl.*, **8**, 376 (1969).

30. K. Vrieze and G. van Koten, *J. R. Netherlands Chem. Soc.*, **99**, 145 (1980).

31. W. H. Kirchoff, *J. Am. Chem. Soc.*, **91**, 2437 (1969).

32. B. Beagly, S. J. Chantrell, R. G. Kirby and D. G. Schmidling, *J. Mol. Struct.*, **25**, 319 (1975).

33. K. O. Christe, M. Gerken, R. Haiges, S. Schneider, T. Schroer, F. S. Tham and A. Vij, *Solid State Sci.*, **4**, 1529 (2002).

34. M. Herberhold and W. Jellen, *Z. Naturforsch.*, **41B**, 144 (1986).

35. T. Maaninen, R. Laitinen and T. Chivers, *Chem. Commun.*, 1812 (2002).

36. T. Maaninen, T. Chivers, R. Laitinen, G. Schatte and M. Nissinen, *Inorg. Chem.*, **39**, 5341 (2000).

37. T. Chivers and G. Schatte, Abstract P-40, *10th International Symposium on Inorganic Ring Systems*, Burlington, Vermont, U.S.A. (2003).

38. A. F. Hill, G. R. Clark, C. E. F. Richard, W. R. Roper and M. Herberhold, *J. Organomet. Chem.*, **401**, 357 (1991).

39. (a) S. J. Laplaca and J. A. Ibers, *Inorg. Chem.*, **5**, 405 (1966); (b) L. Vaska and S. S. Bath, *J. Am. Chem. Soc.*, **88**, 1333 (1966).

40. G. Hartmann, R. Hoppenheit and R. Mews, *Inorg. Chim. Acta*, **76**, L201 (1981).

41. F. Pauer and D. Stalke, *J. Organomet. Chem.*, **418**, 127 (1991).

42. J. K. Brask, T. Chivers, M. Parvez and G. Schatte, *Angew. Chem., Int. Ed. Engl.*, **36**, 1986 (1997).

43. J. K. Brask, T. Chivers, B. McGarvey, G. Schatte, R. Sung and R. T. Boeré, *Inorg. Chem.*, **37**, 4633 (1998).

44. J. K. Brask, T. Chivers and M. Parvez, *Angew. Chem., Int. Ed. Engl.*, **39**, 958 (2000).

45. M. Bartberger, K. N. Houk, S. C. Powell, J. D. Mannion, K. Y. Lo, J. S. Stamler, and E. J. Toone, *J. Am Chem. Soc.*, **122**, 5889 (2000).

46. G. E. Carnahan. P. G. Lenhert and R. Ravichandran, *Acta Crystallogr.*, **B34**, 2645 (1978).

47. N. Arulsamy, D. S. Bohle, J. A. Butt, G. I. Irvine, P. A. Jordan and E. Sagan, *J. Am. Chem. Soc.*, **121**, 7115 (1999).

48. J-M. Lü, J. M. Wittbrodt, K. Wang, Z. Wen, H. B. Schegel, P. G. Wang and J-P. Cheng, *J. Am. Chem. Soc.*, **123**, 2903 (2001).

49. D. Bartberger, J. D. Mannion, S. C. Powell, J. S. Stamler, K. N. Houk and E. J. Toone, *J. Am. Chem. Soc.*, **123**, 8868 (2001).

50. (a) K. Goto, Y. Hino, Y. Takahashi, T. Kawashima, G. Yamamoto, N. Takagi and S. Nagase, *Chem. Lett.*, 1204 (2001);   (b) K. Goto, K. Shimada and T. Kawashima, Abstract IL-3, *IXth International Conference on the Chemistry of Selenium and Tellurium*, Bombay, India, 2004; (c) C. Wismach, W-W. du Mont, P. G. Jones, L. Ernst, U. Papke, G. Mugesh, W. Kaim, M. Wanner and K. D. Becker, *Angew. Chem., Int. Ed. Engl.*, **43**, 3970 (2004).

51. B. K. Oh and M. E. Meyerhoff, *J. Am. Chem. Soc.*, **125**, 9552 (2003).

52. A. R. Butler, S. Elkins-Daukes, D. Parkin and D. L. H. Williams, *Chem. Commun.*, 1732 (2001).

53. (a) H. W. Roesky and H. Wiezer, *Angew. Chem., Int. Ed. Engl.*, **14**, 258 (1975); (b) H. W. Roesky and M. Witt, *Inorg. Synth.*, **25**, 49 (1989).

54. H. W. Roesky, M. Kuhn and J. W. Bats, *Chem. Ber.*, **115**, 3025 (1982).

55. E. Rodek, N. Amin and H. W. Roesky, *Z. Anorg. Allg. Chem.*, **457**, 127 (1979).

56. H. W. Roesky, W. Schaper, O. Petersen and T. Müller, *Chem. Ber.*, **110**, 2695 (1977).

57. H. W. Roesky, M. Witt, B. Krebs and H. J. Korte, *Angew. Chem., Int. Ed. Engl.*, **18**, 415 (1979).

58. M. Witt, H. W. Roesky, M. Noltemeyer, W. Clegg, M. Schmidt and G. M. Sheldrick, *Angew. Chem., Int. Ed. Engl.*, **20**, 974 (1981).

59. T. Chivers, A. W. Cordes, R. T. Oakley and W. T. Pennington, *Inorg. Chem.*, **22**, 2429 (1983).

60. H. W. Roesky, M. Witt, J. Schimkowiak, M. Schmidt, M. Noltemeyer and G. M. Sheldrick, *Angew. Chem., Int. Ed. Engl.*, **21**, 538 (1982).

61. C. Leben and M. Jansen, *Z. Naturforsch.*, **54B**, 757 (1999).

62. V. Kocman and J. Rucklidge, *Acta Crystallogr.*, **B30**, 6 (1974).

63. R. Steudel, P. Luger and H. Bradaczek, *Angew. Chem., Int. Ed. Engl.*, **12**, 316 (1973).

64. R. T. Boeré, R. T. Oakley and M. Shevalier, *J. Chem. Soc., Chem. Commun.*, 110 (1987).

Chapter 10

Acyclic Organo Chalcogen–Nitrogen Compounds

10.1 Introduction

In this chapter the chemistry of acyclic organic derivatives containing a chalcogen–nitrogen functional group as the central feature will be discussed. The simplest examples are thionitrosyls RNS isovalent with nitroso compounds RNO. Little is known about the Se and Te analogues RNE (E = Se, Te) although the cyclic trimers have been characterized (Section 6.3). Sulfur–nitrogen compounds with the composition RNS_2 exist as N-thiosulfinylamines RN=S=S with unbranched structures rather than the branched structures that are typical of the corresponding nitro compounds RNO_2. Chalcogen diimides RN=E=NR (E = S, Se, Te) represent a very widely studied class of chalcogen-nitrogen compounds for which detailed comparisons between the behaviour of the three chalcogens can be made. These compounds are isoelectronic with SO_2, $(SeO_2)_\infty$ and $(TeO_2)_\infty$. Investigations of their reactivity have led to the synthesis of imido analogues of other sulfur-oxygen species such as SO_3, $[SO_3]^{2-}$ and $[SO_4]^{2-}$, i.e., $S(NR)_3$, $[S(NR)_3]^{2-}$ and $[S(NR)_4]^{2-}$, respectively. Hybrid systems of the type RN=E=O (sulfinylamines) are covered under the rubric of chalcogen-nitrogen oxides in Section 9.6.

The foregoing compounds all contain imido groups (NR) attached to the chalcogen centre. A second, very important class of acyclic organo chalcogen–nitrogen compounds involves those in which an amido group (NR_2) is linked to the chalcogen, e.g., $E(NR_2)_2$ (E = S, Se, Te). These compounds have a well-developed chemistry and notable differences

180

between the behaviour of the heavier chalcogen derivatives and that of sulfur have been observed. For example, the reactivity of the Te–N functionality in tellurium amides has been exploited in organic synthesis.

The third class of compounds to be discussed in this chapter are those in which an RE group (E = S, Se, Te) is attached to a nitrogen centre. This category includes amines of the type $(RE)_3N$ and the related radicals $[(RE)_2N]^•$, as well as organochalcogen(II) azides, REN_3, and nitrenes REN (E = S, Se). Covalent azides of the type $RTe(N_3)_3$ and $R_2Te(N_3)_2$, in which the chalcogen is in the +4 oxidation state, have also been characterized.

10.2   Chalcogenonitrosyls, RNE (E = S, Se)

Unlike reactive diatomic chalcogen-nitrogen species NE (E = S, Se) (Section 5.2.1), the prototypical chalcogenonitrosyls HNE (E = S, Se) have not been characterized spectroscopically, although HNS has been trapped as a bridging ligand in the complex $(HNS)Fe_2(CO)_6$ (Section 7.4). *Ab initio* molecular orbital calculations at the self-consistent field level, with inclusion of electron correlation, reveal that HNS is *ca.* 23 kcal mol^{-1} more stable than the isomer NSH.[1] There is no low-lying barrier that would allow thermal isomerization of HNS to occur in preference to dissociation into H + NS. The most common form of HNS is the cyclic tetramer $(HNS)_4$ (Section 6.2.1).

The thionitrosyl group may be stabilized either by a dimethylamino substituent or by the use of a highly bulky *ortho*-substituted aryl groups. *N,N'*-Dimethylthionitrosoamine M_2NNS is obtained as a low melting, deep purple solid from the reaction of 1,1-dimethylhydrazine with sulfur (Eq. 10.1) or by the reduction of Me_2NNSO with $LiAlH_4$.[2] This thermally unstable derivative is monomeric in solution.

$$Me_2NNH_2 + 1/4S_8 \rightarrow Me_2NNS + H_2S \qquad (10.1)$$

The first evidence for thionitrosoarenes $4\text{-}XC_6H_4N{=}S$ (X = H, Cl, Br) was gleaned from trapping experiments in which the Diels–Alder cycloaddition products of the thionitrosoarenes (generated *in situ* from the thermolysis of the *N*,*N'*-thiodianilines $XC_6H_4N(H)SN(H)C_6H_4X$ with dimethylbutadiene were isolated.[3] 2,4-Di-*tert*-butyl-6-cyanothionitrosobenzene (10.1) has been produced by thermal or photochemical reactions (Scheme 10.1) and detected by UV–visible and infrared spectra in argon or undecane matrices at 12 K.[4]

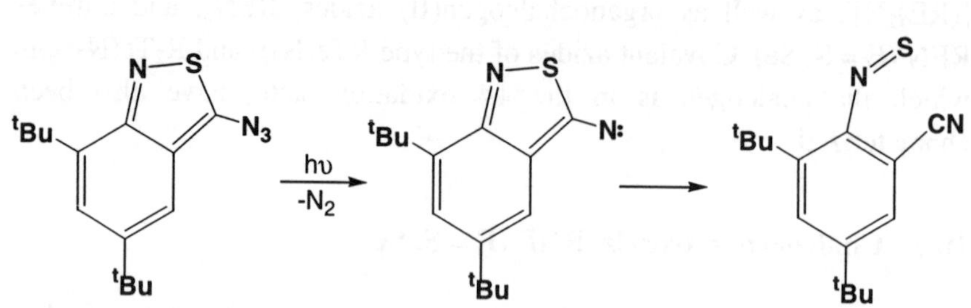

10.1

Scheme 10.1 Thermal or photochemical generation of a nitroso compound

An alternative approach to thionitrosoarenes involves the reaction of amines with SCl_2.[5] This method has also been adapted to the production of selenonitrosoarenes ArN=Se by using the selenium(II) synthon $PhSO_2SeCl$ as the Se source (Scheme 10.2).[6] It is likely that $SeCl_2$, generated *in situ* in THF, could also be used in this process. The Diels–Alder cycloaddition of ArN=Se species with dimethylbutadiene gives 1,2-selenazine derivatives in low yields.

There is no evidence for the tellurium analogues ArN=Te, but the cyclic trimers $(ENtBu)_3$ (E = Se, Te) are stable crystalline solids that have been structurally characterized (Section 6.3).

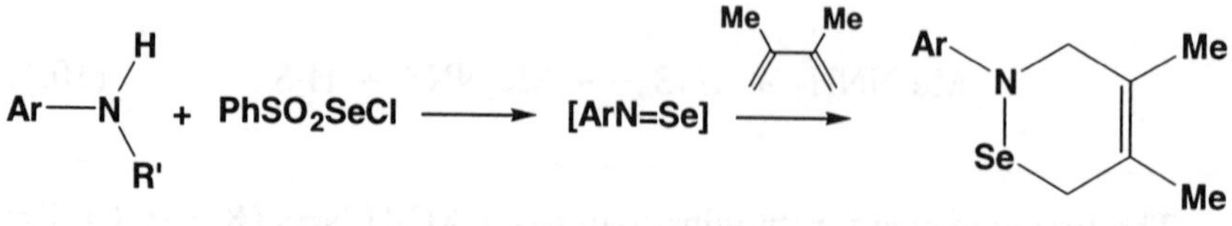

Scheme 10.2 Generation and trapping of selenonitrosoarenes

10.3 *N*-Thiosulfinylamines, RNSS

Dithionitro compounds RNS_2 with a branched structure analogous to that of nitro compounds RNO_2 are unknown. However, a wide range of *N*-thiosulfinylamines RN=S=S have been characterized. *Ab initio* molecular orbital calculations of HNS_2 isomers show that, although the HN=S=S arrangement is the most stable, several other isomers exist within a reasonable energy range.[7] The only plausible options for the oxygen analogue are nitrous acid HO–N=O and the nitro compound HNO_2 (*cf.* $[NO_3]^-$ (branched) and $[NS_3]^-$ (unbranched), Section 5.4.2). The preference for unbranched structures in the S-N systems can be attributed to the relative electronegativities of O, S, and N. In the branched structures of N–O species the more electronegative O atoms carry the negative charge. In S–N systems the unbranched atomic arrangement allows the more electronegative N atom to accommodate some of the negative charge.

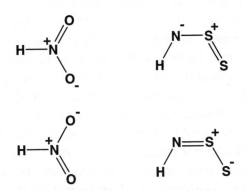

Scheme 10.3 Resonance structures for HNO_2 and HNS_2

The first *N*-thiosulfinylamine $4-Me_2NC_6H_4N=S=S$ (**10.2**) was obtained as a deep violet solid (λ_{max} 510 nm) in low yield by the reaction of phosphorus pentasulfide with *N,N'*-dimethyl-4-nitrosoaniline.[8] Compound **10.2** (M.p. 113-115°C) has much higher thermal stability than the corresponding thionitrosoarenes, but it decomposes to the corresponding azobenzene and sulfur on heating to 200°C.

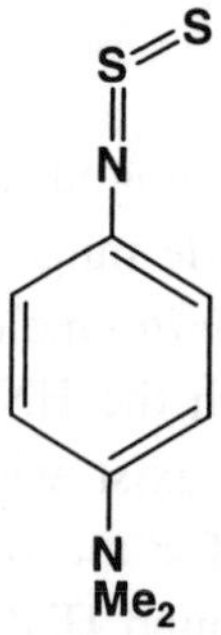

10.2	10.3a (R = Me)	10.3b'
	10.3b (R = ^{t}Bu)	

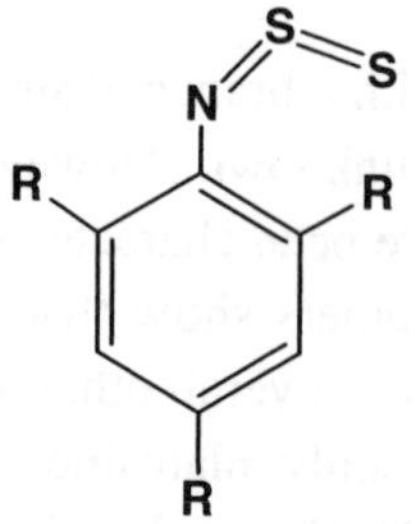

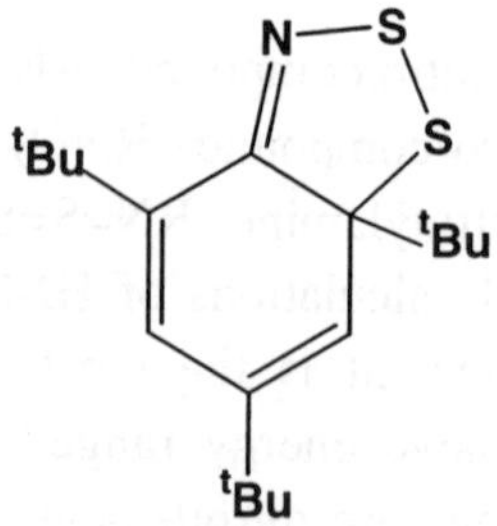

Thermally stable *N*-thiosulfinyl anilines may also be obtained by attaching an aryl group with bulky *ortho* substituents to the nitrogen atom. The derivatives **10.3a** and **10.3b** are obtained in 70-80% yields from the reaction of the corresponding aniline with S_2Cl_2 in diethyl ether in the presence of NEt_3. The structure of **10.3a** has been determined by X-ray crystallography. Interestingly, **10.3a** is a purple compound (λ_{max} *ca.* 540 nm) whereas **10.3b** is yellow in the solid state. The derivative **10.3b** forms red solutions, however, and the ^{1}H NMR spectrum reveals the presence of two compounds, identified as **10.3b** and **10.3b'**, in solution. In the solid state this RNSS derivative exists as the ring-closed isomer **10.3b'**, as demonstrated by an X-ray structural analysis.[9] Compound **10.3a** does not undergo an analogous isomerism in solution. NMR studies have been employed to determine the thermodynamic parameters for the transformation of **10.3b** into **10.3b'**. It is estimated that the π–bond of the –N=S=S group is weaker than that of the –N=S=O group by >25 kcal mol^{-1}.[10]

The photolysis of **10.3a** in pentane or the reaction of **10.3a** with PPh_3 generate the corresponding sulfur diimide ArN=S=NAr (Ar = 2,4-^{t}Bu$_2$-6-MeC$_6$H$_2$) as the major product, presumably via the intermediate formation of the thionitrosoarene.[11,12] By contrast, thermolysis in arene solvents results in ring closure to give the 2,1-benzisothiazole **10.4** and the corresponding aniline.[11]

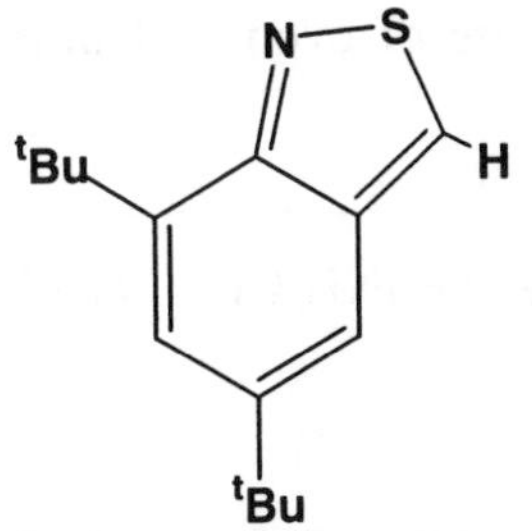

10.4 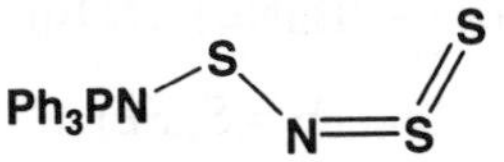   **10.5** 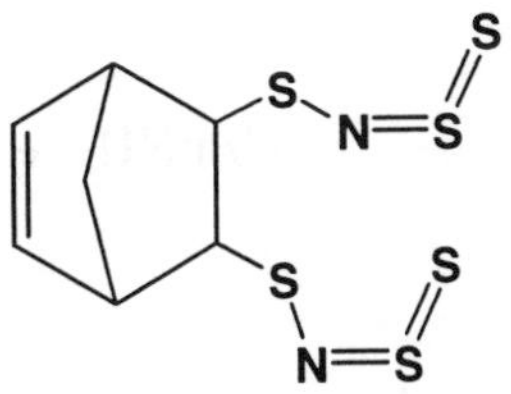   **10.6**

Several compounds containing the SNSS functionality have been prepared and structurally characterized.[13] These include $Ph_3PNSNSS$ (**10.5**)[14] and the norbornenyl derivative **10.6**.[15] The former is obtained by heating the six-membered ring $Ph_3P=N–S_3N_3$ in boiling acetonitrile. The bifunctional derivative **10.6** was isolated as a minor product from the reaction of S_4N_2 and norbornadiene. These *C*- and *N*-bonded derivatives of the SNSS⁻ anion adopt planar *cis, trans* structures with short (*ca.* 1.91 Å) terminal S–S bonds implying, as expected, substantial double bond character. Strong Raman bands are observed in the 570-590 cm⁻¹ region for the S–S bonds in **10.5** and **10.6** consistent with the shortness of these bonds. The SNSS chromophore exhibits a strong visible absorption band in the range 400-500 nm, attributed to a $\pi^* \rightarrow \pi^*$ transition.

10.4   Chalcogen Diimides RN=E=NR (E = S, Se Te)

10.4.1 *Synthesis*

The reactions of primary amines with sulfur(IV) or selenium(IV) halides provide a facile route to the corresponding chalcogen diimides RN=E=NR (E = S, Se), which are of interest from the structural viewpoint, as reagents in organic synthesis and as ligands for transition metals.[16] The first sulfur(IV) diimide $^tBuN=S=N^tBu$ was prepared *ca.* 50 years ago by the reaction of *tert*-butylamine with SCl_4 generated *in situ* (Eq. 10.2).[17] A similar approach can be used to generate $^tBuN=Se=N^tBu$.[18] Selenium(IV) diimides are markedly less thermally robust than their sulfur analogues. For example, $^tBuN=Se=N^tBu$

decomposes at room temperature to give a mixture of cyclic selenium imides and $^tBuN=N^tBu$. (Fig. 3.3).[19]

$$6^tBuNH_2 + ECl_4 \rightarrow {}^tBuN=E=N^tBu + 4^tBuNH_3Cl \qquad (10.2)$$

$$(E = S, Se)$$

The corresponding tellurium diimide $^tBuNTe(\mu\text{-}N^tBu)_2TeN^tBu$ (**10.7**) may be obtained in good yields from the reaction of lithium *tert*-butylamide with $TeCl_4$ in THF (Eq. 10.3).[20a] In toluene solution this reaction also produces the cyclic tellurium(II) imide $(TeN^tBu)_3$.[20b] The dimer **10.7** is obtained as an orange solid, which can be purified by vacuum sublimation at *ca.* 90°C.

$$2TeCl_4 + 8LiNH^tBu \rightarrow {}^tBuNTe(\mu\text{-}N^tBu)_2TeN^tBu$$

$$+ 8LiCl + 4^tBuNH_2 \qquad (10.3)$$

Several methods for the preparation of unsymmetrical sulfur diimides RN=S=NR′ have been developed. One approach involves the addition of a catalytic amount of an alkali metal to a mixture of two symmetrical sulfur diimides, RN=S=NR and R′N=S=NR′.[21] A second method makes use of alkali-metal derivatives of [RNSN]⁻ anions.[22] For example, derivatives in which one of the substituents is a fluoroheteroaryl group can be prepared by the reaction of the anionic nucleophile [RN=S=N]⁻ with pentafluoropyridine.[22b] Sulfur diimides of the type RN=S=NH (R = 2,4,6-$^tBu_3C_6H_2S$) have also been prepared.[22c]

10.4.2 *Structures*

Three different conformations are possible for monomeric chalcogen diimides (Fig. 10.1). Variable-temperature NMR spectra indicate that the *cis,trans* isomer of $S(NR)_2$ is most stable in solution for small organic groups (R = Me, ^tBu).[23] With bulkier organic substituents, small amounts of the *trans,trans* isomer exist in equilibrium with the *cis,cis* isomer.[24,25] The *cis,cis* isomer is observed in solutions of certain sulfur diimides with aryl groups attached to nitrogen $S(NAr)_2$ (Ar = 2,4,6-$C_6H_2Br_3$, 2,6-$C_6H_3Me_2$, C_6F_5).[26] The *syn,anti* or *E,Z* systems of nomenclature may also be used to describe these isomers.

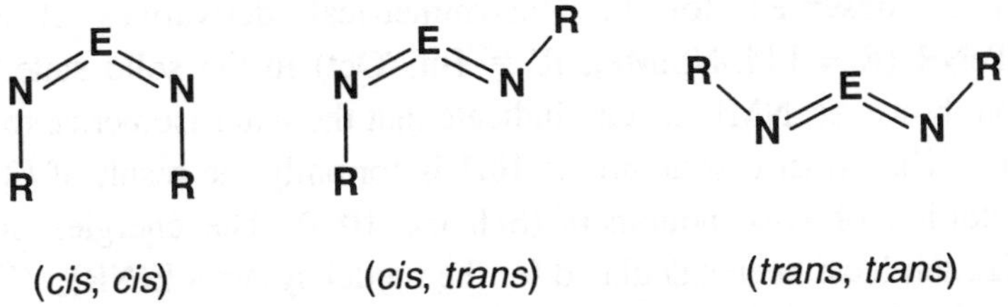

(*cis, cis*)   (*cis, trans*)   (*trans, trans*)

Fig. 10.1   Conformational isomers of chalcogen diimides

X-ray structural determinations of sulfur diimides in the solid state have shown that *cis,trans* or *cis,cis* isomers may exist in the solid state depending on the organic substituent R.[26,27] Although the former is more common, the *cis,cis* conformation is found for R = C_6F_5, 2,6-$Me_2C_6H_3$. Gas phase electron-diffraction studies of the important synthon $Me_3SiNSNSiMe_3$ reveal a *cis,cis* conformation with S=N bond lengths of 1.54 Å,[28] within the typical range for sulfur diimides. Electron density studies for $^tBuN=S=N^tBu$ have shown that the bonding is predominantly electrostatic (S^+–N^-) rather than a covalent double bond (S=N), i.e. valence expansion utilizing d orbitals on sulfur does not occur.[29]

The only selenium diimide to be structurally characterized in the solid state, $Se(NAd)_2$ (Ad = adamantyl),[30] adopts the *cis,trans* conformation consistent with conclusions based on ^{1}H and ^{13}C NMR studies for $Se(N^tBu)_2$.[31]

The geometries and relative energies of the different conformations of model chalcogen diimides $E(NR)_2$ (E = S, Se; R = H, Me, ^tBu and $SiMe_3$) have been investigated by using *ab initio* and DFT molecular orbital methods.[32a] The *cis,trans* conformation is predicted to be most stable with the exception of the parent molecules $E(NH)_2$[32a] and the unsymmetrical systems RNSNH,[32b] for which the *cis,cis* conformation is slightly more stable than the *cis,trans* isomer.

In contrast to the monomeric structures of $E(NR)_2$, the tellurium analogues adopt dimeric structures. The *tert*-butyl derivative **10.7** has a *cis,endo,endo* arrangement of terminal ^{t}Bu groups with respect to the Te_2N_2 ring,[20b] whereas a *trans,exo,exo* arrangement of the exocyclic groups is observed for the unsymmetrical derivatives $RTeN(\mu-NR')_2TeNR$ (R = PPh_2NSiMe_3; R' = ^{t}Bu, ^{t}Oct) in the solid state.[20c] In solution, however, NMR spectra indicate that the latter isomerize to a *cis* isomer.[33] The dimeric structure of **10.7** is formally the result of [2 + 2] cycloaddition of two monomers (Scheme 10.4). The energies of this dimerization have been calculated for the model systems $E(NR)_2$ (E = S, Se, Te; R = H, Me, ^{t}Bu, $SiMe_3$) by a variety of theoretical methods.[34,35] In general, it was found that this process is strongly endothermic for sulfur diimides, approximately thermoneutral for selenium diimides and strongly exothermic for tellurium diimides, consistent with experimental observations. These differences can be attributed to the expected trend to lower π-bond energies for chalcogen-nitrogen $(np-2p)\pi$-bonds along the series S (n = 3), Se (n = 4) and Te (n = 5).

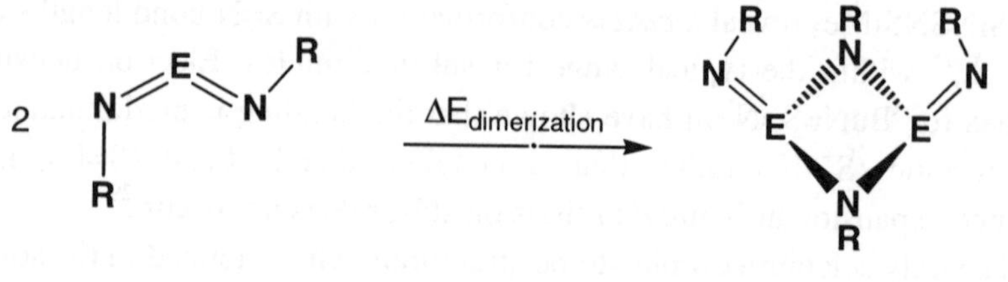

$$2 \quad \begin{array}{c} \text{N} \\ | \\ \text{R} \end{array}\!\!=\!\!E\!\!=\!\!\begin{array}{c} \text{N} \\ | \\ \text{R} \end{array}\!\!-\!\!\text{R} \qquad \xrightarrow{\;\Delta E_{\text{dimerization}}\;} \qquad $$

10.7 (E = Te, R = ^{t}Bu)

Scheme 10.4   Dimerization of chalcogen diimides

10.4.3 *Metal complexes*

Monomeric sulfur diimides have an extensive coordination chemistry as might be anticipated from the availability of three potential donor sites and two π-bonds.[16] In addition, they are prone to fragmentation to produce thionitroso and, subsequently, sulfido and imido ligands. Under mild conditions with suitable coordinatively unsaturated metal complexes sulfur diimides may coordinate without rupture of the – N=S=N– unit. Four modes of coordination have been identified or invoked as intermediates in fluxional processes (Fig. 10.2).

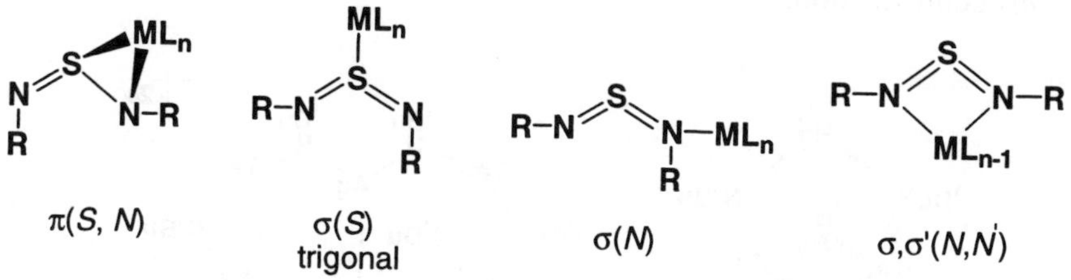

$\pi(S, N)$ $\sigma(S)$ trigonal $\sigma(N)$ $\sigma,\sigma'(N,N)$

Fig. 10.2 Coordination modes for monomeric chalcogen diimides

The bonding mode is dependent on the nature of the metal centre as well as the steric or electronic properties of the imino substituent (*cf.* M-RNSO complexes, Section 9.6). For example, ^{1}H NMR studies indicate that the M(CO)$_5$ unit in the $\sigma(N)$-trigonal complex [W(CO)$_5${S(NMe)$_2$}] undergoes a 1,3-shift between the two nitrogen donors, presumably via a $\pi(N,S)$ intermediate.[36] The corresponding *tert*-butyl derivative readily loses CO to give [W(CO)$_4${S(N^{t}Bu)$_2$}] in which the sulfur diimide is *N,N'*-chelated to the metal.[36,37] This bonding mode is also observed for complexes with main-group metal halides, *e.g.*, [SnCl$_4$·{E(N^{t}Bu)$_2$}] (E = S, Se).[38,39] The effect of metal basicity on the coordination modes is illustrated by the complexes of Pt(II) and Pt(0), [PtCl$_2$(PPh$_3$){η^1-S(NR)$_2$}] (R = alkyl, aryl)[40] and [Pt(PPh$_3$)$_2$(η^2-ArNSNAr)] (Ar = C$_6$H$_4$X-4; X = Me, Cl).[41] An increase in metal electron density favours $\pi(S,N)$ coordination.

The dimeric structure of tellurium diimides enables these versatile ligands to act in a bridging or a chelating bonding mode. In the *cis,exo,exo* conformation the formation *N,N'*-chelated complexes, *e.g.*,with HgCl$_2$ in **10.8**, is observed.[42] With Ag$^+$, however, two metal ions bridge two tellurium diimide ligands in the *cis,exo,exo* conformation in the complex **10.9**.[43] The non-linear (*ca.* 163°) N–Ag–N arrangement is indicative of a metallophilic attraction so that each tellurium diimide ligand is *N,N'*-chelated to an Ag$_2^{2+}$ unit. The *trans* isomer may also bridge metal centres. For example, the interaction of **10.7** with copper(I) trifluoromethanesulfonate brings about a *cis→trans* isomerization of the ligand to give complex **10.10** in which two Cu^+ ions form linear bridges between three tellurium diimide ligands and the central ligand is in the *trans* conformation.[43]

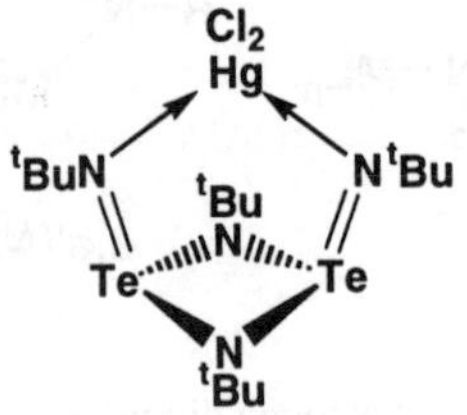

10.8

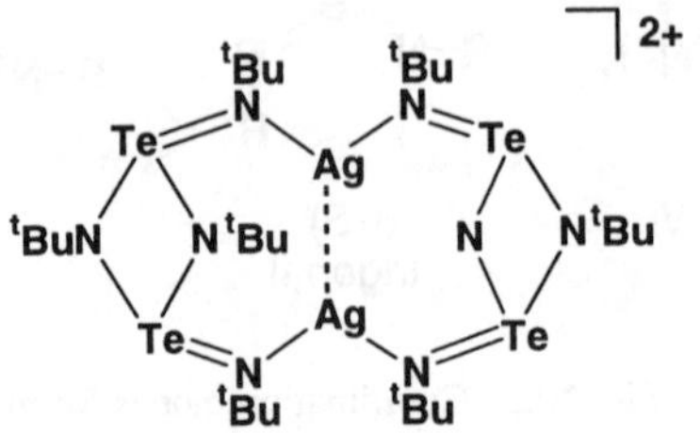

10.9

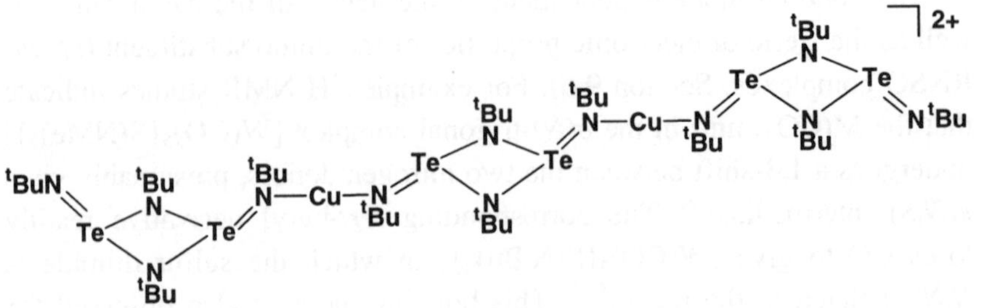

10.10

The most important fragmentation reactions of sulfur diimides occur with polynuclear metal carbonyls, *e.g.*, $M_3(CO)_{12}$ (M = Fe, Ru),[44,45] in which a thionitrosyl RNS is stabilized by π-coordination to a dimetallic fragment in complexes of the type $[M_2(\mu{:}\pi\text{-}SNR)(CO)_6]$ (M = Fe, R = $SiMe_3$; M = Ru, R = ^tBu) (Section 7.4). These fragmentation processes are dependent on the metal. For example, further bond rupture occurs in reactions with Fe, Co or Ni carbonyl complexes to give both imido and sulfido complexes.

10.4.4 *Reactions*

Sulfur diimides may be reduced chemically (by alkali metals)[46] or electrochemically[47] to the corresponding radical anions $[S(NR)_2]^{-•}$, which exhibit five-line (1:2:3:2:1) EPR spectra consistent with coupling of the unpaired electron with two equivalent nitrogen centers.[46] These anion radicals are isoelectronic with the sulfur dioxide radical anion $[SO_2]^{-\cdot}$, which is known to dimerize to the the dithionite dianion $[S_2O_4]^{2-}$ in the solid state. A mechanism involving the centrosymmetric association and rearrangement of two sulfur diimide radical anions has been proposed to account for the scrambling process that occurs when a small amount of an alkali metal is added to a mixture of two symmetrical sulfur diimides (Scheme 10.5).[21]

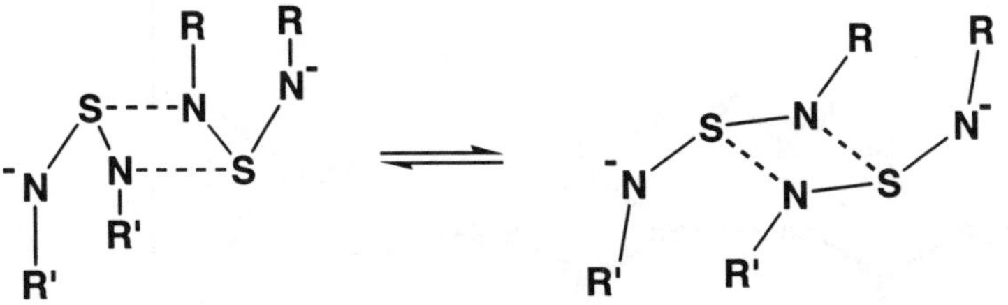

Scheme 10.5 Skeletal scrambling of sulfur diimide radical anions

Redox reactions occur between sulfur diimides and reagents containing Si–Si bonds to give diaminosulfanes (Eq. 10.4).[48]

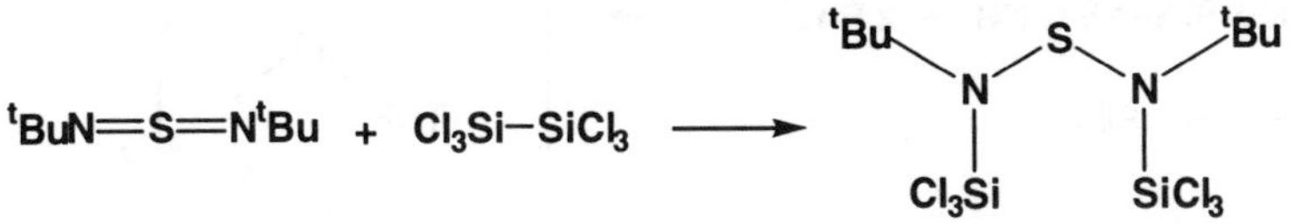

$$^tBuN{=}S{=}N^tBu \; + \; Cl_3Si{-}SiCl_3 \; \longrightarrow$$

$$(10.4)$$

A redox process also occurs in the reaction of selenium diimides with bis(amino)stannylenes. For example, the cyclic stannylene $Me_2Si(\mu\text{-}N^tBu)_2Sn$ reacts in a 1:1 molar ratio with $^tBuN{=}Se{=}N^tBu$ to give a spirocyclic tin complex, which reacts with a second equivalent of the stannylene to generate a Sn–Sn bond [d(Sn–Sn) = 2.85 Å, $^1J(^{119}Sn{-}^{117}Sn)$ = 13,865 Hz)] (Scheme 10.6).[49]

Sulfur diimides react readily and quantitatively with organolithium reagents at the sulfur centre to produce lithium sulfinimidinates of the type $\{Li[RS(NR')_2]\}_x$.[50] The lithium derivatives may be hydrolyzed by water to give the corresponding sulfinimidamide R'NS(R)NHR' which, upon treatment with MH (M = Na, K) or the metal (M = Rb, Cs) in THF, produces the heavier alkali-metal derivatives of these interesting bidentate, monoanionic ligands (Scheme 10.7).[51]

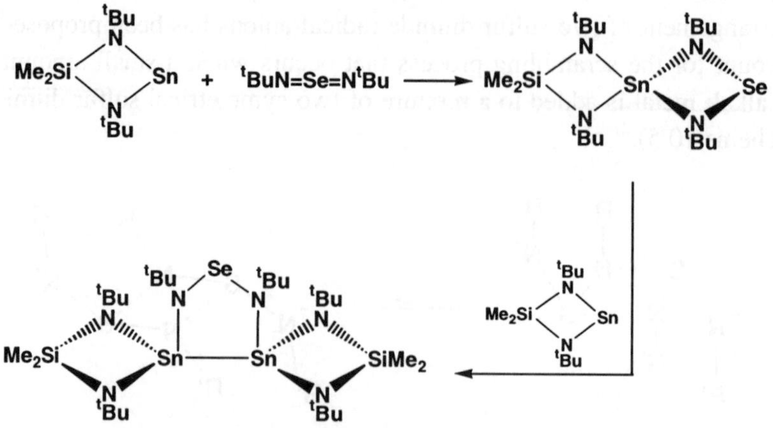

Scheme 10.6　Reaction of a selenium diimide with a bis(amino)stannylene

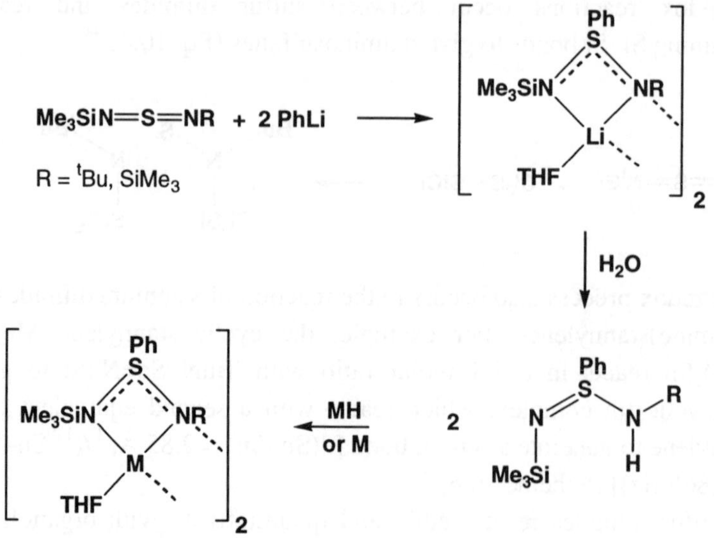

Scheme 10.7　Synthesis of alkali metal sulfinimidinates

The structures of these alkali-metal complexes are influenced by (a) the size and electronic properties of the R′ group (b) the size of the alkali metal cation and (c) solvation of the alkali-metal cation. Several structural types have been established including step-shaped ladders, eight-membered rings and ion-solvated complexes (Fig. 10.3).[51] In contrast to the reaction of silylated sulfur diimides with organolithium reagents, the corresponding reaction with $Me_3SnN=S=NSnMe_3$ yields the N-lithio salt $Me_3SnN=S=NLi$.[52]

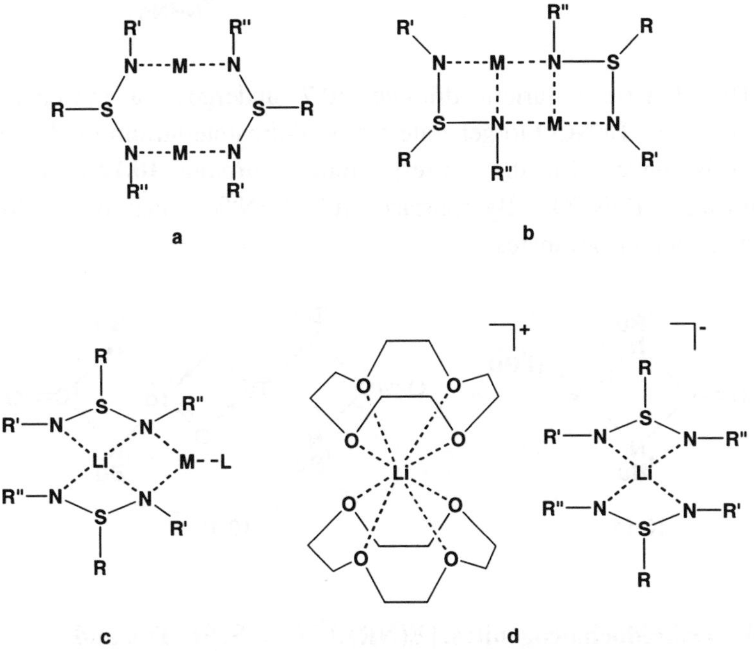

Fig. 10.3   Coordination modes for $[RS(NR')(NR'')]^-$ anions

Selenium diimides react in a manner similar to SeO_2 with unsaturated organic compounds. The first report of $^tBuN=Se=N^tBu$ described its use as an *in situ* reagent for allylic amination of olefins and acetylenes.[18] Improved procedures for this process (Eq. 10.5) and for the diamination of 1,3-dienes (Eq. 10.6) have been developed using the reagents $RN=Se=NR$ [R = *para*-toluenesulfonyl (Ts), *ortho*-nitrobenzenesulfonyl (Ns)].[53]

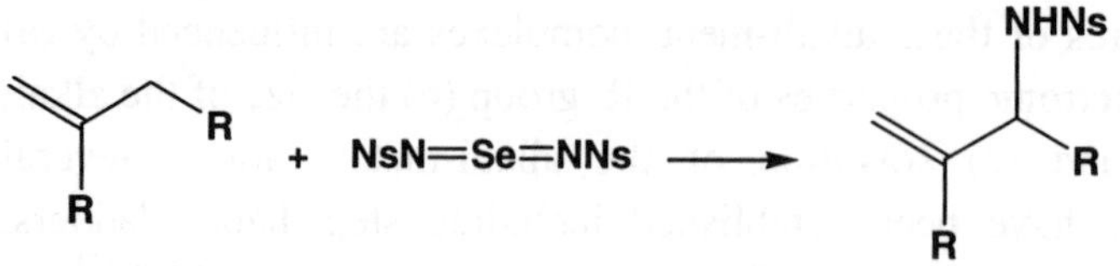

$$\text{(10.5)}$$

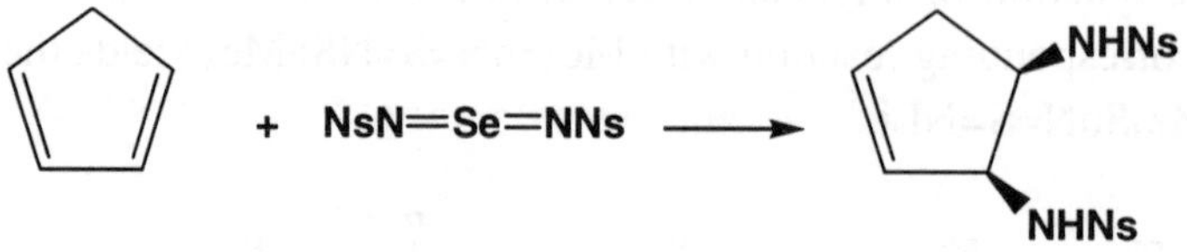

$$\text{(10.6)}$$

The dimeric tellurium diimide **10.7** undergoes a cycloaddition reaction with ^{t}BuNCO to generate the N,N'-ureatotellurium imide **10.11**, which is converted to the corresponding telluroxide **10.12** by reaction with excess ^{t}BuNCO.[54] By contrast, ^{t}BuN=S=N^{t}Bu undergoes exchange reactions with isocyanates.

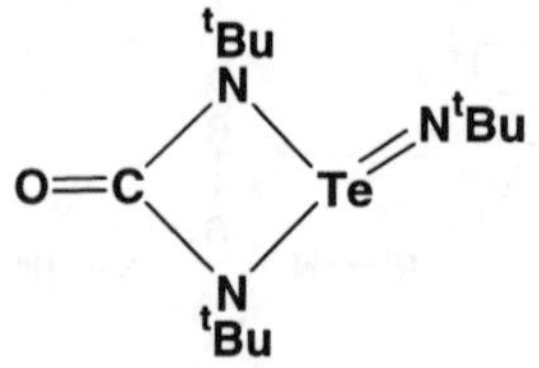

10.11 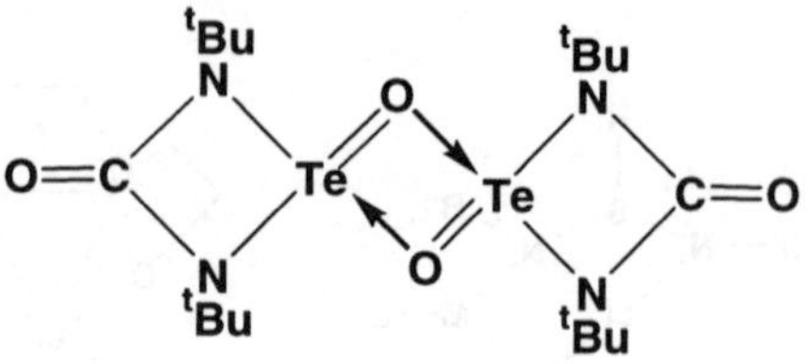   **10.12**

10.5 Triimidochalcogenites, [E(NR)₃]²⁻ (E = S, Se, Te), and Sulfur Triimides, S(NR)₃ [55,56]

Dianions of the type [E(N^{t}Bu)₃]²⁻, isoelectronic with chalcogenite ions EO_3^{2-} (E = S,[57] Se,[58] Te[59]), are easily made by the nucleophilic addition of LiNH^{t}Bu to the chalcogen diimide followed by deprotonation with a second equivalent of LiNH^{t}Bu (Eq. 10.7). In this one-pot reaction the lithium reagent behaves both as a nucleophile and as a base. In the analogous reaction of N,N'-di-*tert*-butylcarbodiimide with LiNH^{t}Bu, the intermediate amidodiimido complex [(thf)Li₂{C(N^{t}Bu)₂(NH^{t}Bu)}₂] has been isolated.[60] In that case, however, the stronger base ^{n}BuLi is necessary for the deprotonation step to give the triimidocarbonate anion [Li$_2${C(N^{t}Bu)$_3$}]$_2$

$$4LiNH^tBu + 2E(N^tBu)_2 \rightarrow [Li_2\{E(N^tBu)_3\}]_2 + 2H_2N^tBu \quad (10.7)$$

$$(E = S, Se, Te)$$

The dilithium triimidochalcogenites [Li$_2${E(N^{t}Bu)$_3$}]$_2$ form dimeric structures in which two pyramidal [E(N^{t}Bu)$_3$]$^{2-}$ dianions are bridged by four lithium cations to form distorted, hexagonal prisms of the type **10.13**. A fascinating feature of these cluster systems is the formation of intensely coloured [deep blue (E = S) or green (E = Se)] solutions upon contact with air. The EPR spectra of these solutions (Section 3.4), indicate that one-electron oxidation of **10.13a** or **10.13b** is accompanied by removal of one Li$^+$ ion from the cluster to give neutral radicals in which the dianion [E(N^{t}Bu)$_3$]$^{2-}$ and the radical monoanion [E(N^{t}Bu)$_3$]$^{-•}$ are bridged by three Li$^+$ ions.[61,62]

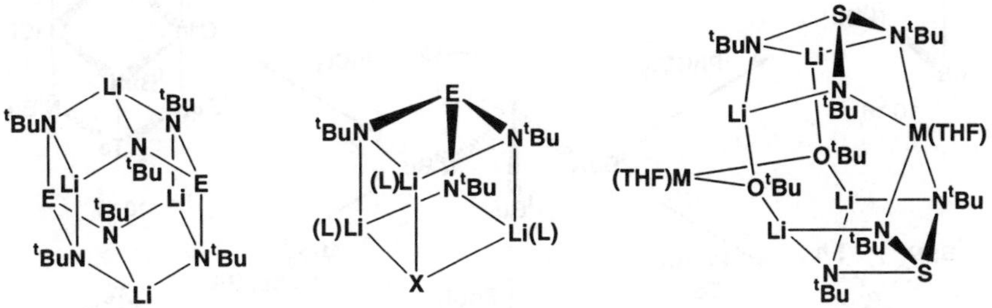

10.13a (E = S)
10.13b (E = Se)
10.13c (E = Te)

10.14 (E = S, Se)
(X = Cl, Br, I)
(L = THF)

10.15

In view of the facile oxidation of **10.13a-c** it is not surprising that some metathetical reactions with metal halides result in redox behaviour.[63] Interestingly, lithium halides disrupt the dimeric structures of **10.13a** or **10.13c** to give distorted cubes of the type **10.14**, in which a molecule of the lithium halide is entrapped by a Li$_2$[E(N^{t}Bu)$_3$] monomer.[61,64] Similar structures are found for the MeLi, LiN$_3$ and LiOCH=CH$_2$ adducts of **10.13a**.[65] In the LiN$_3$ adduct, the terminal nitrogen of the azido ligand bridges to one lithium ion of a neighbouring cube in the crystal lattice.[65b] The OCH=CH$_2$ group in the LiOCH=CH$_2$ adduct is derived from cleavage of THF solvent molecules.[65c]

In the reactions of **10.13a** with alkali metal *tert*-butoxides cage expansion occurs to give the sixteen-atom cluster **10.15,** in which two molecules of MO^{t}Bu (M = Na, K) are inserted into the dimeric structure.[66] The cluster **10.13a** also undergoes transmetallation reactions with coinage metals. For example, the reactions with silver(I) or copper(I) halides produces complexes in which three of the Li$^+$ ions are replaced by Ag$^+$ or Cu$^+$ ions and a molecule of lithium halide is incorporated in the cluster.[67]

Although redox processes are sometimes observed in metathetical reactions with metal halides, the pyramidal dianion [Te(NtBu)$_3$]$^{2-}$ has a rich coordination chemistry (Scheme 10.8).[68] For example, the reaction

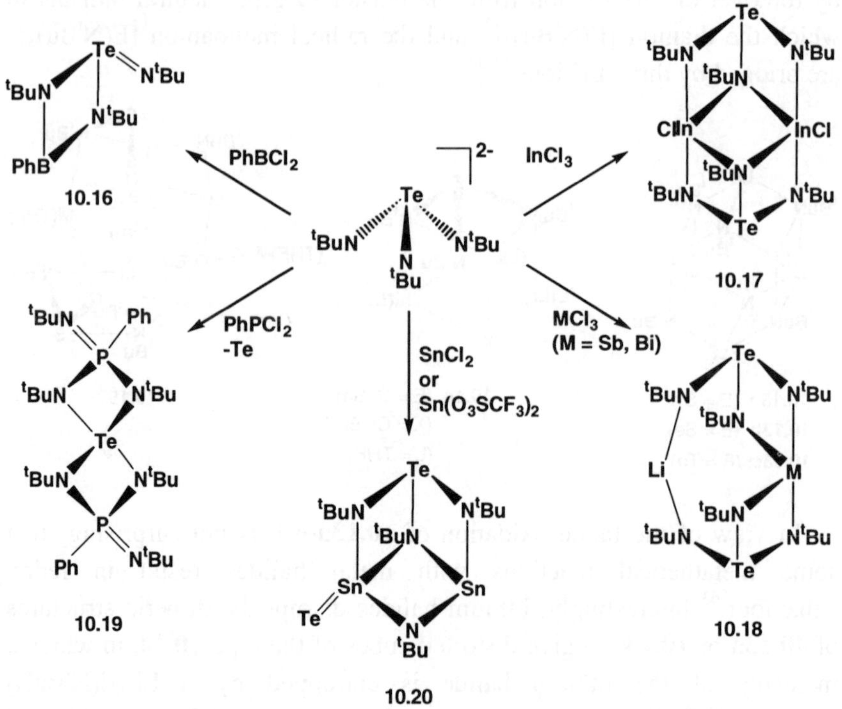

Scheme 10.8　Reactions of the [Te(N^{t}Bu)$_3$]$^{2-}$ dianion

of **10.13c** with $PhBCl_2$ gives the N,N'-chelated complex **10.16**, in which a monomeric tellurium imide, *cf*. **10.11**, is stabilized by the boraamidinato ligand $[PhB(N^tBu)_2]^{2-}$.[59] An alternative approach to **10.16** is the reaction of $PhB(\mu-N^tBu)_2TeCl_2$ with two equivalents of $LiNH^tBu$.[69] The dimeric cluster **10.17**, obtained from the reaction of **10.13c** with $InCl_3$, involves a second chelation of the $[Te(N^tBu)_3]^{2-}$ dianion to the other monomeric unit to give five-coordinate indium centres in an arm-chair structure.[70] In **10.18** two $[Te(N^tBu)_3]^{2-}$ dianions bis-chelate to a heavier group 15 centre to give a spirocyclic monoanion, which is linked to the Li^+ counter-ion by the two terminal N^tBu groups.[71] By contrast, redox behaviour is observed in the reaction of **10.13c** with $PhPCl_2$ to give the neutral spirocyclic complex **10.19** in which the tellurium(IV) centre is chelated by two $[PhP(N^tBu)_3]^{2-}$ ligands.[59] An especially intriguing outcome of such redox processes is the formation of the stannatellone **10.20** in the reaction of **10.13c** with Sn(II) salts.[72]

The first sulfur triimide $S(NSiMe_3)_3$ was prepared by the reaction of NSF_3 with $LiN(SiMe_3)_2$ (Section 8.3).[73a] The derivatives $S(NR)_3$ (R = $SiMe_3$, ^tBu) have trigonal planar structures, *cf*. SO_3, with S–N bond lengths in the range 1.50-1.52 Å and an NSN bond angle of 120°.[73b,c] An alternative preparation of $S(N^tBu)_3$ is the oxidation of the trisimidosulfite **10.13a** with halogens.[61] The SN vibrations in the Raman spectrum of $S(N^tBu)_3$ occur at much lower wave numbers (640-920 cm^{-1}) than expected for a covalent S=N bond indicating that the bonding is predominantly electrostatic $(S^+–N^-)$.[74] This conclusion is supported by electron density measurements.[29] Monomeric sulfonyl diimides $(RN)_2SO$ or sulfonyl imides $RNSO_2$ have not been characterized.[75]

10.6 Tetraimidosulfate, $[S(N^tBu)_4]^{2-}$, and Methylenetrimidosulfate, $[H_2CS(N^tBu)_3]^{2-}$

The tetraimidosulfate anion $[S(NtBu)_4]^{2-}$, isoelectronic with SO_4^{2-}, is prepared by a methodology similar to that employed for the synthesis of triimidosulfites. The reaction of the sulfur triimide $S(N^tBu)_3$ with two equivalents of $LiNH^tBu$ produces the solvated monomeric complex $[(thf)_4Li_2S(N^tBu)_4]$ (**10.21**) (Eq. 10.8).[76] The nucleophilic addition of sterically undemanding alkyl-lithium reagents, *e.g.*, MeLi, to $S(N^tBu)_3$ produces the dimeric complex $[(thf)_2Li_2\{(N^tBu)_3SMe\}_2]$ (**10.22**).[74] Treatment of **10.22** with $[^tBuNH_3]Cl$ gives the parent acid $MeS(N^tBu)_3H$. *N,N'*-chelated complexes of the $[MeS(N^tBu)_3]^-$ anion are obtained by the nucleophilic addition of Me_2Zn or Me_3Al to $S(N^tBu)_3$.[77]

$$S(N^tBu)_3 + 2LiNH^tBu \rightarrow [(thf)_4Li_2S(N^tBu)_4] + H_2N^tBu \quad (10.8)$$

In the monomeric structure of **10.21** the solvation of both Li^+ ions by two THF molecules prevents further aggregation. The four S–N bond lengths are equal at *ca.* 1.60 Å indicating that the negative charge is delocalized over the $S(N^tBu)_4$ unit. In the dimer **10.22** one $[MeS(N^tBu)_3]^-$ anion is coordinated to both Li^+ ions, one of which is bis-solvated by THF, while the other is chelated only to the unsolvated Li^+ ion.

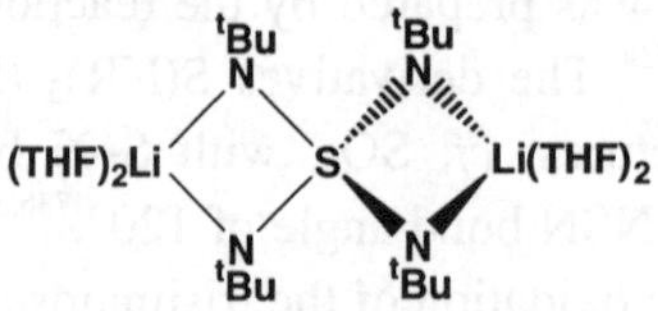

10.21

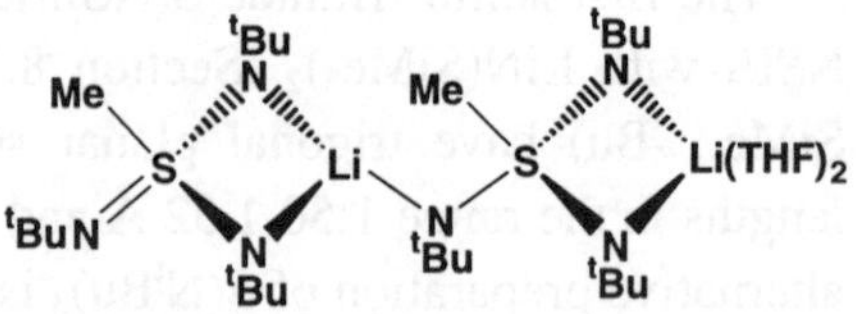

10.22

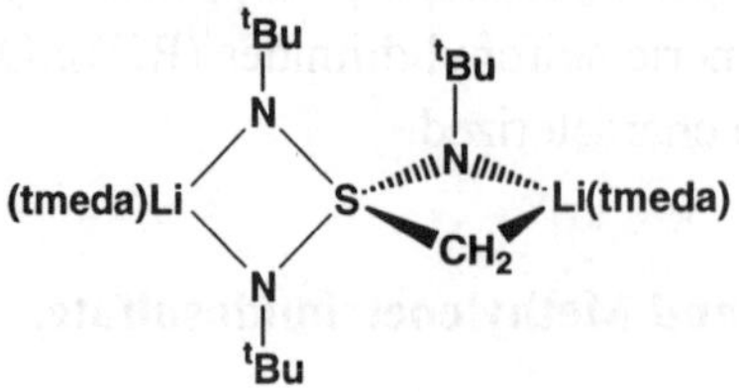

10.23

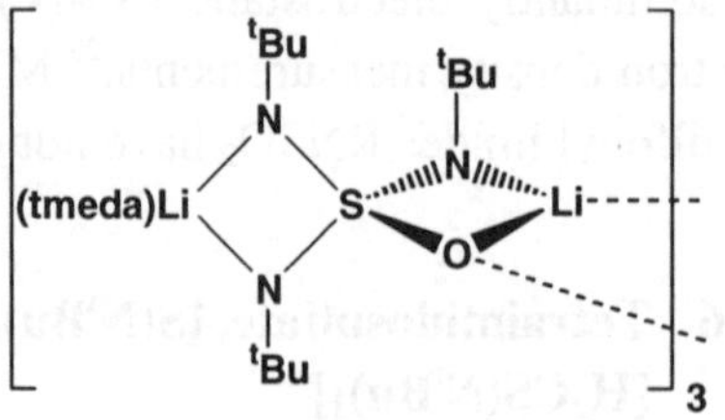

10.24

The methylenetriimido sulfite dianion $[CH_2S(N^tBu)_3]^{2-}$, isoelectronic with both $[S(N^tBu)_4]^{2-}$ and $[OS(N^tBu)_3]^{2-}$, is prepared by the treatment of $[CH_3S(N^tBu)_3]^-$ with methyllithium in the presence of TMEDA (Eq. 10.9).[78] The structure of $[(tmeda)_2Li_2\{H_2CS(N^tBu)_3\}]$ (**10.23**) resembles that of **10.21** with the TMEDA ligands replacing the two THF ligands on each Li centre. The bond length of the ylidic S^+–C^- bond in **10.23** is *ca.* 0.07 Å shorter than the mean S–CH_3 bond distance in **10.22**. Consistent with this ylidic character, **10.23** is preferentially hydrolyzed by water at the S–C bond to give the triimidosulfate dianion $[OS(N^tBu)_3]^{2-}$. The monosolvated dilithium complex **10.24** forms a trimer via Li–O interactions. The dianion $[OS(NtBu)_3]^{2-}$ has also been obtained in a monomeric cubic structure, *i.e.*, **10.14a** (E = S(O), X = I), with an exocyclic S=O double bond (1.46 Å).[61]

$$[(thf)_2Li_2\{(N^tBu)_3SMe\}_2] + MeLi \rightarrow$$

$$[(tmeda)_2Li_2\{H_2CS(N^tBu)_3\}] + CH_4 \qquad (10.9)$$

Reaction of **10.23** with $S(N^tBu)_3$ and subsequent protonation with $[^tBuNH_3]Cl$ yields $H_2C[S(N^tBu)_2(NH^tBu)]_2$, an imido analogue of methane disulfonic acid.[79]

10.7 Chalcogen Diamides $E_x(NR_2)_2$, (E = S, Se, Te; x = 1-4)

Sulfur diamides (diaminosulfanes) are obtained by the action of sulfur chlorides with an aliphatic secondary amine (Eq. 10.10).[80] The monoselanes $Se(NR_2)_2$ (R = Me, Et) have also been prepared.[81] Polyselanes $Se_x(NR_2)_2$ (x = 2 - 4, NR_2 = morpholinyl; x = 4, NR_2 = piperidinyl) are formed in the reaction of elemental selenium with the boiling amine in the presence of Pb_3O_4.[82]

$$4R_2NH + S_xCl_2 \rightarrow R_2N–S_x–NR_2 + 2[R_2NH_2]Cl \qquad (10.10)$$

$$(x = 1- 4)$$

The acyclic tellurium(II) diamide $Te(NMe_2)_2$ is prepared by the reaction of $TeCl_4$ with $LiNMe_2$.[83] Derivatives containing one Te(II)–N linkage are obtained according to Eq. 10.11.[84]

$$ArTeI + LiN(SiMe_3)(SiMe_2R) \rightarrow ArTeN(SiMe_3)(SiMe_2R) + LiI \qquad (10.11)$$

$$(Ar = Ph, Naph) \qquad \qquad (R = Me, {}^tBu)$$

Silylated amino derivatives of the chalcogens of the type $E[(N(SiMe_3)_2]_2$ (E = S, Se. Te) are useful for the synthesis of other chalcogen–nitrogen compounds via reactions with element halides and elimination of Me_3SiCl. These reagents are made by the reactions of $LiN(SiMe_3)_2$ with SCl_2, Se_2Cl_2 (Eq. 10.12) or $TeCl_4$, respectively.[85,86] The corresponding polysulfanes and polyselanes $E_x[N(SiMe_3)_2]$ (E = S, Se; x = 2-4) are obtained as inseparable mixtures by adding the elemental chalcogen to a mixture of these reagents.[87] They have been characterized by mass spectra and by ^{13}C and ^{77}Se NMR spectra. The reaction of $HN(SiMe_3)_2$ with S_2Cl_2 provides a route to the pure trisulfane $S_3[N(SiMe_3)_2]_2$, which is a useful reagent for the preparation of S_4N_2 by treatment with a mixture of S_2Cl_2 and $SOCl_2$ (Section 5.2.5).[87]

$$2LiN(SiMe_3)_2 + Se_2Cl_2 \rightarrow Se[N(SiMe_3)_2] + 2LiCl + 1/8Se_8 \quad (10.12)$$

The series $E[N(SiMe_3)_2]_2$ (E = S, Se, Te)[85,86] displays E–N bond lengths of 1.72, 1.87 and 2.05 Å, respectively, indicative of single chalcogen–nitrogen bonds. The <NEN bond angles of 109.6°, 108.0° and 105.8°, respectively, reflect the increasing s character of the lone pair on the chalcogen. In contrast to the monomeric structure of $Te[N(SiMe_3)_2]_2$, the dimethylamino derivative $[Te(NMe_2)_2]_\infty$ has a polymeric structure with intermolecular Te•••N contacts of 2.96 Å that give rise to trapezoidal Te_2N_2 rings (Section 3.1).[83] The X-ray structures of the polyselanes $Se_x(NR_2)_2$ (x = 2-4, NR_2 = morpholino; x = 4, NR_2 = piperidino) exhibit typical single Se–N bond lengths in the range 1.82-1.85 Å.[82,88]

The polar tellurium(II)–nitrogen bond is readily susceptible to protolysis by weakly acidic reagents. For example, the reaction of $[Te(NMe_2)_2]_\infty$ with two equivalents of Ph_3CSH produces the monomeric thiolato derivative $Te(SCPh_3)_2$.[83] Alkynyl tellurides may be prepared by the reaction of terminal acetylenes with arenetellurenamides (Eq. 10.13).[89]

$$C_6H_5TeN^iPr_2 + RC\equiv CH \rightarrow RC\equiv CTeC_6H_5 + HN^iPr_2 \quad (10.13)$$

Dialkynyl tellurides are obtained in moderate yield by the one-pot reaction of $TeCl_4$ with $LiN(SiMe_3)_2$, followed by the addition of a terminal acetylene. Alkenyl tellurium(II) derivatives may also be prepared by the addition of arenetellurenamides or $Te[N(SiMe_3)_2]_2$ to the $-C\equiv C-$ bond of dimethylacetylene dicarboxylate.[90]

The intriguing radical cation $[Te\{N(SiMe_3)_2\}_2]^{+\bullet}$ is formed (as the blue AsF_6^- salt) by oxidation of $Te[N(SiMe_3)_2]_2$ with AsF_5. This deep blue salt is monomeric in the solid state with $d(Te–N) = 1.97$ Å, consistent with multiple bonding. The broad singlet in the EPR spectrum indicates that the unpaired electron is located primarily on the tellurium atom.[91]

10.8 Organochalcogenyl Azides and Nitrenes[92]

Organosulfenyl azides RSN_3 are of interest as sources of the corresponding sulfenyl nitrenes RSN, which may be used in the construction of S–N heterocycles. Sulfenyl nitrenes are also accessible by the oxidation of sulfenamides, *e.g.*, with lead tetraacetate. An alternative approach involves the thermal extrusion of bridging nitrogen from 1,4-dihydro-1,4-iminonaphthalenes followed by trapping with an alkene to give an aziridine (Scheme 10.9).[93] Sulfenyl azides cannot be isolated. The trifluoromethyl derivative CF_3SN_3, prepared from CF_3SCl and trimethylsilyl azide, acts as a source of monomeric CF_3SN which may be trapped by hexachlorocyclopentadiene to give $C_5Cl_6=NSCF_3$; NSCl reacts in a similar manner to give $C_5Cl_6=NSCl$.[94] In the case of selenenyl azides, $RSeN_3$, stabilization has been achieved by intramolecular coordination in $2-Me_2NC_6H_4SeN_3$.[95] The thermal instability of sulfenyl azides has been exploited in the synthesis of benzo-1,3,2-dithiazolyl derivatives from the reaction of 1,2-arenedisulfenyl dichlorides with trimethylsilyl azide (Section 11.3.5).[96]

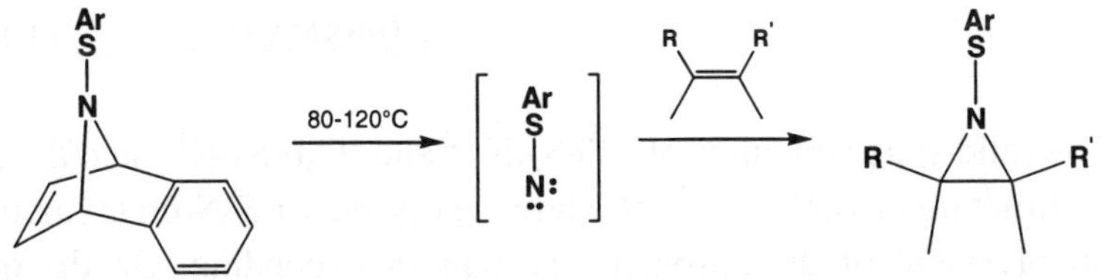

Scheme 10.9   Formation and trapping of an arenesulfenylnitrene (Ar = Ph, 2,4-$(NO_2)_2C_6H_3$)

Sulfinyl azides have higher thermal stability than their divalent counterparts. For example, benzenesulfinyl azides $ArS(O)N_3$ (Ar = Ph, 4-MeC$_6$H$_4$, 4-NO$_2$C$_6$H$_4$), which are obtained by the metathetical reaction of the sulfinyl chloride and sodium azide in acetonitrile, can be stored in the solid state at low temperatures.[97]

Organotellurium(IV) azides with two or three azido groups attached to tellurium, $R_2Te(N_3)_2$ (R = alkyl, Ph, C$_6$F$_5$) and $RTe(N_3)_3$ (R = alkyl, 2,4,6-Me$_3$C$_6$H$_2$), have sufficient thermal stability to allow structural characterization in the solid state.[98] These azides are prepared by the reaction of organotellurium(IV) fluorides with trimethylsilyl azide (Eq. 10.14). In the solid state organotellurium(IV) azides form polymeric networks via weak Te•••N interactions in the range 2.8–3.3 Å resulting in seven- or eight-coordination at the tellurium atom.[98] Covalent organoselenium(IV) azides have not been characterized, but the structures of ionic selenonium azides [R$_3$Se]N$_3$, obtained from [R$_3$Se]I and AgN$_3$, have been determined.[99]

$$RTeF_3 + 3Me_3SiN_3 \rightarrow RTe(N_3)_3 + 3Me_3SiF \qquad (10.14)$$

10.9 Trisulfenamides, (RS)$_3$N, and the Radical [(PhS)$_2$N]•

Tribenzenesulfenamide (PhS)$_3$N is obtained as a pale yellow solid by the treatment of the sodium salt of dibenzenesulfenamide, generated *in situ*, with acetic anhydride (Eq. 10.15).[100] The perfluorinated analogue (C$_6$F$_5$S)$_3$N is prepared by the reaction of (C$_6$F$_5$S)$_2$NH and C$_6$F$_5$SCl in diethyl ether.[101]

$$2(PhS)_2NH + Ac_2O + 2NaH \rightarrow (PhS)_3N + NaOAc$$
$$+ [PhSNAc]Na \qquad (10.15)$$

The solid-state structures of (PhS)$_3$N[102] and (C$_6$F$_5$S)$_3$N[101] and the gas-phase structure of (CF$_3$S)$_3$N[103] all show nearly planar S$_3$N units implying the involvement of the nitrogen lone pair in π–bonding. On the other hand, the S–N bond lengths of *ca.* 1.80 Å are longer than typical single bond values. Tribenzenesulfenamide decomposes at *ca.* 80°C to give

PhSSPh and N_2 via the intermediate formation of the purple radical $[(PhS)_2N]^{\bullet}$, which is readily detected by its five-line (1:2:3:2:1) EPR spectrum. This radical is also generated by the oxidation of $(PhS)_2NH$ with lead dioxide.[100]

10.10 Metal Complexes of Sulfimido, $[Ph_2S=N]^-$, and Sulfenamido, $[RSNR']^-$, Anions

Although the sulfimide ion $[NSPh_2]^-$ is electronically similar to the widely studied $[NPPh_3]^-$ monoanion, only a few metal complexes of this potentially interesting ligand are known. Some transition-metal complexes, *e.g.*, $WF_4(NSPh_2)_2$, $Cl_2VO(NSPh_2)$ and $[Cl_2Fe(NSPh_2)]_2$, are prepared by the reaction of metal halides with $Me_3SiNSPh_2$.[104,105] The ion-separated uranium(VI) complex $[Ph_4P][UOCl_4(NSPh_2)]$ is obtained in a similar manner.[106] Treatment of $Cp*_2UCl_2$ with two equivalents of $LiNSPh_2$ produces the uranium(IV) complex $Cp*_2U(NSPh_2)_2$.[107] Short U–N bond lengths indicative of bond orders ≥ 2 are observed for both the U(IV) and the U(VI) complexes. The S–N bond lengths are in the range 1.55–1.60 Å, close to the double bond value. The ability of the $[Ph_2SN]^-$ ligand to adjust to different bonding environments is reflected in the wide range of MNS bond angles (130-172°) observed in metal complexes.[104-107]

Complexes of the neutral ligand with Cu(II), Co(II), Pd(II) and Pt(II) have been characterized.[108,109] Both square planar and *pseudo*-tetrahedral geometries are observed for *trans*-$[CuCl_2(Ph_2SNH)_2]$.[108] Although the reactions of metal halides with Ph_2SNH in acetonitrile are generally straightforward, a metal-assisted addition of the sulfimide to MeCN occurs with $[Pt(MeCN)_2Cl_2]$ to give the cyclic platinum(II) complex **10.25**.[110]

Complexes of the sulfenamido anion $[RSNR']^-$ with several transition metals [Zr(IV), Ti(IV), Mo(VI), W(VI), Ni(II), and U(IV)] are known.[111,112] They are prepared by the reaction of the lithium derivative of the sulfenamido anion with a metal halide complex. A selenium complex $W(N^tBu)_2(^tBuNSePh)_2$ has been obtained in a similar manner.[112]

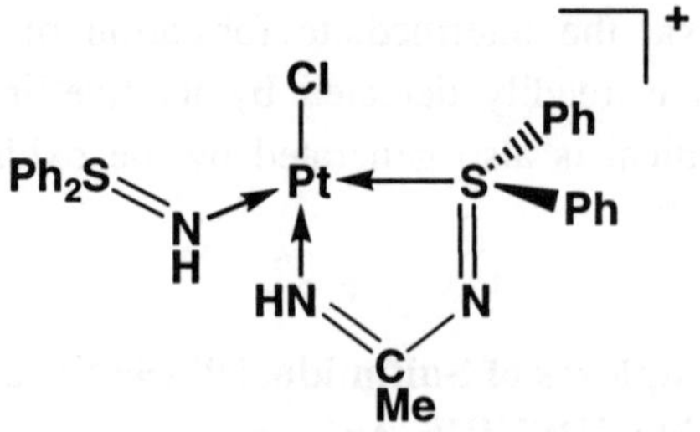

10.25

The solvated sulfenamides $[Li_2(^tBuNSC_6H_4Me\text{-}4)_2(THF)_n]$ (n = 2,4) have dimeric structures with a central Li_2N_2 ring.[113] The coordination mode is determined by the extent of solvation of the Li^+ ions; monosolvation allows for η^2-N,S coordination whereas disolvation restricts the coordination mode to η^1-N. Variable temperature NMR studies indicated that a dynamic exchange between these two structural types occurs in THF solution (Scheme 10.10). The dihapto coordination mode is observed exclusively in transition-metal complexes and the geometry at nitrogen is planar.[111,112]

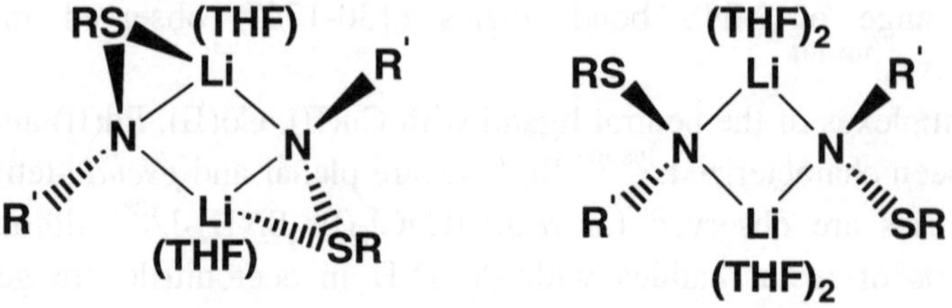

Scheme 10.10　η^1 and η^2 Coordination modes for the [RSNR']⁻ anion in solvated lithium complexes

References

1.　J. Wasilewski and V.Staemmler, *Inorg. Chem.*, **25**, 4221 (1986).

2.　W. J. Middleton, *J. Am. Chem. Soc.*, **88**, 3842 (1966).

3.　P. Tavs, *Angew. Chem., Int. Ed. Engl.*, **5**, 1048 (1966).

4.   M. Takahashi, R. Okazaki, N. Inamoto, T. Sugawara and H. Iwamura, *J. Am. Chem. Soc.*, **114**, 1830 (1992).

5.   (a) M. R. Bryce, J. N. Heaton, P. C. Taylor and M. Anderson, *J. Chem. Soc., Perkin Trans. 1*, 1935 (1994); (b) M. R. Bryce, J. Becher and B. Fält-Hansen, *Adv, Heterocycl. Chem.*, **1**, 55 (1992).

6.   M. R. Bryce and A. Chesney, *J. Chem. Soc., Chem. Commun.*, 195 (1995).

7.   S. Nakamura, M. Takahashi, R. Okazaki and K. Morokuma, *J. Am. Chem. Soc.*, **109**, 4142, (1987).

8.   D. H. R. Barton and M. J. Robson, *J. Chem. Soc., Perkin Trans. 1*, 1245 (1974).

9.   Y. Inagaki, R. Okazaki and N. Inamoto, *Bull. Chem. Soc. Jpn.*, **52**, 1998 (1979).

10. Y. Inagaki, R. Okazaki, N. Inamoto. K. Yamada and H. Kawazura, *Bull. Chem. Soc. Jpn.*, **52**, 2008 (1979).

11. Y. Inagaki, R. Okazaki and N. Inamoto, *Bull. Chem. Soc. Jpn.*, **52**, 2002 (1979).

12. Y. Inagaki, T. Hosogai, R. Okazaki and N. Inamoto, *Bull. Chem. Soc. Jpn.*, **53**, 205 (1980).

13. T. Chivers, *Sulfur Reports*, **7**, 89 (1986).

14. T. Chivers, A. W. Cordes, R. T. Oakley and P. N. Swepston, *Inorg. Chem.*, **20**, 2376 (1981).

15. A. W. Cordes, H. Koenig, M. C. Noble and R. T. Oakley, *Inorg. Chem.*, **22**, 3375 (1983).

16. A. F. Hill, *Adv. Organomet. Chem.*, **36**, 159 (1994).

17. W. Goehring and G. Weis, *Angew. Chem.*, **68**, 687 (1956).

18. K. B. Sharpless, T. Hori, L. K. Truesdale and C. O. Dietrich, *J. Am. Chem. Soc.*, **98**, 269 (1976).

19. T. Maaninen, T. Chivers, R. Laitinen, G. Schatte and M. Nissinen, *Inorg. Chem.*, **39**, 5341 (2000).

20. (a) T. Chivers, N. Sandblom and G. Schatte, *Inorg. Synth.*, **34**, 42 (2004); (b) T. Chivers, X. Gao and M. Parvez, *J. Am. Chem. Soc.*, **117**, 2359 (1995); (c) T. Chivers, X. Gao and M. Parvez, *J. Chem. Soc., Chem. Commun.*, 2149 (1994).

21. K. Bestari, R. T. Oakley and A. W. Cordes, *Can. J. Chem.*, **69**, 94 (1991).

22. (a) E. Lork, R. Mews, M. M. Shakirov, P. G. Watson and A. V. Zibarev, *Eur. J. Inorg. Chem.*, 2123 (2001); (b) E. Lork, R. Mews, M. M. Shakirov, P. G. Watson and A. V. Zibarev, *J. Fluorine Chem.*, **115**, 165 (2002); (c) I. Y. Bagryanskaya, Y. V. Gatilov, M. A. Shakirov and A. V. Zibarev, *Mendeleeev Commun.*, 167 (2002).

23. J. R. Grunwell, C. F. Hoyng and J. A. Rieck, *Tetrahedron Lett.*, **26**, 2421 (1973).

24. J. Kuyper and K. Vrieze, *J. Organomet. Chem.*, **86**, 127 (1975).

25. W. Sicinska, L. Stefaniak, M. Witanowski and G. A. Webb, *J. Mol. Struct.*, **158**, 57 (1987).

26. I. Yu. Bagryanskaya, Y. Gatilov, M. M. Shakirov and A. V. Zibarev, *Mendeleev Commun.*, 136 (1994).

27. I. Yu. Bagryanskaya, Y. Gatilov, M. M. Shakirov and A. V. Zibarev, *Mendeleev Commun.*, 167 (1994).

28. D. G. Anderson, H. E. Robertson, D. W. H. Rankin and J. D. Woollins, *J. Chem. Soc., Dalton Trans.*, 859 (1989).

29. D. Leusser, J. Henn, J. N. Kocher, B. Engels and D. Stalke, *J. Am. Chem. Soc.*, **126**, 1781 (2004).

30. T. Maaninen, R. Laitinen and T. Chivers, *Chem. Commun.*, 1812 (2002).

31. B. Wrackmeyer, B. Distler, S. Gerstmann and M. Herberhold, *Z. Naturforsch.*, **48B**, 1307 (1993).

32. (a) H. M. Tuononen, R. J. Suontamo, J. U. Valkonen, R. S. Laitinen and T. Chivers, *Inorg. Chem.*, **42**, 2447 (2003); (b) S. Shahbazian, M. Zahedi and S. W. Ng, *J. Mol. Struct. – Theochem.*, **672**, 211 (2004).

33. T. Chivers, X. Gao and M. Parvez, *Inorg. Chem.*, **35**, 9 (1996).

34. N. Sandblom, T. Ziegler and T. Chivers, *Inorg. Chem.*, **37**, 354 (1998).

35. T. Maaninen, H. M. Tuononen, G. Schatte, R. Suontamo, J. Valkonen, R. Laitinen and T. Chivers, *Inorg. Chem.*, **43**, 2097 (2004).

36. R. Meij, J. Kuyper, D. Stufkens and K. Vrieze, *J. Organomet. Chem.*, **110**, 219 (1976).

37. E. Lindsell and G. R. Faulds, *J. Chem. Soc., Dalton Trans.*, 40 (1975).

38. H. W. Roesky, H-G. Schmidt, M. Noltemeyer and G. M. Sheldrick, *Chem. Ber.*, **116**, 1411 (1983).

39. J. Gindl, M. Björgvinsson, H. W. Roesky, C. Freire-Erdbrügger and G. M. Sheldrick, *J. Chem. Soc., Dalton Trans.*, 811 (1993).

40. J. Kuyper and K. Vrieze, *J. Organomet. Chem.*, **74**, 289 (1974).

41. R. Meij, D. J. Stufkens, K. Vrieze, E. Rosendaal and H. Schenk, *J. Organomet. Chem.*, **115**, 323 (1978).

42. T. Chivers and G. Schatte, *Can. J. Chem.*, **81**, 1307 (2003).

43. T. Chivers, M. Parvez and G. Schatte, *Angew. Chem., Int. Ed. Engl.*, **38**, 2217 (1999).

44. M. Herberhold and W. Buhlmeyer, *Angew. Chem., Int. Ed. Engl.*, **23**, 80 (1984).

45. M. Herberhold, W. Buhlmeyer, A. Gieren, T. Hubner and J. Wu, *Z. Naturforsch.*, **42B**, 65 (1987).

46. J. A. Hunter, B. King, W. E. Lindsell and M. A. Neish, *J. Chem. Soc., Dalton Trans.*, 880 (1980).

47. G. Brands and A. Golloch, *Z. Naturforsch.*, **37B**, 1137 (1982).

48. B. Wrackmeyer, S. M. Frank, M. Herberhold, A. Simon and H. Borrmann, *J. Chem. Soc., Dalton Trans.*, 2607 (1991).

49. B. Wrackmeyer, C. Köhler, W. Milius and M. Herberhold, *Z. Anorg. Allg. Chem.*, **621**, 1625 (1995).

50. O. J. Scherer and R. Schmitt, *J. Organomet. Chem.*, **16**, P11 (1969).

51. F. Pauer and D. Stalke, *J. Organomet. Chem.*, **418**, 127 (1991).

52. D. Hanssgen and R. Steffens, *J. Organomet. Chem.*, **236**, 53 (1982).

53. M. Bruncko, T-A. V. Khuong and K. B. Sharpless, *Angew. Chem., Int. Ed. Engl.*, **35**, 454 (1996).

54. G. Schatte, T. Chivers, C. Jaska and N. Sandblom, *Chem. Commun.*, 1657 (2000).

55. (a) R. Fleischer and D. Stalke, *Coord. Chem. Rev.*, **176**, 431 (1998); (b) D. Stalke, *Proc. Ind. Acad. Sci.*, **112**, 155 (2000).

56. J. K. Brask and T. Chivers, *Angew. Chem., Int. Ed. Engl.*, **40**, 3988 (2001).

57. R. Fleischer, S. Freitag, F. Pauer and D. Stalke, *Angew. Chem., Int Ed. Engl.*, **35**, 204 (1996).

58. T. Chivers, M. Parvez and G. Schatte, *Inorg. Chem.*, **35**, 4094 (1996).

59. T. Chivers, X. Gao and M. Parvez, *Angew. Chem., Int. Ed. Engl.*, **34**, 2549 (1995).

60. T. Chivers, M. Parvez and G. Schatte, *J. Organomet. Chem.*, **550**, 213 (1998).

61. R. Fleischer, S. Freitag and D. Stalke, *J. Chem. Soc., Dalton Trans.*, 193 (1998).

62. J. K. Brask, T. Chivers, B. McGarvey, G. Schatte, R. Sung and R. T. Boeré, *Inorg. Chem.*, **37**, 4633 (1998).

63. R. Fleischer and D. Stalke, *Organometallics*, **17**, 832 (1998).

64. T. Chivers, M. Parvez and G. Schatte, *Inorg. Chem.*, **40**, 540 (2001).

65. (a) B. Walfort, L. Lameyer, W. Weiss, R. Herbst-Irmer, R. Bertermann, J. Rocha and D. Stalke, *Chem. Eur. J.*, **7**, 1417 (2001); (b) R. Fleischer and D. Stalke, *Chem. Commun.*, 343 (1998); (c) B.Walfort, S. K. Pandey and D. Stalke, *Chem. Commun.*, 1640 (2001).

66. D. Ilge, D. S. Wright and D. Stalke, *Chem. Eur. J.*, **4**, 2275 (1998).

67. B. Walforth, T. Auth, B. Degel, H. Helten and D. Stalke, *Organometallics*, **21**, 2208 (2002).

68. T. Chivers, *Can. J. Chem.*, **79**, 1841 (2001).

69. T. Chivers, C. Fedorchuk, G. Schatte and J. K. Brask, *Can. J. Chem.*, **80**, 821 (2002).

70. T. Chivers and G. Schatte, *Eur. J. Inorg. Chem.*, 2266 (2002).

71. T. Chivers, M. Parvez, G. Schatte and G. P. A. Yap, *Inorg. Chem.*, **38**, 1380 (1999).

72. T. Chivers and G. Schatte, *Chem. Commun.*, 2264 (2001).

73. (a) O. Glemser and J. Wegener, *Angew. Chem., Int. Ed. Engl.*, **9**, 309 (1970); (b) O. Glemser, S. Pohl, F-M. Tesky and R. Mews, *Angew. Chem., Int. Ed. Engl.*, **16**, 789 (1977); (c) S. Pohl, B. Krebs, U. Seyer and G. Henkel, *Chem. Ber.*, **112**, 1751 (1979).

74. R. Fleischer, B. Walfort, A. Gbureck, P. Scholz, W. Kiefer and D. Stalke, *Chem. Eur. J.*, **4**, 2266 (1998).

75. R. Mews, P. G. Watson and E. Lork, *Coord. Chem. Rev.*, **158**, 233 (1997).

76. R. Fleischer, A. Rothenberger and D.Stalke, *Angew. Chem., Int. Ed. Engl.*, **36**, 1105 (1997).

77. B. Walfort, A. P. Leedham, C. A. Russell and D. Stalke, *Inorg. Chem.*, **40**, 5668 (2001).

78. B. Walfort and D. Stalke, *Angew. Chem., Int. Ed. Engl.*, **40**, 3846 (2001).

79. D. Leusser, B. Walfort and D. Stalke, *Angew. Chem., Int. Ed. Engl.*, **41**, 2079 (2002).

80. Q. E. Thompson, *Quart. Rep. Sulfur Chem.*, **5**, 251 (1970).

81. R. Paetzold and E. Rönsch, *Z. Anorg. Allg. Chem.*, **338**, 22 (1965).

82. O. Foss and V. Janickis, *J. Chem. Soc., Dalton Trans.*, 620 (1980).

83. R. E. Allan, H. Gornitzka, J. Karcher, M. A. Paver, M-A. Rennie, C. A. Russell, P. R. Raithby, D. Stalke, A. Steiner and D. S. Wright, *J. Chem. Soc., Dalton Trans.*, 1727 (1996).

84. T. Murai, K. Kimurai and S. Kato, *Chem. Letters*, 2017 (1989).

85. G. Schubert, G. Kiel and G. Gattow, *Z. Anorg. Allg. Chem.*, **575**, 129 (1989).

86. M. Björgvinsson, H. W. Roesky, F. Paner, D. Stalke and G. M. Sheldrick, *Inorg. Chem.*, **29**, 5140 (1990).

87. J. Siivari, A. Maaninen, E. Haapiniemi, R. S. Laitinen and T. Chivers, *Z. Naturforsch.*, **50B**, 1575 (1995).

88. O. Foss and V. Janickis, *J. Chem. Soc., Dalton Trans.*, 628 (1980).

89. T. Murai, K. Nonomura, K. Kimura and S. Kato, *Organometallics*, **10**, 1095 (1991).

90. T. Murai, K. Imaeda, S. Kajita, K. Kimura, H. Ishihara and S. Kato, *Phosphorus, Sulfur, and Silicon*, **67**, 239 (1992)..

91. M. Björgvinsson, T. Heinze, H. W. Roesky, F. Pauer, D. Stalke and G. M. Sheldrick, *Angew. Chem., Int. Ed. Engl.*, **30**, 1677 (1991).

92. A. Senning, *Sulfur Letters*, **12**, 211 (1991).

93. R. S. Atkinson, M. Lee and J. R. Malpass, *J. Chem. Soc., Chem Commun.*, 919 (1984).

94. A. Haas and T. Mischo, *Can. J. Chem.*, **67**, 1902 (1989).

95. B. Krumm, T. M. Klapötke and K. Polborn, *J. Am. Chem. Soc.*, **126**, 710 (2004).

96. G. Wolmershäuser, M. Schnauber and T. Wilhelm, *J. Chem. Soc., Chem. Commun.*, 573 (1984).

97. T. J. Maricich and V. L. Hoffman, *J. Am. Chem. Soc.*, **96**, 7770 (1974).

98. (a) T. M. Klapötke, B. Krumm, P. Mayer, H. Piotrowski, O. P. Ruscitti and A. Schiller, *Inorg. Chem.*, **41**, 1184 (2002); (b) T. M. Klapötke, B. Krumm, P. Mayer and O. P. Ruscitti, *Inorg. Chem.*, **39**, 5426 (2000).

99. T. M. Klapötke, B. Krumm, P. Mayer, H. Piotrowski. K. Polborn and I. Schwab, *Z. Anorg. Allg. Chem.*, **628**, 1831 (2002).

100. D. H. R. Barton, I. A. Blair, P. D. Magnus and R. K. Norris, *J.Chem. Soc., Perkin Trans. 1*, 1032 (1973).

101. A. Biehl, R. Boese, A. Haas, C. Klare and M. Peach, *Z. Anorg. Allg. Chem.*, **622**, 1262 (1996).

102. J. R. Carruthers, K. Prout and D. J. Watkin, *Cryst. Struct. Comm.*, **10**, 1217 (1981).

103. C. J. Marsden and L. S. Bartell, *J. Chem. Soc., Dalton Trans.*, 1582 (1977).

104. H. W. Roesky, M. Zimmer, H. G. Schmidt, M. Noltemeyer and G. M. Sheldrick, *Chem. Ber.*, **121**, 1377 (1988).

105. H. W. Roesky, M. Zimmer, H. G. Schmidt, and M. Noltemeyer, *Z. Naturforsch.*, **43B**, 1490 (1988).

106. V. C. Williams, M. Müller, M. A. Leech, R. G. Denning and M. L. H. Green, *Inorg. Chem.*, **39**, 2538 (2000).

107. K. A. N. S. Ariyaratne, R. E. Cramer and J. W. Gilje, *Organometallics*, **21**, 5799 (2002).

108. P. F. Kelly, A. M. Slawin and K. W. Waring, *J. Chem. Soc., Dalton Trans.*, 2853 (1997).

109. P. F. Kelly, A. M. Slawin and K. W. Waring, *Inorg. Chem. Commun.*, **1**, 249 (1998).

110. P.F. Kelly and A. M. Z. Slawin, *Chem. Commun.*, 1081 (1999).

111. A. A. Danopoulos, D. M. Hankin, S. M. Cafferkey and M. B. Hursthouse, *J. Chem. Soc., Dalton Trans.*, 1613 (2000)

112. D. M. Hankin, A. A. Danopoulos, G. Wilkinson, T. K. N. Sweet and M. B. Hursthouse, *J. Chem. Soc., Dalton Trans.*, 1309 (1996).

113. A. H. Mahmoudkhani, S. Rauscher, B. Grajales and I. Vargas-Baca, *Inorg. Chem.*, **42**, 3849 (2003).

Chapter 11

Five-membered Carbon–Nitrogen–Chalcogen Ring Systems: from Radicals to Functional Materials

11.1 Introduction

An RC unit is isolobal with S^+ as a substituent in a sulfur–nitrogen system. Consequently, it is not surprising that a number of C–N–S ring systems are known that are isoelectronic with the cyclic binary S–N cations discussed in Chapter 5. The replacement of one or more S^+ substituents by RC groups confers greater stability on the ring system and provides a link between these inorganic heterocycles and benzenoid compounds. Activity in this area of chemistry was stimulated in the 1980s by the theoretical prediction that polymers of the type $(RCNSN)_x$ will have conducting properties similar to those of $(SN)_x$ (Section 14.2).[1] Subsequently, there have been extensive investigations of C–N–S heterocycles as potential precursors for such polymers. Although this goal has not been realized, a new area of radical chemistry based on $[RCN_2S_2]^{\cdot}$ (dithiadiazolyl) and related radical systems has emerged.[2] These heterocycles and their selenium analogues exhibit considerable potential for the development of novel one-dimensional metals or materials with unique magnetic properties.

Chapters 11 and 12 both deal with C–N–E (E = S, Se, Te) heterocycles. The coverage is limited to ring systems in which the number of hetero atoms exceeds the number of carbon atoms. The topics are organized according to ring size. After a brief section on four-

212

membered rings, Chapter 11 will focus on the extensively studied five-membered ring systems. Within each section heterocycles that incorporate two-coordinate sulfur will be discussed followed, where they are known, by ring systems that contain three- or four-coordinate sulfur. Comparisons will be made between C–N–S and C–N–Se compounds where possible. The chemistry of C–N–Te hetrocycles that fall within the scope of this monograph (see Preface) is virtually unexplored, with the exception of telluradiazoles (Section 11.3.5). In addition to a number of general reviews,[3-7] the following specific aspects of the chemistry of C–N–E heterocycles (E = S, Se) have been the subject of comprehensive surveys: coordination chemistry,[8] electrochemistry,[9] radical chemistry[10,11] and magnetic properties.[12] The chemistry of six-membered and larger C–N–E heterocycles is covered in Chapter 12.

11.2 Four-membered Rings

Four-membered C–N–E rings are known only for E = Te. The cyclocondensation reaction of $TeCl_4$ with $PhCN_2(SiMe_3)_3$ produces a CN_2Te ring incorporating five-coordinate tellurium (Eq. 11.1).[13] Similar cyclocondensation reactions of amidinates with sulfur or selenium halides generate either five-membered rings containing two-coordinate chalcogen atoms (Section 11.3) or eight-membered rings involving either two- or three-coordinate chalcogen atoms (Section 12.4.1).

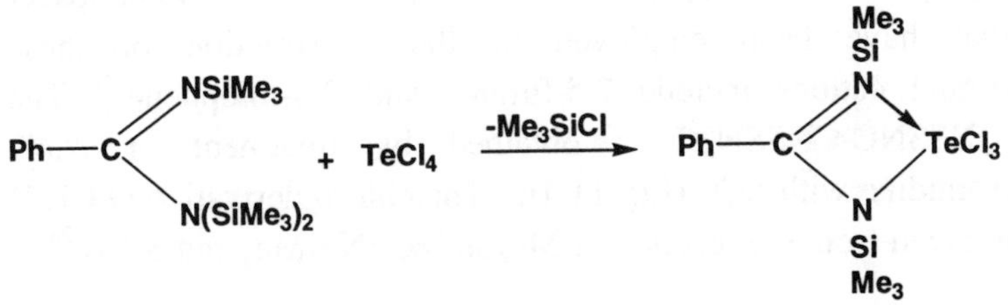

$$Ph-C{\overset{NSiMe_3}{\underset{N(SiMe_3)_2}{\big\langle}}} \ + \ TeCl_4 \ \xrightarrow{-Me_3SiCl} \ Ph-C{\overset{\underset{N}{\overset{Me_3}{Si}}}{\underset{\underset{Me_3}{Si}}{\overset{N}{\big\langle}}}}TeCl_3 \qquad (11.1)$$

11.3   Five-membered Rings

11.3.1   *1,2,3,5-Dichalcogenadiazolium [RCNEEN]$^+$ and dichalcogenadiazolyl [RCNEEN]$^•$ rings (E = S, Se)*

The first examples of the 1,2,3,5-dithiadiazolium cation (**11.1**) were obtained from the treatment of organic nitriles with $(NSCl)_3$.[14] This reaction was subsequently shown to proceed by the intermediate formation of six-membered ring systems of the type $RCN_3S_2Cl_2$ which, upon thermolysis, yield **11.1** (Eq. 11.2).[15]

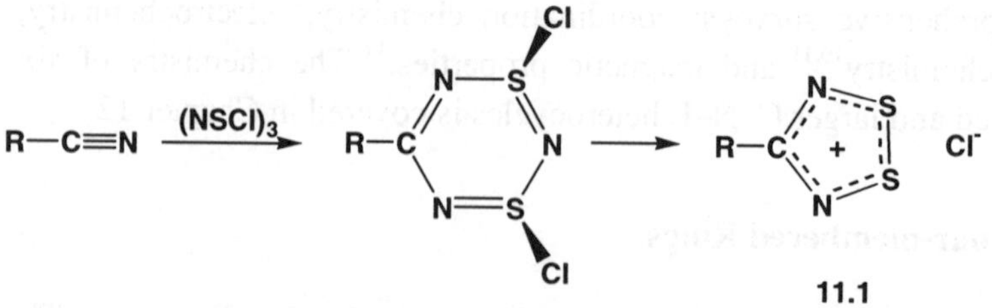

$$R-C{\equiv}N \xrightarrow{(NSCl)_3} \quad \quad \longrightarrow \quad \quad Cl^- \quad (11.2)$$

11.1

A more versatile synthesis of **11.1** (and the selenium analogue) involves the cyclocondensation of trisilylated amidines with sulfur dichloride or $SeCl_2$ generated *in situ* (Eq. 11.3).[16-18] This route can be used to prepare the prototypal systems [HCNEEN]$^+$ (E = S, Se).[19] It is also readily extended to the synthesis of multi-dichalcogenadiazolium cations such as 1,3- or 1,4-$C_6H_4(CNEEN)_2$]$^{2+}$ (**11.2**, E = S, Se),[16,20] [1,3,5-$C_6H_3(CNSSN)_3$]$^{3+}$ [21] 1,3,5-$C_3N_3(CNSSN)$]$^{3+}$ (**11.3**).[22] Other spacer groups that have been employed in the construction of these multifunctional cations include 2,5-furan[23] and 2,5-thiophene.[24] The dication [NSSNC-CNSSN]$^{2+}$ is obtained by treatment of *N,N'*-diaminooxamidine with SCl_2 (Eq. 11.4).[25] The chloro derivative (**11.1**, R = Cl) is prepared from the reaction of $Me_3SiN=C=NSiMe_3$ and S_2Cl_2.[26]

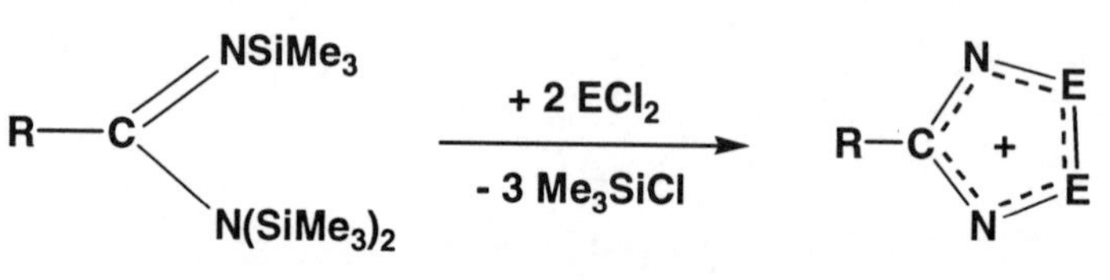

$$R-C{\begin{array}{c} NSiMe_3 \\ N(SiMe_3)_2 \end{array}} \xrightarrow[- 3\ Me_3SiCl]{+\ 2\ ECl_2} \quad R-C \quad (11.3)$$

11.2 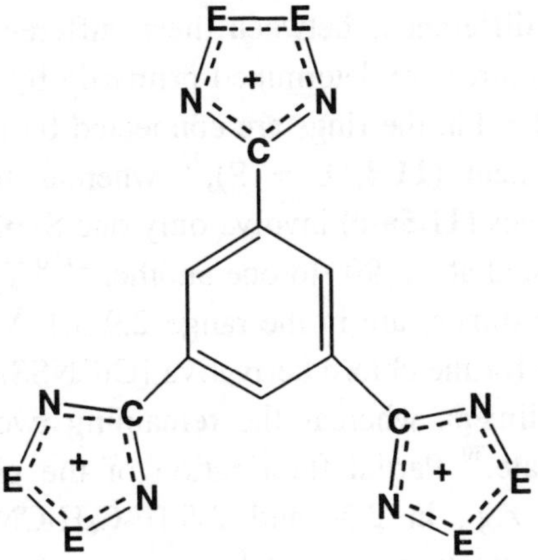   **11.3**

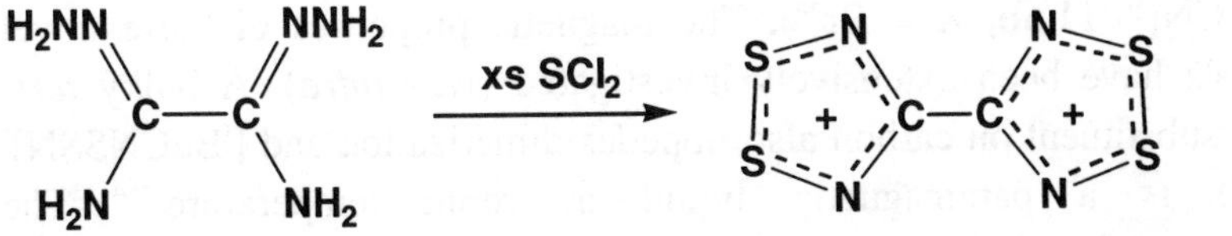

(11.4)

The colour of salts containing cations of the type **11.1** depends markedly on charge-transfer effects. Weak donor (hard) anions, *e.g.*, [AsF$_6$]$^-$, give yellow salts, whereas softer anions, *e.g.*, Br$^-$ or I$^-$, bestow darker (burgundy or black) colours on the salts.[2] The unperturbed CN$_2$S$_2$ ring in salts of the former type is planar. In simple aryl derivatives the the phenyl and dithiadiazolyl rings are almost co-planar, but there is some dependence on both steric effects (especially for *ortho*-substituted phenyl derivatives) and/or packing effects.[2]

The [RCN$_2$S$_2$]$^+$ ring is a six π-electron system (isoelectronic with [S$_3$N$_2$]$^{2+}$, Section 5.3.4), which is readily reduced, *e.g.*, by Ph$_3$Sb or Zn/Cu couple in SO$_2$, to the corresponding seven π-electron radical [RCN$_2$S$_2$]$^{\bullet}$. The structures of these radicals are markedly dependent on the nature of R. By changing R it is possible to generate a monomer, a variety of *cis*, *trans* and staggered dimers, or polymeric structures. The energy differences between these different arrangements are small and the structures are determined primarily by packing effects. For example, when R = Ph, the rings are connected by two S•••S interactions in a *cis* arrangement (**11.4**, E = S),[27] whereas the Me_2N,[28] CF_3[26] and Me[29] derivatives (**11.5a-c**) involve only one S•••S contact and the CN_2S_2 rings are twisted at *ca.* 90° to one another.[29,30] Typical S•••S contact distances in these dimers are in the range 2.9-3.1 Å. Four polymorphs have been isolated for the chloro derivative [ClCNSSN]•; two of them crystallize as *cisoid* dimers, whereas the remaining two form twisted dimers in the solid state.[30] Partial fluorination of the phenyl substituent attached to carbon, *e.g.*, in 2,3- and 2,5-[$F_2C_6H_3$CNSSN]•, results in a twisted dimeric structure, presumably as a result of the dipole generated in the aromatic ring by the unsymmetrical fluorine substitution.[31] The polyfluorinated aryl derivatives 4-XC_6F_4(CNSSN)• are of especial interest, since they adopt monomeric structures in the solid state (**11.6a**, X = CN;[32] **11.6b**, X = Br[33]). The magnetic properties of these novel radicals have been extensively investigated (*vide infra*). A bulky *tert*-butyl substituent on carbon also impedes dimerization and [^{t}BuCNSSN]• (**11.7**) is a paramagnetic liquid at room temperature.[34a] The trifluoromethyl derivative **11.5b** melts at 35°C with a dramatic increase in volume to give a paramagnetic liquid.[34b] Crystalline forms of dithiadiazolyl radicals are usually obtained by vacuum sublimation. When this process is carried out under a partial atmosphere of N_2, CO_2, SO_2 or Ar, the dimer [$CF_3C_6H_3$FCNSSN]$_2$ forms inclusion structures in which the guest molecules are intercalated between layers of the dithiadiazolyl radicals.[35]

The selenium analogue [PhCNSeSeN]• and cyano-functionalized diselenadiazolyl radicals adopt cofacial dimeric structures, *e.g.*, **11.4** (E = Se), with unequal Se•••Se interactions of *ca.* 3.15 and 3.35 Å. In the latter case the radical dimers are linked together by electrostatic CN•••Se contacts.[36,37] Tellurium analogues of dithiadiazolyl radicals (or the corresponding cations) are unknown, but calculations predict that the radical dimers, *e.g.*, **11.4** (E = Te), will be more strongly associated than the sulfur or selenium analogues.[38]

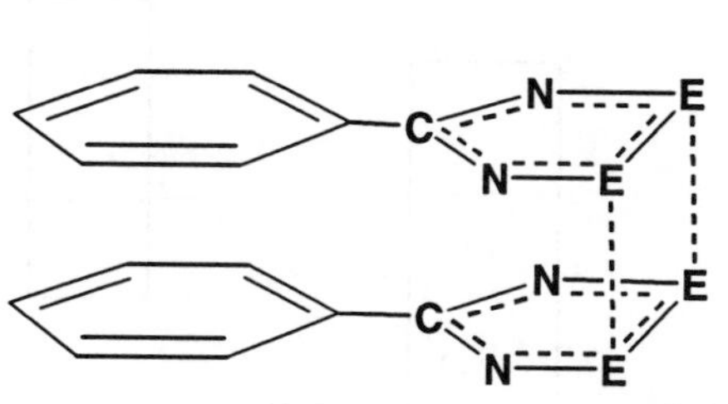

11.4

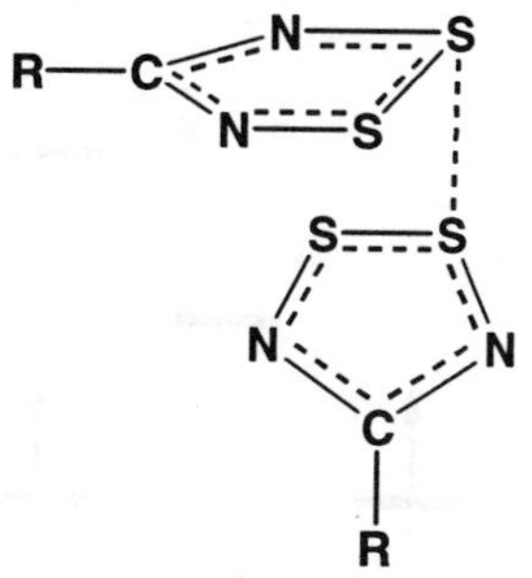

11.5a, R = NMe$_2$

11.5b, R = CF$_3$

11.5c, R = Me

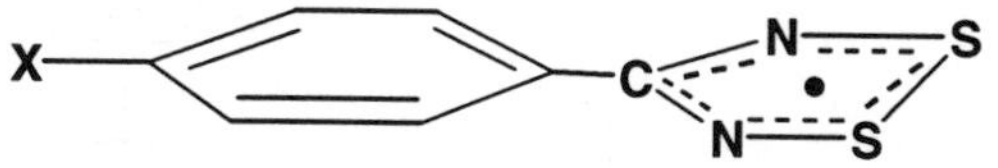

11.6a, X = CN

11.6b, X = Br

11.6c, X = NO$_2$

11.6d, X = 4-NCC$_6$F$_4$

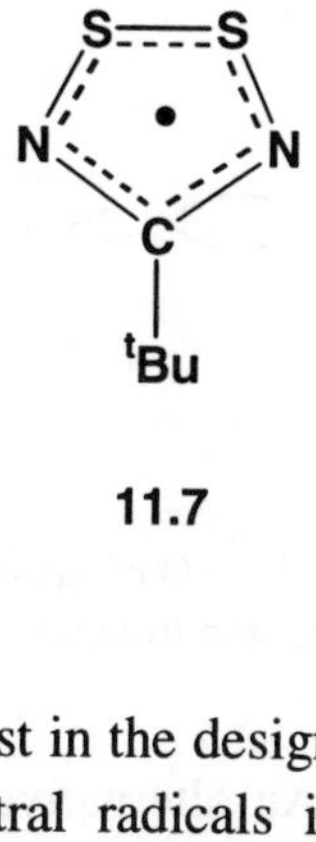

11.7

1,2,3,5-Dithiadiazolyl radicals are of especial interest in the design of molecular conductors.[6,10] The stacking of these neutral radicals in a column may be predicted to produce a material with a half-filled conducting band analogous to that of an alkali metal such as sodium (Fig. 11.1). A wide variety of mono-, bi- and trifunctional radicals have been prepared by reduction of the corresponding cations and their solid-state structures have been examined by a combination of experimental techniques and theoretical methods. Several of the selenium derivatives, *e.g.*, 1,3- and 1,4-C$_6$H$_4$(CNSeSeN˙)$_2$, are small band gap semiconductors.[17,20] Metallic behaviour has not been realized, however, as a result of the Peierls instability associated with an exactly half-filled energy band, *i.e.*, the radicals associate into closed-shell dimers.

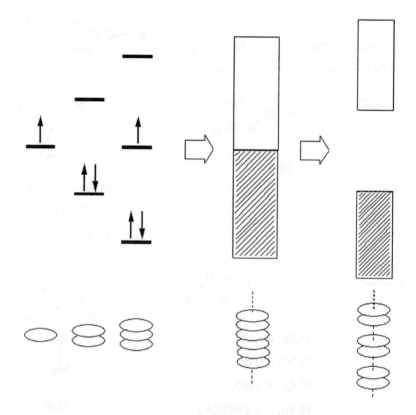

Fig. 11.1   Qualitative band energy diagrams for a column of radicals (Reproduced with permission from reference 6. Copyright NRC (Canada) Research Press 1993)

An alternative approach to stabilizing the metallic state involves p-type doping. For example, partial oxidation of neutral dithiadiazolyl radicals with iodine or bromine will remove some electrons from the half-filled level. Consistently, doping of biradical systems with halogens can lead to remarkable increases in conductivity and several iodine charge transfer salts exhibiting metallic behaviour at room temperature have been reported.[39-43] However, these doped materials become semi-conductors or even insulators at low temperatures.

An intrinsic requirement in the design of neutral radical conductors is a low disproportionation energy for reaction 11.5. Electrochemical investigations of the redox behaviour of 1,2,3,5-dithia and diselenadiazoles,[44] combined with experimentally determined gas-phase ionization potentials,[45] indicate that these systems have higher than desirable disproportionation energies. Consequently, recent attention in this area has been diverted to studies of chalcogen-nitrogen heterocycles with lower disproportionation energies, *e.g.*, benzo-fused dithiazolyl radicals (Section 11.3.5).

$$2 \text{ radical} \leftrightarrow \text{cation}^+ + \text{anion}^- \qquad (11.5)$$

The magnetic properties of the monomeric dithiadiazolyl **11.6a** are of particular interest. This compound crystallizes in two forms. Both form a chain-like motif through electrostatic CN•••S interactions. The α-phase is centric, however, whereas the β-phase is polar. The β-phase undergoes a magnetic phase transition below 36 K to a weakly ferromagnetic state.[12] Studies of its magnetic behaviour under pressure have revealed an enhancement of the ordering temperature up to 65.5 K.[46] The related derivative **11.6c** also forms a chain-like motif through NO_2•••S interactions. This compound has been shown to order as a ferromagnet below 1.3 K.[47] The radical **11.6b** is also monomeric,[33] but it does not undergo long range magnetic order. The spin density in **11.6a–d** has been shown by both experimental and theoretical methods to be mainly concentrated on the heterocyclic ring, with nearly negligible spin density on the fluorinated aromatic ring.[48] The difference in magnetic behaviour of these derivatives is due to their different packing arrangements.

The reaction chemistry of the highly reactive 1,2,3,5-dithiadiazolyl radical dimer **11.4** (E = S) has been investigated extensively. Electrochemical investigations provide evidence for the formation of the corresponding monoanion [PhCNSSN]⁻, which was isolated as the [Na(18-crown-6)]⁺ salt.[49] Dithiadiazolyls may also act as a source of chelating ligands in oxidative-addition reactions with low-valent metal complexes.[8] For example, the reaction of [PhCNSSN]₂ with $Pt(PPh_3)_3$ produces the square planar complex **11.8**, which can be depicted either as a sixteen-electron Pt(II) complex with the dianionic heterocyclic ligand acting as a two-electron donor or as a seventeen-electron Pt(I) complex in which the monoanionic radical ligand contributes three electrons.[50]

The EPR spectrum of **11.8** reveals coupling of the unpaired electron to two ^{14}N and two ^{31}P nuclei accompanied by ^{195}Pt satellites.[51] The dithiadiazolyl ring may also act as a bridging ligand, as illustrated in the binuclear complex **11.9** obtained from the reaction of **11.4** (E = S) with [CpNi(CO)]$_2$.[52] In some cases, *e.g.*, in the formation of the di-iron complex **11.10**, the heterocyclic ligand becomes protonated.[53] This behaviour is attributed to the very high reactivity of the corresponding radical, which results from the accommodation of the unpaired electron in a σ^* orbital rather than a delocalized π-orbital. The trinuclear palladium complex **11.11** is obtained from the reaction of Pd(PPh$_3$)$_4$ and [ArCNSSN]$^{\bullet}$ (Ar = 4-pyridyl).[54]

The neutral radical **11.12** is an interesting heterocyclic analogue of 2,2'-bipyridyl. This paramagnetic (spin-bearing) ligand forms an *N,N'*-chelated complex with bis(hexafluoroacetylacetonato)cobalt(II).[55]

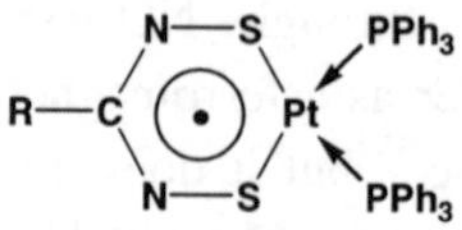

11.8 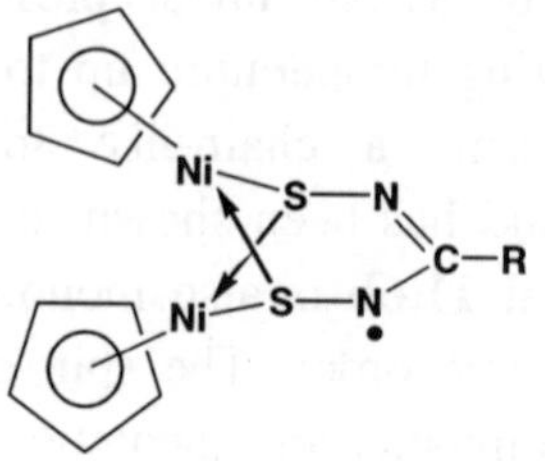   11.9 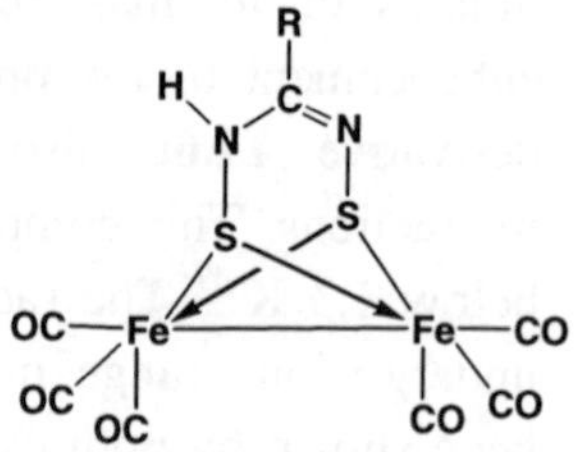   11.10

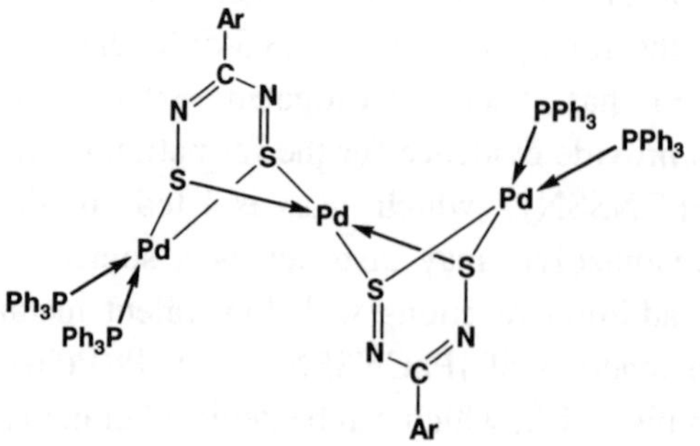

11.11 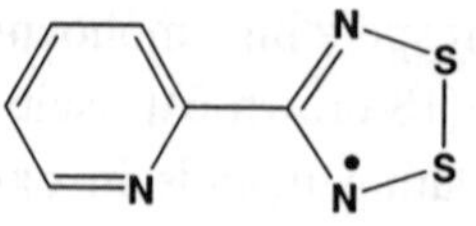   11.12

The reactions of binary cyclic S–N cations with the radical dimer **11.4** (E = S) result in ring contraction. For example, the combination of **11.4** (E =S) with [S_5N_5]Cl produces the six-membered [S_3N_3]⁻ anion (Section 5.4.6) as the [(PhCNSSN)$_2$Cl]⁺ salt.[56] In this cation a Cl⁻ anion bridges two [PhCNSSN]⁺ cations with four S•••Cl distances in the range 2.9-3.0 Å. The radical dimer **11.4** (E = S) also behaves as a dechlorinating agent in reactions with reagents containing P–Cl, Si–Br or activated C–X (X = Br, Cl) bonds. Thus **11.4** (E = S) may be used as a coupling reagent to convert Ph$_2$PCl to Ph$_2$PPPh$_2$ or Me$_3$SiBr to Me$_3$SiSiMe$_3$.[57a] The five-membered ring in **11.4** (E = S) undergoes ring expansion upon treatment with atomic nitrogen, under plasma conditions, to give the dithiatrazine dimers [ArCN$_3$S$_2$]$_2$ (Ar = Ph, 4-ClC$_6$H$_4$).[57b] Alternative methods to these six-membered hetrocycles are discussed in Section 12.2.2.

11.3.2 *1,3,2,4-Dithiadiazolium, [RCNSNS]⁺, and dithiadiazolyl, [RCNSNS]⁻ rings*

The 1,3,2,4-dithiadiazolium cation (**11.13**) is prepared by the cycloaddition of the [S$_2$N]⁺ cation to the C≡N triple bond of organic nitriles (Section 5.3.2 and Eq. 11.6).[58] This methodology may also be applied to the synthesis of molecules containing more than one 1,3,2,4-dithiadiazolyl ring, *e.g.*, by the reaction of 1,3-, 1,4-, or 1,5-C$_6$H$_4$(CN)$_2$ with two equivalents of [S$_2$N][AsF$_6$] to give the corresponding dications 1,3-, 1,4-, or 1,5-[C$_6$H$_4$(CNSNS)$_2$]$^{2+}$.[59] The double addition of [S$_2$N]⁺ to cyanogen produces the dication [(SNSNC)–(CNSNS)]$^{2+}$ [60] and the triple addition of [S$_2$N]⁺ to the tricyanomethide anion [C(CN)$_3$]⁻ proceeds in a stepwise fashion to give the dication [C(CNSNS)$_3$]$^{2+}$.[61] The versatility of this approach is further illustrated by the cycloaddition of [S$_2$N]⁺ to SF$_5$CN[62] and Hg(CN)$_2$[63] to give the corresponding 1,3,2,4-dithiadiazolium mono- and di-cations, respectively. In contrast to the latter reaction, the combination of PhHgCN with [S$_2$N]⁺ produces the cationic sulfur-nitrogen chain [PhS$_4$N$_3$Ph]⁺ (Section 14.3).[63b] A mixed 1,3,2,4-/1,2,3,5-dithiadiazolium dication **11.14** is produced by reaction of [4-CNC$_6$H$_4$CNSSN]AsF$_6$ with [S$_2$N]AsF$_6$.[64]

Although the reagent $[Se_2N]^+$ has been generated *in situ* from $[N(SeCl)_2]^+$ and $SnCl_2$ and shown to undergo cycloaddition with $CF_3C{\equiv}CCF_3$,[65] it has not been used to generate selenium analogues of **11.13**.

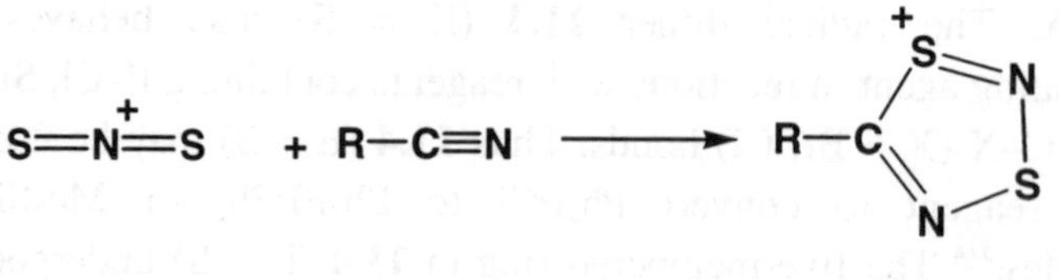

$$S{=}\overset{+}{N}{=}S \;\; + \;\; R{-}C{\equiv}N \;\; \longrightarrow \;\; R{-}C \quad (11.6)$$

11.13

11.14

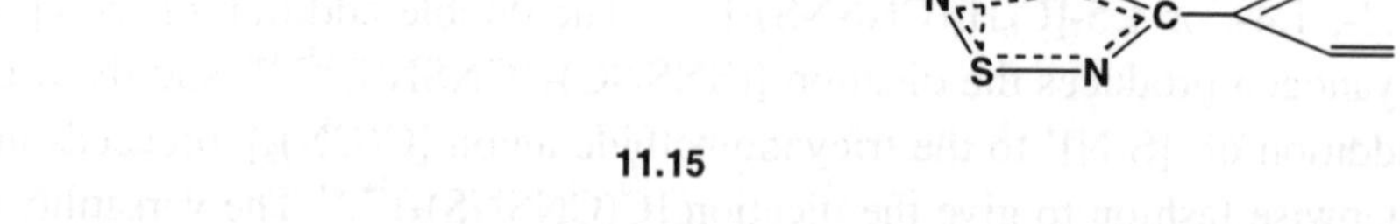

11.15

1,3,2,4-Dithiadiazolyl radicals are typically prepared by reduction of the corresponding cations with $SbPh_3$. They are unstable with respect to isomerization to the 1,2,3,5-isomers both in solution[34] and in the solid state.[66] The isomerization is a photochemically symmetry-allowed process, which is thermally symmetry forbidden. A bimolecular head-to-tail rearrangement has been proposed to account for this isomerization (Scheme 11.1).[67] This rearrangement process is conveniently monitored by EPR spectroscopy (Section 3.4). The 1,3,2,4-dithiadiazoles exhibit a 1:1:1 triplet, whereas the 1,2,3,5-isomers give rise to a 1:2:3:2:1 quintet. In addition to this isomerization, the derivatives [RCNSNS]$^{\bullet}$ (R = ^{t}Bu, Ph) undergo a unimolecular photolysis to RCN and [SNS]$^{\bullet}$.[67]

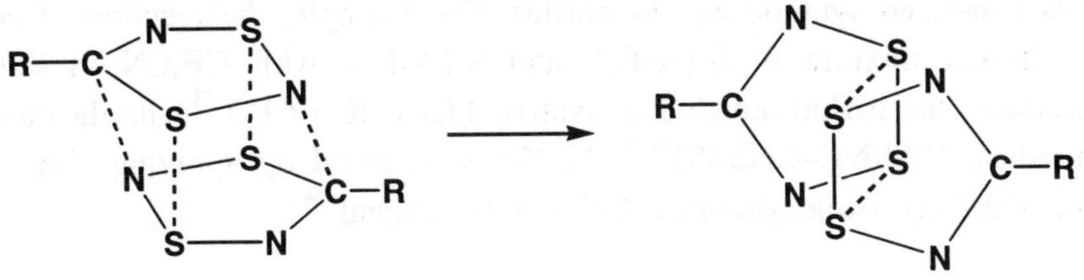

Scheme 11.1   Rearrangement of 1,3,2,4-dithiadiazoles into 1,2,3,5-isomers

As a consequence of their instability, the properties of 1,3,2,4-dithiadiazolyls have been less well studied than those of the 1,2,3,5-isomers. However, several derivatives have been structurally characterized. An interesting example is the phenylene-bridged, diradical 1,4-[C$_6$H$_4$(CNSNS)]$_2$$^{\bullet\bullet}$, (11.15) which has a polymeric structure.[59] The related diradical [SNSNC-CNSNS]$^{\bullet\bullet}$ exhibits a large increase in paramagnetism (from 1.03 to 2.55 BM) upon mechanical grinding, apparently as a result of a pressure-induced phase change.[68]

11.3.3   5-Oxo-1,3,2,4-dithiadiazole

The cyclic ketone 5-oxo-1,3,2,4-dithiadiazole, S$_2$N$_2$CO, may be prepared by two methods (Eq. 11.7 and 11.8).[69,70]

$$Me_2SnN_2S_2 + COF_2 \rightarrow S_2N_2CO + Me_2SnF_2 \tag{11.7}$$

$$ClC(O)SCl + Me_3SiNSNSiMe_3 \rightarrow S_2N_2CO + 2Me_3SiCl \tag{11.8}$$

X-Ray structural data and recent high level theoretical calculations confirm that this neutral, diamagnetic dithiadiazole is an aromatic six π-electron ring system.[71] The gas-phase infra-red and photoelectron spectra of S$_2$N$_2$CO have also been reported.[72]

11.3.4 1,2,3,4-Trithiazolium [RCNSSS]$^{+•}$ rings

1,2,3,4-Trithiazolium radical cations [RCNSSS]$^{+•}$ (**11.16**) are isoelectronic with the corresponding dithiadiazolyl radicals [RCN$_2$S$_2$]$^•$. These seven π-electron systems are prepared by cycloaddition of the *in situ* generated synthon S$_3$$^{+•}$ to nitriles. For example, the reaction of an equimolar mixture of S$_4$[AsF$_6$]$_2$ and S$_8$[AsF$_6$]$_2$ with CF$_3$CN in SO$_2$ produces the trifluoromethyl derivative **11.16** (R = CF$_3$).[73] The dication diradical [SSSNC–CNSSS]$^{2+2•}$ (**11.17**) is obtained by applying similar methodology using cyanogen (CN)$_2$ as the reagent.[74]

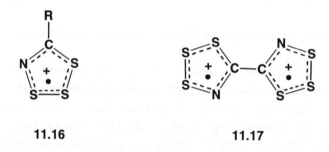

11.16   **11.17**

The solid-state structure of the [AsF$_6$]$^-$ salt of **11.16** (R = CF$_3$) consists of alternating layers of monomeric, planar heterocyclic rings and the counter-anions. The magnetic moment of this salt exhibits normal Curie-Weiss behaviour with a magnetic moment of 1.65 BM between 25 and 200 K.[73] The EPR spectrum of **11.16** (R = CF$_3$) shows no coupling with the ^{14}N nucleus, consistent with calculations that reveal the SOMO for the model system **11.16** (R = H) resides primarily on the three sulfur atoms. The [AsF$_6$]$^-$ salt of the dication diradical **11.17** also consists of non-interacting dications and anions.[74] The room temperature magnetic moment of this salt is 2.81 BM, consistent with the presence of two unpaired electrons.

11.3.5 1,2,3- and 1,3,2-Dithiazolium and dithiazolyl radicals

The 1,2,3- and 1,3,2-dithiazolium ions [(RC)$_2$NS$_2$]$^+$ (**11.18** and **11.19**, respectively) may be regarded as six π-electron heterocycles, isoelectronic with dithiadiazolium rings [RCN$_2$S$_2$]$^+$ via the formal replacement of a nitrogen atom by an RC group. The primary interest in these ring systems is as a source of the corresponding seven π-electron dithiazolyl radicals via one-electron reduction.[11] Although the structures and properties of these radicals will be the main focus of this section, a brief account of the synthesis of their cationic precursors is also included. Comprehensive coverage of the chemistry of dithiazolium cations may be found elsewhere.[75]

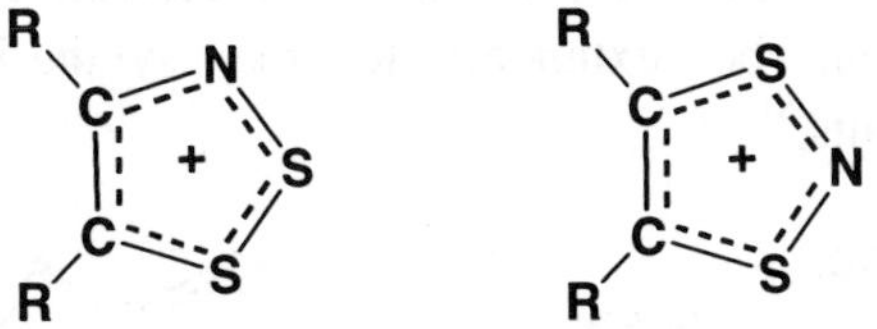

11.18   **11.19**

1,2,3-Dithiazolium salts were first prepared by Herz from the reaction of aromatic amines with an excess of S_2Cl_2 (Eq. 11.9). In this reaction the aromatic ring usually becomes chlorinated *para* to the amine N atom. When aromatic amines with bulky *ortho* substituents are employed, this reaction produces acyclic thiosulfinylamino compounds ArN=S=S (Section 10.3).[76] An alternative to the Herz reaction involves the condensation reaction between thionyl choride and an *ortho*-aminothiophenol (Eq. 11.10).[77] This approach has been exploited to generate a benzo-bis(1,2,3-dithiazolium) salt.[78.79] It may also be extended to generate ring systems in which either (or both) of the sulfur atoms is replaced by selenium atoms by employing *ortho*-aminoselenophenols in condensation reactions with $SOCl_2$ or, in the case of selenium, selenous acid or selenium tetrachloride.[77]

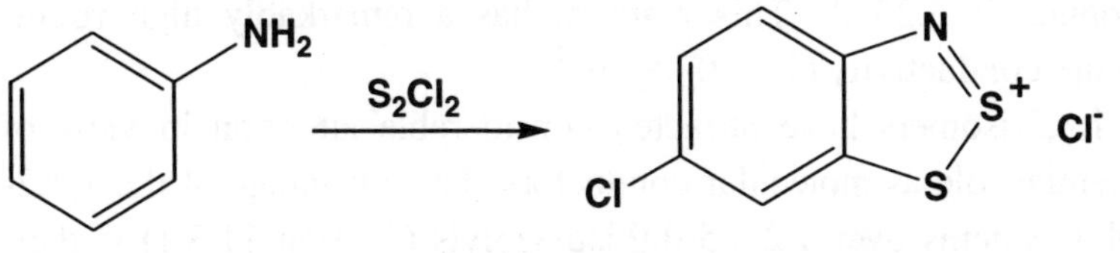

$$(11.9)$$

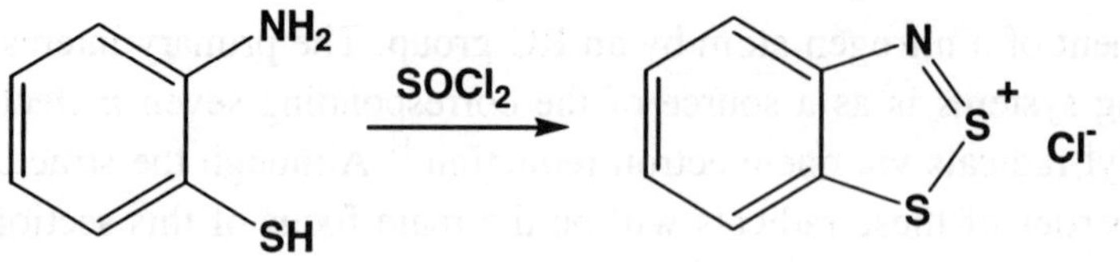

$$(11.10)$$

The primary route to 1,3,2-dithiazolium salts involves the cyclocondensation of 1,2-bis(sulfenyl chlorides) with trimethylsilyl azide, which is illustrated for benzo-fused derivatives in Eq. 11.11.[80] This method can be extended to the synthesis of bis(1,3,2-dithiazolylium) salts.[81]

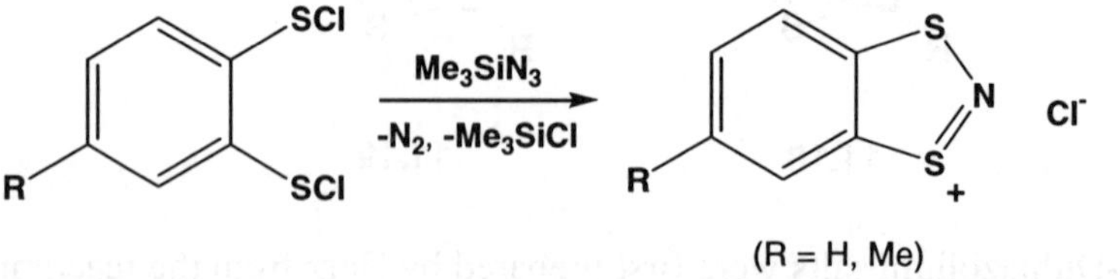

$$(11.11)$$

(R = H, Me)

1,2,3- and 1,3,2-Dithiazolyl radicals are generated from the corresponding cations by reduction with reagents such as $SbPh_3$, sodium dithionite or Zn/Cu couple.[11] Although 1,2,3-dithiazolyls have been characterized in solution, *e.g.*, by EPR and UV–visible spectroscopy, these isomers have limited stability and rearrangement processes involving C–C bond formation may occur in the absence of steric protection.[82] However, the pentafluorophenyl derivative $(C_6F_5C_2S_2NCl)_2$ has been structurally characterized in the solid state.[83] In this dimer the C_6F_5 group is twisted (by 58°) away from the plane of the C_2S_2N ring and the heterocyclic rings are linked by a single S•••S contact (3.30 Å). The structure of the tricyclic system **11.20** has also been determined.[84] It consists of slipped stacks of head-to-tail π-dimers. The closest intradimer S•••S contact is 3.23 Å. This material has a remarkably high room-temperature conductivity (1×10^{-4} S cm^{-1}).

The 1,3,2-isomers have attracted considerable attention in view of their potential role as molecular conductors. The advantage of the 1,3,2-dithiazolyl systems over 1,2,3,5-dithiadiazolyls (Section 11.3.1) is their relatively low disproportionation energies (IP – EA) (Eq. 11.5). This parameter provides an estimate of the Coulombic barrier to electron transfer in a neutral radical conductor. The monocyclic derivatives **11.21** (R = R' = CN,[85] CF$_3$[34b]) crystallize as cofacial, diamagnetic dimers linked by weak S•••S contacts; the dimers are further associated into tetrameric units. The benzo derivative **11.22** (X = H) exhibits a rare centrosymmetric mode of association,[86] whereas the naphthalene compound **11.23** consists of discrete radicals locked into herringbone arrays.[87] The thiadiazole[88] and thiadiazolopyrazine-based[89] systems, **11.24** and **11.25**, are of particular interest. These materials adopt slipped π-stack arrays in the solid state.

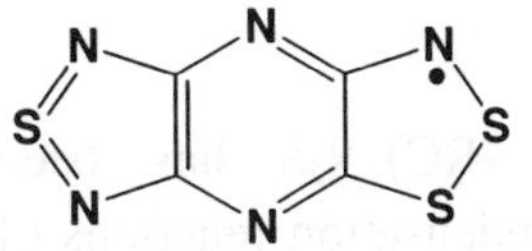

| **11.20** | **11.21** | **11.22** (X = H, Me) |

| **11.23** | **11.24** | **11.25** |

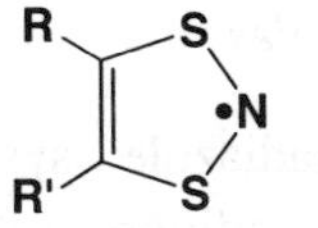

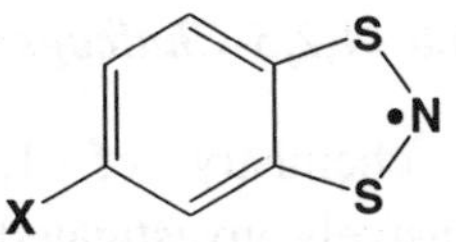

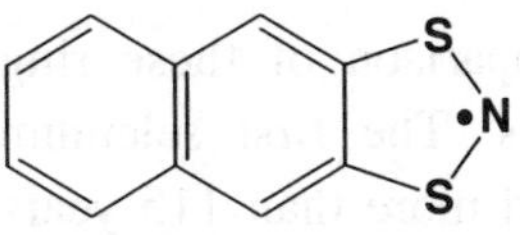

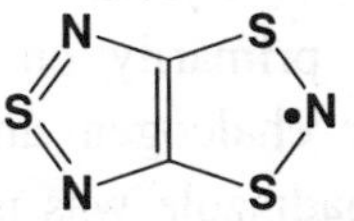

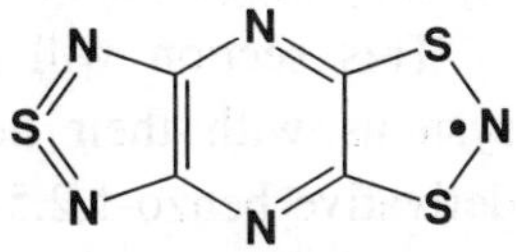

The magnetic properties of 1,3,2-dithiazolyl radicals have also attracted considerable attention. Some derivatives retain their paramagnetism at room temperature, *e.g.*, methylbenzodithiazolyl **11.22** (X = Me)[90] and the naphthaleno-dithiazolyl **11.22**.[87] Several members of the dithiazolyl family exhibit polymorphism. This property gives rise to the phenomenon of bistability, *i.e.*, the possibility of switching between two different polymorphs of the same compound. For example, pristine samples of **11.22** (X = H) undergo an unusual solid–liquid–solid transformation upon heating.[87] The resultant solid is strongly paramagnetic and orders as an antiferromagnet below 11 K. Both **11.24** and **11.25** exhibit a form of bistability in which a paramagnetic phase (undimerized π-stacks) coexists over a specific temperature range with a diamagnetic phase (π-dimer stacks).[91] Thermodynamic studies on the trithiatriazapentalenyl radical **11.24** show that the entropy change for the transition is equivalent to the unpairing of the two electrons and the gain in entropy upon breaking the π^*-π^* dimers compensates for the slightly lower lattice enthalpy of the monomeric structure.[92] This phase transition is particularly interesting since the region of bistability (230-320 K) encompasses room temperature. The possible applications of these and related bistable radical-based materials in molecular switching devices is being considered.[93]

11.3.6 *1,2,5-Chalcogenadiazoles*

The chemistry of 1,2,5-thiadiazole systems $(RC)_2N_2S$ has been extensively investigated.[94] In addition to the condensation reactions of sulfur halides with 1,2-diaminobenzenes, this ring system is obtained in high yields by the reaction of S_4N_4 with acetylenes (Section 5.2.6).[95] For example, the reaction of S_4N_4 with diphenylacetylene produces 3,4-diphenyl-1,2,5-thiadiazole in 87% yield.

This section will focus primarily on a comparison of these ring systems with their heavier chalcogen analogues. The first selenium derivative benzo-1,2,5-selenadiazole was prepared more than 115 years ago by the condensation reaction of selenium dioxide with 1,2-diaminobenzene (Eq. 11.12) and other benzo derivatives may be prepared in a similar manner.[96] The parent 1,2,5-selenadiazole has also been reported.[97] This reagent has been employed to make the tellurium analogue via treatment with ethylmagnesium bromide followed by the addition of tellurium tetrachloride (Eq. 11.13).[98]

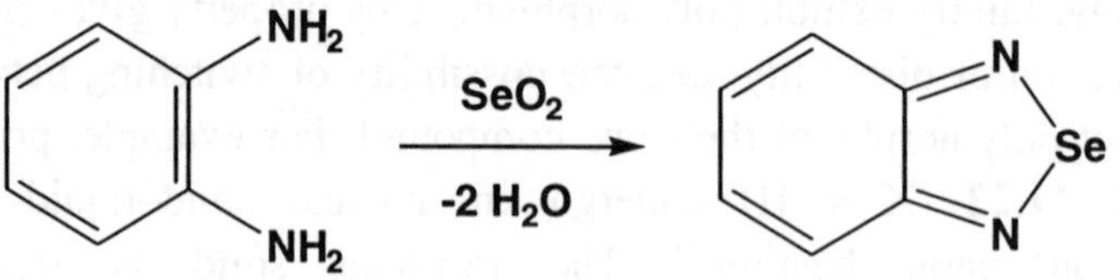

$$\text{(11.13)}$$

Two benzo-1,2,5-telluradiazoles have been reported. The tetrafluoro derivative **11.27** is obtained by the condensation of tellurium tetrachloride with 1,2-diaminobenzene upon heating in a high boiling solvent.[99] The telluradiazole [$^tBu_2C_6H_2N_2Te$]$_2$ (**11.28**) is the unexpected product of the reaction depicted in Scheme 11.2.[100] Apparently C(aryl)–C(CH$_3$)$_3$ bond cleavage occurs during the course of this reaction resulting in the loss of a ^tBu group from the aromatic ring. A similar ring closure has been observed to occur in the reaction of SeOCl$_2$ with 2,4,6-$^tBu_3C_6H_2NH_2$.[101]

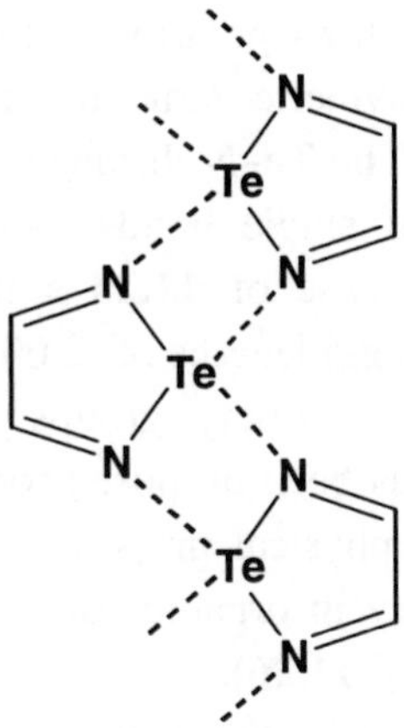

11.26 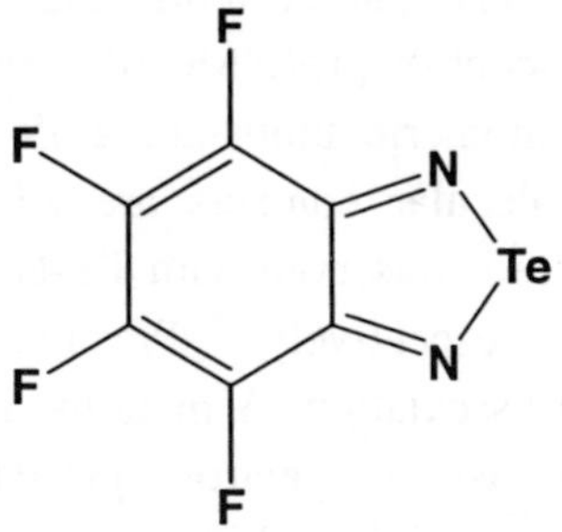   **11.27**

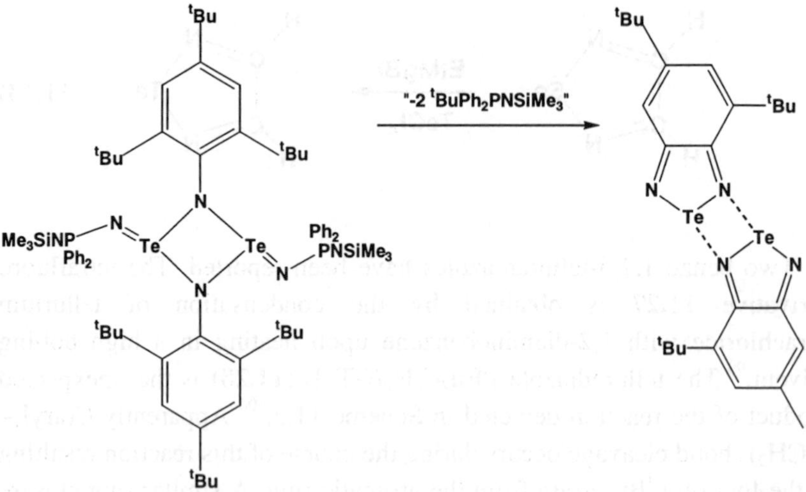

Scheme 11.2   Formation of the 1,2,5-telluradiazole **11.28** via C–C bond cleavage

In contrast to thiadiazoles, the heavier chalcogen analogues tend to form associated structures as a result of the high polarity of the E–N bonds. The parent telluradiazole has a polymeric structure **11.26**[102] reminiscent of $[Te(NMe_2)_2]_\infty$ (Section 10.7). The Te–N distances within the monomeric units are 2.05 Å, typical of single bonds, while the intermolecular contacts are 2.96 Å. In the case of **11.28** a dimeric structure is observed with Te–N and Te•••N bond lengths of 2.00 Å and 2.63 Å, respectively.[100] Presumably, the bulky *tert*-butyl groups prevent further association. A monomeric structure has been proposed for **11.27** on the basis of spectroscopic data,[99] but the physical properties of this derivative (high melting point and insolubility in organic solvents) are more consistent with an associated structure (*cf.* **11.26**).

In addition to intermolecular interactions in the solid state, the relative importance of the two resonance structures **A** and **B** is an important issue in the structural determinations of selena- and telluradiazoles.[103] The Se–N bond lengths fall within the range 1.78–1.81 Å and the Te–N bond lengths are 2.00–2.05 Å compared to single bond values of 1.86 and 2.05 Å, respectively. It can be concluded that resonance structure **A** is more important than **B** for the Se and, especially, the Te systems consistent with weaker E–N π-bonds for the heavier chalcogens. X-ray structural investigations of 1,2,5-thiadiazoles indicate extensive π-delocalization in the heterocyclic ring and, in the case of the benzo derivative, quinonoid character for the benzene ring, suggesting contributions from both resonances structures **A** and **B**.[104] These conclusions are supported by ^{1}H NMR, microwave, and photoelectron spectra, and by *ab initio* molecular orbital calculations.[105] The reactions of the parent 1,2,5-thiadiazole indicate that ionic resonance forms such as **C** are also important contributors to the resonance hybrid.[106]

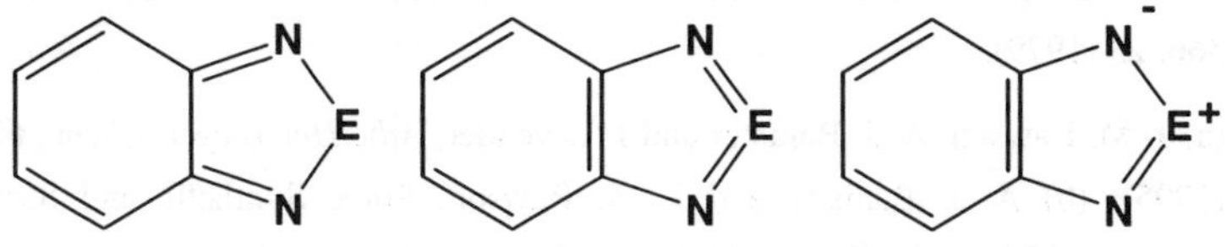

A   **B**   **C**

An *N*-bonded 1:1 adduct of benzo-1,2,5-thiadiazole with the Lewis acid AsF$_5$ has been structurally characterized.[107] The coordination of AsF$_5$ to one of the nitrogens introduces asymmetry in the heterocyclic ring [d(S–N) = 1.63 and 1.58 Å, *cf.* 1.60 Å in the free ligand]. The quinonoid character of the benzene ring is still apparent in the adduct. A 1:1 complex with Pt(II) and 1:1 or 1:2 complexes with Cr(0), Mo(0) and W(0) have been reported.[108-110] These complexes are considered to be *N*-monodentate or *N,N'*-chelated, respectively, on the basis of spectroscopic evidence. 1,2,5-Selenadiazole also forms mono-adducts with M(CO)$_5$ (M= Cr, Mo, W).[110] Se–N bond insertion occurs in the reaction of the selenadiazole (CN)$_2$C$_2$N$_2$Se with Pt(C$_2$H$_4$)(PPh$_3$)$_2$ to give the six-membered ring **11.29**.[111]

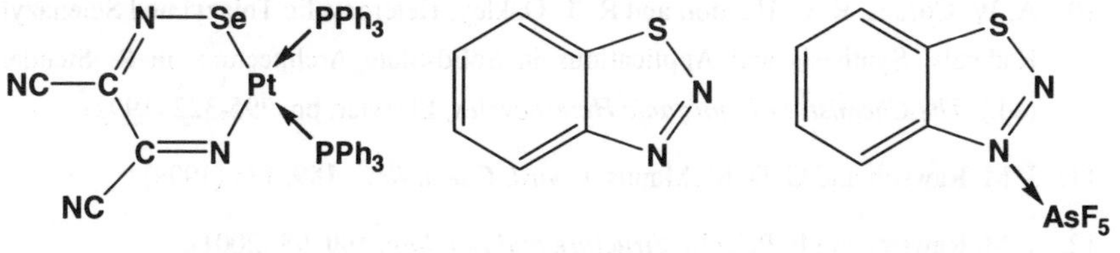

11.29   **11.30**   **11.31**

The structure of the isomeric benzo-1,2,3-thiadiazole **11.30** is unknown,[112] but the 1:1 adduct with AsF_5 (**11.31**) has been structurally characterized. The AsF_5 molecule is coordinated to the carbon-bonded nitrogen atom.[107] Cycloocteno-1,2,3-selenadiazole is an effective source of selenium for the production of semi-conductors such as cadmium selenide.[113]

References

1. M-H. Whangbo, R. Hoffman and R. B. Woodward, *Proc. R. Soc. London, Ser. A*, **366**, 23 (1979).

2. (a) J. M. Rawson, A. J. Banister and I. Lavender, *Adv. Heterocycl. Chem.*, **62**, 137 (1995); (b) A. J. Banister and J. M. Rawson, Some Synthetic and Structural Aspects of Dithiadiazoles, RCN_2S_2, and Related Compounds, in R. Steudel (ed.) *The Chemistry of Inorganic Heterocycles*, Elsevier, pp. 323-348 (1992).

3. J. L. Morris and C. W. Rees, *Chem. Soc. Rev.*, **15**, 1 (1986).

4. T. Chivers, Sulfur-Nitrogen Heterocycles, in I. Haiduc and D. B. Sowerby (ed.) *The Chemistry of Inorganic Homo- and Heterocycles*, Academic Press, London, Vol. 2, pp. 793-870 (1987).

5. R. T. Oakley, *Prog. Inorg. Chem.*, **36**, 1 (1988).

6. R. T. Oakley, *Can. J. Chem.*, **71**, 1775 (1993).

7. T. Chivers, Sulfur-Nitrogen Compounds, in R. B. King (ed.) *Encyclopedia of Inorganic Chemistry*, 2nd Edition, John Wiley & Sons Ltd., in press (2004).

8. A. J. Banister, I. May, J. M. Rawson and J. N. B. Smith, *J. Organomet. Chem.*, **550**, 241 (1998).

9. R. T. Boeré and T. L. Roemmele, *Coord. Chem. Rev.*, **210**, 369 (2000).

10. A. W. Cordes, R. C. Haddon and R. T. Oakley, Heterocyclic Thiazyl and Selenazyl Radicals: Synthesis and Applications in Solid-State Architecture, in R. Steudel (ed.) *The Chemistry of Inorganic Heterocycles*, Elsevier, pp. 295-322 (1992).

11. J. M. Rawson and G. D. McManus, *Coord. Chem. Rev.*, **189**, 135 (1999).

12. J. M. Rawson and F. Palacio, *Structure and Bonding*, **100**, 93 (2001).

13. E. Hey, C. Ergezinger and K. Dehnicke, Z. *Naturforsch.*, **44B**, 205 (1984).

14. (a) G. G. Alange, A. J. Banister, B. Bell and P. W. Millen, *J. Chem. Soc., Perkin Trans. 1*, 1192 (1979); (b) A. J. Banister, N. R. M. Smith and R. G. Hey, *J. Chem. Soc., Perkin Trans. 1*, 1181 (1983).

15. A. Apblett and T. Chivers, *Inorg. Chem.*, **28**, 4544 (1989).

16. P. Del Bel Belluz, A. W. Cordes, E. M. Kristof, P. V. Kristof, S. W. Liblong and R. T. Oakley, *J. Am. Chem. Soc.*, **111**, 9276 (1989).

17. A. W. Cordes, R. C. Haddon, R. T. Oakley, L. F. Schneemeyer, J. V. Waszczak, K. M. Young and N. M. Zimmerman, *J. Am. Chem. Soc.*, **113**, 582 (1991).

18. A. Amin and C. W. Rees, *J. Chem. Soc., Perkin Trans. 1*, 2495 (1989).

19. A. W. Cordes, C. D. Bryan, W. M. Davis, R. H. de Laat, S. H. Glarum. J. D. Goddard. R. C. Haddon. R. G. Hicks, D. K. Kennepohl, R. T. Oakley, S. R. Scott and N. P. C. Westwood, *J. Am. Chem. Soc.*, **115**, 7232 (1993).

20. M. P. Andrews, A. W. Cordes, D. C. Douglass, R. M. Fleming, S. H. Glarum, R. C. Haddon, P. Marsh, R. T. Oakley, T. T. M. Palstra, L. F. Schneemeyer, G. W. Trucks, R. R. Tycko, J. V. Waszczak, W. W. Warren, K. M. Young and N. M. Zimmerman, *J. Am. Chem. Soc.*, **113**, 3559 (1991).

21. A. W. Cordes, R. C. Haddon, R. G. Hicks, R. T. Oakley, T. T. M. Palstra, L. F. Schneemeyer and J. F. Waszczak, *J. Am. Chem. Soc.*, **114**, 5000 (1992).

22. A. W. Cordes, R. C. Haddon, R. G. Hicks, D. K. Kennepohl, R. T. Oakley, L. F. Schneemeyer and J. F. Waszczak *Inorg. Chem.*, **32**, 1554 (1993).

23. A. W. Cordes, C. M. Chamchoumis, R. G. Hicks, R. T. Oakley, K. M. Young and R. C. Haddon, *Can. J. Chem.*, **70**, 919 (1992).

24. A. W. Cordes, R. C. Haddon, C. D. MacKinnon, R. T. Oakley, G. W. Patenaude, R. W. Reed, T. Rietveld and K. E. Vajda, *Inorg. Chem.*, **35**, 7626 (1996).

25. C. D. Bryan, A. W. Cordes, R. C. Haddon, R. G. Hicks, R. T. Oakley, T. M. Palstra and A. J. Perel, *J. Chem. Soc., Chem. Commun.*, 1447 (1994).

26. H.-U. Höfs, J. W. Bats, R. Gleiter, G. Hartmann, R. Mews, M. Eckert-Maksić, H. Oberhammer and G. M. Sheldrick, *Chem. Ber.*, **118**, 3781 (1985).

27. A. Vegas, A. Pérez-Salazar, A. J. Banister and R. G. Hey, *J. Chem. Soc., Dalton Trans.*, 1812 (1980).

28. A. W. Cordes, J. D. Goddard, R. T. Oakley and N. P. C. Westwood, *J. Am. Chem.. Soc.*, **111**, 6147 (1989).

29. A. J. Banister, M. I. Hansford, Z. V. Hauptman, S. T. Wait and W. Clegg, *J. Chem. Soc., Dalton Trans.*, 1705 (1989).

30. (a) A. D. Bond, D. A. Haynes, C. M. Pask and J. M. Rawson, *J. Chem. Soc., Dalton Trans.*, 2522 (2002); (b) C. S. Clarke, S. I. Pascu and J. M. Rawson, *Cryst. Eng. Commun.*, 79 (2004).

31. A. J. Banister, A. S. Batsanov, O. G. Dawe, P. L. Herbertson, J. A. K. Howard, S. Lynn, I. May. J. N. B. Smith, J. M. Rawson, T. E. Rogers, B. K. Tanner, G. Antorrena and F. Palacio, *J. Chem. Soc., Dalton Trans.*, 2539 (1997).

32. A J. Banister, N. Bricklebank, I. Lavender, J. M. Rawson, C. I. Gregory, B. K. Tanner, W. Clegg, M. R. J. Elsegood and F. Palacio, *Angew. Chem., Int. Ed. Engl.*, **35**, 2533 (1996).

33. G. Antorrena, J. E. Davies, M. Hartley, F. Palacio, J. M. Rawson, J. N. B. Smith and A. Steiner, *Chem. Commun.*, 1394 (1999).

34. (a) W. V. F. Brooks, N. Burford, J. Passmore, M. J. Schriver and L. H. Sutcliffe, *J. Chem. Soc., Chem. Commun.*, 69 (1987); (b) H. Du, R. C. Haddon. I. Krossing, J. Passmore, J. M. Rawson and M. J. Schriver, *Chem. Commun.*, 1836 (2002).

35. C. S. Clarke, D. A. Haynes, J. M. Rawson and A. D. Bond, *Chem. Commun.*, 2774 (2003).

36. A. W. Cordes, R. C. Haddon, R. G. Hicks, R. T. Oakley and T. T. M. Palstra, *Inorg. Chem.*, **31**, 1802 (1992).

37. W. M. Davis, R. G. Hicks, R. T. Oakley, B. Zhao and N. J. Taylor, *Can. J. Chem.*, **71**, 180 (1993).

38. W. M. Davis and J. D. Goddard, *Can. J. Chem.*, **74**, 810 (1996).

39. C. D. Bryan, A. W. Cordes, R. M. Fleming, N. A. George, S. H. Glarum, R. C. Haddon, R. T. Oakley, T. T. M. Palstra, A. S. Perel, L. F. Schneemeyer and J. V. Waszczak, *Nature*, **365**, 821 (1993).

40. C. D. Bryan, A. W. Cordes, R. C. Haddon, R. G. Hicks, D. K. Kennepohl, C. D. MacKinnon, R. T. Oakley, T. T. M. Palstra, A. S. Perel, S. R. Scott, L. F. Schneemeyer and J. V. Waszczak, *J. Am. Chem. Soc.*, **116**, 1205 (1994).

41. C. D. Bryan, A. W. Cordes, R. M. Fleming, N. A. George, S. H. Glarum, R. C. Haddon, C. D. MacKinnon, R. T. Oakley, T. T. M. Palstra and A. S. Perel, *J. Am. Chem. Soc.*, **117**, 6880 (1995).

42. C. D. Bryan, A. W. Cordes, J. D. Goddard, R. C. Haddon, R. G. Hicks, C. D. MacKinnon, R. C. Mawhinney, R. T. Oakley, T. T. M. Palstra and A. S. Perel, *J. Am. Chem. Soc.*, **118**, 330 (1996).

43. A. W. Cordes, N. A. George, R. C. Haddon, D. K. Kennepohl, R. T. Oakley, T. T. M. Palstra and R. W. Reed, *Chem. Mater.*, **8**, 2774 (1996).

44. R. T. Boeré and K. H. Moock, *J. Am. Chem. Soc.*, **117**, 4755 (1995).

45. J. Campbell, D. Klapstein, P. F. Bernath, W. M. Davis, R. T. Oakley and J. D. Goddard, *Inorg. Chem.*, **35**, 4264 (1996).

46. M. Mito, T. Kawae, K. Takeda, S. Takagi, Y. Matsushita, H. Deguchi, J. M. Rawson and F. Palacio, *Polyhedron*, **20**, 1509 (2001).

47. A. Alberola, R. J. Lees, C. M. Pask, J. M. Rawson, F. Palacio, P. Oliete, C. Paulsen, A. Yamaguchi, R. D. Farley and D. M. Murphy, *Angew. Chem., Int. Ed. Engl.*, **42**, 4782 (2003).

48. P. J. Alonso, G. Antorrena, J. I. Martinez, J. J. Novoa, F. Palacio, J. M. Rawson and J. N. B. Smith. *Appl. Mag. Reson.*, **20**, 231 (2001).

49. C. M. Aherne, A. J. Banister, I. B. Gorrell, M. I. Hansford, Z. V. Hauptman, A. W. Luke and J. M. Rawson, *J. Chem. Soc., Dalton Trans.*, 967 (1993).

50. A. J. Banister, I. B. Gorrell, J. A. K. Howard, S. E. Lawrence, C. W. Lehman, I. May, J. M. Rawson, B. K. Tanner, C. I. Gregory, A. J. Blake and S. P. Fricker, *J. Chem. Soc., Dalton Trans.*, 377 (1997).

51. A. J. Banister, I. B. Gorrell, S. E. Lawrence, C. W. Lehmann, I. May, G. Tait, A. J. Blake and J. M. Rawson, *J. Chem. Soc., Chem. Commun.*, 1779 (1994).

52. A. J. Banister, I. B. Gorrell, W. Clegg and K. A. Jorgensen, *J. Chem. Soc., Dalton Trans.*, 1105 (1991).

53. R. T. Boeré, K. H. Moock, V. Klassen, J. Weaver, D. Lentz, and H. Michael-Schultz, *Can. J. Chem.*, **73**, 1444 (1995).

54. W-K. Wong, C. Sun, W-Y. Wong, D. W. J. Kwong and W-T. Wong, *Eur. J. Inorg. Chem.*, 1045 (2000).

55. N. G. R. Hearns, K. E. Preuss, J. F. Richardson and S. Bin-Salamon, *J. Am. Chem., Soc.*, **126**, Published on-line July 23 (2004).

56. A. J. Banister, W. Clegg, Z. V. Hauptmann, A. W. Luke and S. T. Wait, *J. Chem. Soc., Chem. Commun.*, 351 (1989).

57. (a) N. Adamson, A. J. Banister, I. B. Gorrell, A. W. Luke and J. M. Rawson, *J. Chem. Soc., Chem. Commun.*, 919 (1993); (b) A. J. Banister, M. I. Hansford, Z. V. Hauptmann, S. T. Wait and W. Clegg, *J. Chem. Soc., Chem. Commun.*, 1705 (1989).

58. (a) G. K. MacLean, J. Passmore, M. N. S. Rao, M. J. Schriver, P. S. White, D. Bethell, R. S. Pilkington and L. H. Sutcliffe, *J. Chem. Soc., Dalton Trans.*, 1405 (1985); (b) S. Parsons and J. Passmore, *Acc. Chem. Res.*, **27**, 101 (1994).

59. A. J. Banister, J. M. Rawson, W. Clegg and S. L. Birkby, *J. Chem. Soc., Dalton Trans.*, 1099 (1991).

60. S. Parsons, J. Passmore, M. J. Schriver and P. S. White, *J. Chem. Soc., Chem. Commun.*, 369 (1991).

61. A. J. Banister, I. Lavender, J. M. Rawson and W. Clegg, *J. Chem. Soc., Dalton Trans.*, 859 (1992).

62. J. Jacobs, S. E. Ulic, H. Willner, G. Schatte, J. Passmore, S. V. Sereda and T. S. Cameron, *J. Chem. Soc., Dalton Trans.*, 383 (1996).

63. (a) A. J. Banister, I. Lavender, S. E. Lawrence, J. M. Rawson and W. Clegg, *J. Chem. Soc., Chem. Commun.*, 29 (1994); (b) C. M. Aherne, A. J. Banister, I. Lavender, S. E. Lawrence and J. M. Rawson, *Polyhedron*, **15**, 1877 (1996).

64. A. J. Banister, I. Lavender, J. M. Rawson and R. J. Whitehead, *J. Chem. Soc., Dalton Trans.*, 1449 (1992).

65. K. B. Borisenko, M. Broschag. I. Hargittai. T. M. Klapötke, D. Schröder, A. Schulz, H. Schwarz, I. C. Tornieporth-Oetting and P. S. White, *J. Chem. Soc., Dalton Trans.*, 2705 (1994).

66. C. Aherne, A. J. Banister, A. W. Luke, J. M. Rawson and R. J. Whitehead, *J. Chem. Soc., Dalton Trans.*, 1277 (1992).

67. J. Passmore and X. Sun, *Inorg. Chem.* **35**, 1313 (1996).

68. G. Antorrena, S. Brownridge, T. S. Cameron, R. Palacio, S. Parsons, J. Passmore, L. K. Thompson and F. Zarlaida, *Can. J. Chem.*, **80**, 1568 (2002).

69. H. W. Roesky and E. Wehner, *Angew. Chem., Int. Ed. Eng.*, **14**, 498 (1975).

70. R. Neidlein, P. Leinberger, A. Gieren and B. Dederer, *Chem. Ber.*, **111**, 698 (1978).

71. (a) H. W. Roesky, E. Wehner, E. J. Zehnder, H. J. Deiseroth and A. Simon, *Chem. Ber.*, **111**, 1670 (1978); (b) J. Van Droogenbroeck, K. Tersago. C. Van Alsenoy, S. M. Aucott, H. L. Milton, J. D. Woollins and F. Blockhuis, *Eur. J. Inorg. Chem.*, Published on-line July 22 (2004).

72. R.H. Delaat, L. Durham, E. G. Livingstone and N. P. C. Westwood, *J. Phys. Chem.*, **97**, 11216 (1993).

73. T. S. Cameron, R. C. Haddon, S. M. Mattar, S. Parsons, J. Passmore and A. Ramirez, *Inorg. Chem.*, **31**, 2274 (1992).

74. P. D. Boyle, S. Parsons, J. Passmore and D. J. Wood, *J. Chem. Soc., Chem. Commun.*, 199 (1993).

75. L. I. Khmelnitski and O. A. Rakitin, in A. R. Katritzky, C. W. Rees and E. F. V. Scriven (ed.) *Comprehensive Heterocyclic Chemistry*, Vol. II, Pergamon, Oxford, p. 916 (1996).

76. Y. Inagaki, R. Okazaki, and N. Inamoto, *Bull. Chem. Soc. Jpn.*, **52**, 1998 (1979).

77. L. D. Huestis, M. L. Walsh and N. Hahn, *J. Org. Chem.*, **30**, 2763 (1965).

78. T. M. Barclay, A. W. Cordes, J. D. Goddard, R. C. Mawhinney, R. T. Oakley, K. E. Preuss and R. W. Reed, *J. Am. Chem. Soc.*, **119**, 12136 (1997).

79. L. Beer, J. L. Brusso, A. W. Cordes, R. C. Haddon, M. E. Itkis, K. Kirschbaum, D. S. MacGregor, R. T. Oakley, A. A. Pinkerton and R. W. Reed, *J. Am. Chem. Soc.*, **124**, 9498 (2002).

80. (a) G. Wolmershäuser, M. Schnauber and T. Wilhelm, *J. Chem. Soc., Chem.. Commun.*, 573 (1984); (b) G. Heckmann, R. Johann, G. Kraft and G. Wolmershäuser, *Synth. Met.* **41-43**, 3287 (1991).

81. T. M. Barclay, A. W. Cordes, R. H. de Laat, J. D. Goddard, R. C. Haddon, D. Y. Jeter, R. C. Mawhinney, R. T. Oakley, T. T. M. Palstra, G. W. Patenaude, R. W. Reed, and N. P. C. Westwood, *J. Am. Chem. Soc.*, **119**, 2633 (1997).

82. A. W. Cordes, R. C. Haddon and R. T. Oakley, *Adv. Mater.*, **6**, 798 (1994).

83. T. M. Barclay, L. Beer, A. W. Cordes, R. T. Oakley, K. E. Preuss, N. J. Taylor and R. W. Reed, *J. Chem. Soc., Chem. Commun.*, 531 (1999).

84. T. M. Barclay, A. A. Cordes, R. C. Haddon, M. E. Itkis, R. T. Oakley, R. W. Reed and H. Zhang, *J. Amer. Chem. Soc.*, **121**, 969 (1999).

85. G. Wolmerhäuser and G. Kraft, *Chem. Ber.*, **123**, 881 (1990).

86. (a) E. G. Awere, N. Burford, R. C. Haddon, S. Parsons, J. Passmore, J. Waszczak and P. S. White, *Inorg. Chem.*, **29**, 4821 (1990): (b) E. G. Awere, N. Burford, C. Mailer, J. Passmore, M. J. Schriver, P. S. White, A. J. Banister, H. Oberhammer and L. H. Sutcliffe, *J. Chem. Soc., Chem. Commun.*, 66 (1987).

87. T. M. Barclay, A. W. Cordes, N. A. George, R. C. Haddon, R. T. Oakley, T. T. M. Palstra, G. W. Patenaude, R. W. Reed, J. F. Richardson and H. Zhang, *J. Chem. Soc., Chem. Commun.*, 873 (1997).

88. W. Fujita and K. Awaga, *Science*, **286**, 261 (1999).

89. T. M. Barclay, A. W. Cordes, N. A. George, R. C. Haddon, M. E. Itkis, M. S. Mashuta, R. T. Oakley, G. W. Patenuade, R. W. Reed, J. F. Richardson and H. Zhang, *J. Am. Chem. Soc.*, **120**, 352 (1998).

90. G. D. McManus, J. M. Rawson, N. Feeder, F. Palacio and P. Oliete, *J. Mater. Chem.*, **10**, 2001 (2000).

91. M. E. Itkis, X. Chi, A. W. Cordes and R. C. Haddon, *Science*, **296**, 1443 (2002).

92. G. D. McManus, J. M. Rawson, N. Feeder, J. van Duijn, E. J. L. McInnes, J. J. Novoa, R. Burriel, F. Palacio and P. Oliete, *J. Mater. Chem.*, **11**, 1992 (2001).

93. J. L. Brusso, O. P. Clements, R. C. Haddon, M. E. Itkis, A. A. Leitch, R. T. Oakley, R. W. Reed and J. F. Richardson, *J. Am. Chem. Soc.*, **126**, published on-line June 15 (2004).

94. T.L. Gilchrist, *Heterocyclic Chemistry*, Pitman, London (1985).

95. S. T. A. K. Daley and C. W. Rees, *J. Chem. Soc., Perkin Trans. 1*, 207 (1987).

96. J. F. Alicina and J. A. Kowald, in D. L. Klayman and W. H. Günther (ed) *Organic Selenium Compounds: Their Chemistry and Biology*, John Wiley & Sons, Ch. 17, pp. 1050-1081 (1973).

97. (a) L. M. Weinstock, P. Davis, D. M. Mulvey and J. C. Schaeffer, *Angew. Chem., Int. Ed. Engl.,* **6**, 364 (1967); (b) V. Bertini, *Angew. Chem., Int. Ed. Engl.,* **6**, 563 (1967).

98. V. Bertini and F. Lucchesini, *Synthesis,* 681 (1982).

99. V. N. Kovtonyuk, A. Yu. Makarov, M. M. Shakirov and A. V. Zibarev, *Chem. Commun.,* 1991 (1996).

100. T. Chivers, X. Gao and M. Parvez, *Inorg. Chem.,* **35**, 9 (1996).

101. H. W. Roesky, K.-L. Weber, U. Seseke, W. Pinkert, M. Noltemeyer, W. Clegg and G. M. Sheldrick, *J. Chem. Soc., Dalton Trans.,* 565 (1985).

102. V. Bertini, P. Dapporto, F. Lucchesini, A. Sega and A. De Munno, *Acta Crystallogr.,* **C40**, 653 (1984).

103. M. Björgvinsson and H. W. Roesky, *Polyhedron,* **10**, 2353 (1991).

104. M. Gieren, H. Beta, T. Hübner, V. Lamm, R. Neidlein and D. Droste, *Z. Naturforsch.,* **39B**, 485 (1984).

105. M. H. Palmer and S. M. F. Kennedy, *J. Mol. Struct.* **43**, 33 (1978).

106. W. G. Salmond, *Quart. Rev.,* **22**, 253 (1968).

107. A. Apblett, T. Chivers and J. F. Richardson, *Can. J. Chem.,* **64**, 849 (1986).

108. J. Kuyper and K. Vrieze, *J. Organomet. Chem.,* **86**, 127 (1975).

109. R. Meij, T. A. M. Kaandorp, D. J. Stufkens and K. Vrieze, *J. Organomet. Chem.,* **128**, 203 (1977).

110. W. Kaim and S. Kohlman, *Inorg. Chim. Acta,* **101**, L21 (1985).

111. H. W. Roesky, T. Gries, H. Hofmann, J. Schimkowiak, P. G. Jones, K. Meyer-Bäse and G. M. Sheldrick, *Chem. Ber.,* **119**, 366 (1986).

112. V. A. Bakulev and W. Dehaen, *The Chemistry of 1,2,3-Thiadiazoles. The Chemistry of Heterocyclic Compounds, Vol. 62,* John Wiley and Sons Inc., Hoboken, N. J. (2004).

113. P. K. Khanna, C. P. Morley, R. M. Gorte, R. Gokhale, V. V. V. S. Subbarao and C. Satyanarayana, *Mater. Chem. Phys.,* **83**, 323 (2004).

Chapter 12

Six-membered and Larger Carbon–Nitrogen– Chalcogen Ring Systems

12.1 Introduction

This chapter is a continuation of the preceding discussion of five-membered carbon–nitrogen–chalcogen ring systems, which is now extended to include heterocycles with six or more atoms in the ring, as well as bicyclic systems. The isolobal relationship between an RC unit and S^+ as a substituent in a sulfur-nitrogen ring is still apposite (Section 4.5). Thus, a number of these larger neutral heterocycles have isoelectronic analogues among the cyclic binary S–N cations described in Chapter 5. This class of chalcogen–nitrogen heterocycles embodies neutral radical systems that engage in intermolecular $\pi^*-\pi^*$ interactions, as well as cyclic systems that exhibit weak transannular S•••S ($\pi^*-\pi^*$) interactions. The chapter is sub-divided into sections according to increasing ring size. Within each section, ring systems that incorporate two-coordinate sulfur (or selenium) will be discussed first followed by those which contain three or four-coordinate sulfur. Detailed accounts of some of the early aspects of the chemistry of these heterocycles can be found in several review articles.[1-3]

12.2 Six-membered Rings

12.2.1 *Chalcogenatriazinyl Rings, [(RC)$_2$N$_3$E]$^{\cdot}$ (E = S, Se)*

The chalcogenatriazinyl systems (**12.1**) are prepared by reduction of the corresponding S–Cl or Se–Cl derivatives (**12.2**) with triphenylantimony.[4] These six-membered rings are seven π-electron radicals, which form cofacial diamagnetic dimers in the solid state with E•••E distances of 2.67 Å (E = S) or 2.79 Å (E = Se) (Section 4.7). The solution EPR spectra of a series of thiatriazinyl radicals [(RC)$_2$N$_3$S]$^{\cdot}$ exhibit slightly larger hyperfine couplings to the unique nitrogen atom compared to those involving the symmetrically equivalent nitrogen centres (4.1-4.9 G *vs.* 3.3-3.9 G), except in the case of R = 4-MeOC$_6$H$_4$ for which accidental equivalence of all three nitrogen centres is observed.[5] Oxidation of **12.1** (E = S) with [NO][PF$_6$] generates the corresponding trithiazinium cation, while reduction by sodium in liquid ammonia produces the anion, which has been isolated as the protonated derivative (PhC)$_2$N$_3$(H)S (**12.3**).[6]

12.1 (E = S, Se)   **12.2** (E = S, Se)   **12.3**

Examples of the C$_2$N$_3$S ring system containing three- or four-coordinate sulfur are also well known. The monohalogenated derivatives **12.2** (E = S, Se), are best prepared by the condensation of imidoyl amidines with SCl$_2$ or SeCl$_4$, respectively (Scheme 12.1).[4] In the case of the selenium derivative, the initial product is heated at 60°C and then at 120°C in order to convert it to **12.2** (E = Se) via (PhC)$_2$(NH)N$_2$SeCl$_2$. In the solid state this intermediate is a weakly associated, centrosymmetric dimer with Se–Cl and Se•••Cl distances of 2.42 and 3.39 Å, respectively.[7]

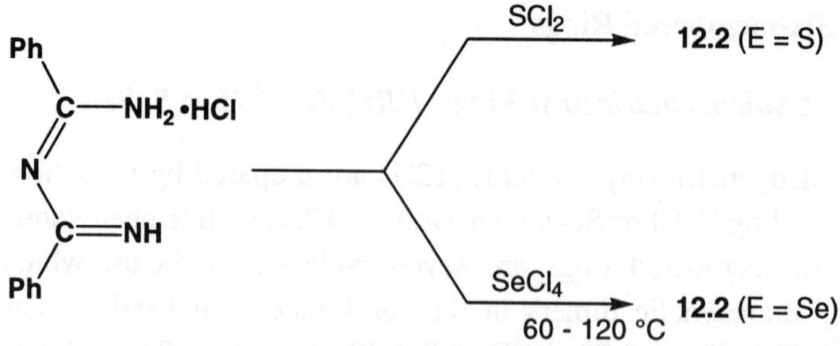

Scheme 12.1   Synthesis of $(PhC)_2N_3ECl$ (E = S, Se)

The related cyanuric–thiazyl system $(ClC)_2N_3SCl$ (**12.4**, X = Cl) is obtained in good yield by the reaction of sodium dicyanamide with thionyl chloride in DMF.[8] This trihalogenated heterocycle reacts with sodium alkoxides or aryloxides with preferential substitution at sulfur.[8b] Trisubstitution can be achieved without ring degradation to give $[(RO)C]_2N_3S(OR)$ (R = Me, Et, CH_2CF_3, Ph). By contrast, the reaction of the fluorinated derivative **12.4** (X = F) with $[(Me_2N)_3S][Me_3SiF_2]$ involves nucleophilic (fluoride) attack at the carbon centre to give the anion $[(F_2CN)(FCN)(FSN)]^-$.[9] The reaction of **12.4** (X = Cl) with trialkylamines results in cleavage of a C–N bond of the tertiary amine and regiospecific substitution of the dialkylamino group on the carbon atoms of the heterocycle.[10] For example, treatment of **12.4** (X = Cl) with tetramethylmethylenediamine produces $[(Me_2N)C]_2N_3SCl$. Unlike the corresponding phosphorus-containing ring system $(Cl_2P)_2N_3SCl$ (Section 14.4), the heterocycle **12.4** (X = Cl) does not undergo ring-opening polymerization cleanly. Thermolysis at 220°C yields a viscous orange gum with some loss of sulfur as volatile sulfur chlorides.[8b]

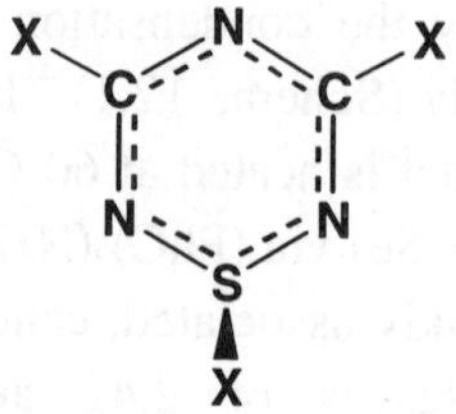

12.4 (X = Cl, F) 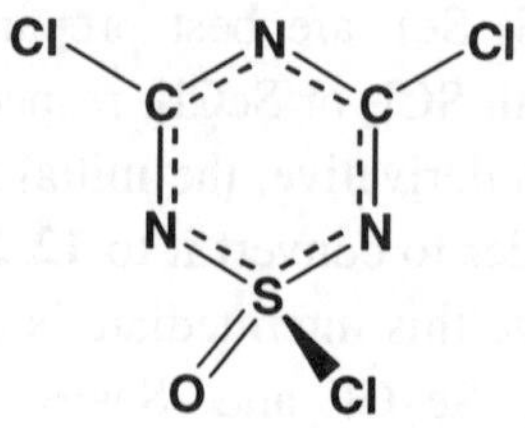   **12.5**

The oxidation of **12.4** (X = Cl) with a mixture of $KMnO_4$ and $CuSO_4.xH_2O$ (x = 4-6) produces the hybrid cyanuric–sulfanuric system **12.5** containing a four-coordinate sulfur centre.[8b] The X-ray structure of this low melting solid reveals a relatively short S–Cl bond length of 1.99 Å and a mean S–N bond length of 1.58 Å [8c] [*cf.* 2.13 and 1.62 Å, respectively for the sulfur (IV) system **12.4** (X = Cl) [11]]. The structure of the corresponding trifluoro derivative **12.4** (X = F) has been determined by electron diffraction [d(S–N) = 1.59 Å].[12] The reaction of **12.5** with $NaOCH_2CF_3$ in a 1:3 molar ratio produces the trialkoxy derivative **12.6** which, upon prolonged thermolysis at 175°C, undergoes isomerization involving migration of a CH_2CF_3 group from oxygen to nitrogen to give the sulfone **12.7** (Eq. 12.1).[8b]

12.6   **12.7**

(12.1)

12.2.2 *Dithiatriazines, RCN_3S_2*

Aryl derivatives of the dithiatriazine ring (**12.8**) are prepared by the reduction of the corresponding *S,S'*-dichloro derivative **12.9** (R = aryl).[13,14] The selenium analogues are unknown. The six-membered rings **12.8** are eight π-electron systems isoelectronic with $[S_3N_3]^+$ (Section 5.3.5). In common with the structure of **12.1**, they exist as cofacial dimers in the solid state with S•••S separations in the range of 2.51-2.53 Å (Section 4.7).[13,14] NMR studies indicate that the dimeric structure is preserved in solution.[14] The six-membered ring **12.8** (R = Ph) is also formed by insertion of excited nitrogen atoms, generated in a low pressure direct current discharge, into the S–S bond of the corresponding dithiadiazolyl ring (Section 11.3.1).[15] Two other dithiatriazines RCN_3S_2 (R = CF_3, NMe_2) are obtained as red solids by reduction of the corresponding S,S'-dichloro derivatives **12.9** using Zn or SbPh$_3$ (R = CF$_3$) and Hg(SiMe$_3$)$_2$ (R = NMe$_2$).[16-18] Although the structures of these derivatives have not been established, they undergo S,S'-cycloaddition with norbornadiene, in a manner typical of the CN$_3$S$_2$ ring system, to give the 1,3-norbornenyl adducts **12.10** (R = Ph, CF$_3$, NMe$_2$) (Section 4.8.1).

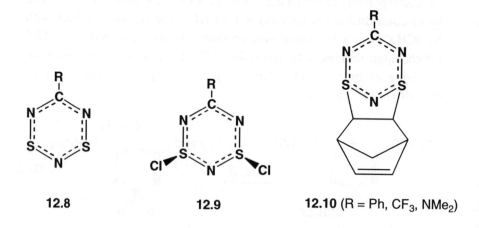

| **12.8** | **12.9** | **12.10** (R = Ph, CF$_3$, NMe$_2$) |

Ab initio molecular orbital calculations for the model systems RCN$_3$S$_2$ (R = H, NH$_2$) show that these dithiatriazines are predicted to be ground state singlets with low-lying triplet excited states (Section 4.4).[19] The singlet state is stabilized by a Jahn–Teller distortion from C_{2v} to C_s symmetry. In this context the observed dimerization of these antiaromatic (eight π-electron) systems is readily understood.

The S,S'-dichloro derivatives **12.9** (R =Ar) are prepared by chlorination of the bicyclic systems ArCN$_5$S$_3$ (Section 12.5) with gaseous chlorine.[13,14] The treatment of organic nitriles with (NSCl)$_3$ at 23°C also gives good yields of **12.9** (R = ^{t}Bu, CCl$_3$, CF$_3$, Ph), but the reactions are slow.[16,20] By contrast, the combination of dialkylcyanamides R$_2$NCN with (NSCl)$_3$ produces **12.9** (R = NMe$_2$, NEt$_2$, N^{i}Pr$_2$) efficiently upon mild heating in CCl$_4$.[20] The chlorinated derivative **12.9** (R = Cl) is obtained in modest yield from ClCN and (NSCl)$_3$.[11] The X-ray structures of **12.9** (R = Cl,[11] CF$_3$,[16] Ph,[21] NMe$_2$,[22] NEt$_2$[23]) reveal a *cis* arrangement of the two chlorine substituents in axial positions with S–Cl bond lengths ranging from *ca.* 2.10 Å (R = Cl, CF_3) to *ca.* 2.21 Å (R = NMe_2, NEt_2). As in the case of cyclotrithiatriazines (Section 8.7), this arrangement is stabilized by the anomeric effect (nitrogen lone pair → S–Cl σ^* bond).[24]

12.2.3 *Trithiadiazines, $R_2CN_2S_3$*

1,3,5,2,4-Trithiadiazines (**12.11**) were first obtained from the reaction of S_4N_4 with diazoalkanes.[25] For example, the parent system **12.11** (R = H) was isolated as a red solid in 40% yield after passing diazomethane into a hot dichloromethane solution of S_4N_4. They are more conveniently synthesized from 1,1-bis(sulfenyl chlorides) and $Me_3SiNSNSiMe_3$.[26] The six-membered ring in **12.11** adopts a half-chair conformation with long S–N bonds (1.67 Å) connecting the –SCH_2S– and –N=S=N–units, in which the sulfur-nitrogen distances are *ca.* 1.55 Å.[25] These structural parameters are reminiscent of those of the binary sulfur nitride S_4N_2 (Section 5.2.5), with which **12.11** (R = H) is isoelectronic. One of the sulfur(II) centres in **12.11** (R = H) is oxidized by *m*-chloroperbenzoic acid or dinitrogen tetroxide to the corresponding sulfoxide.[25,26]

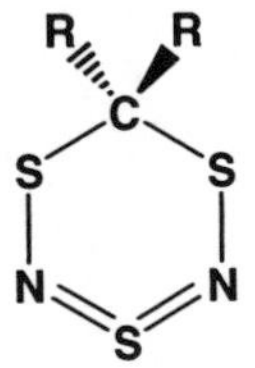

12.11
(R = H, alkyl, aryl)

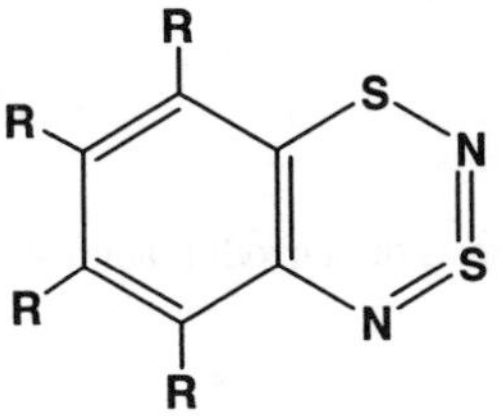

12.12
(R = H, F)

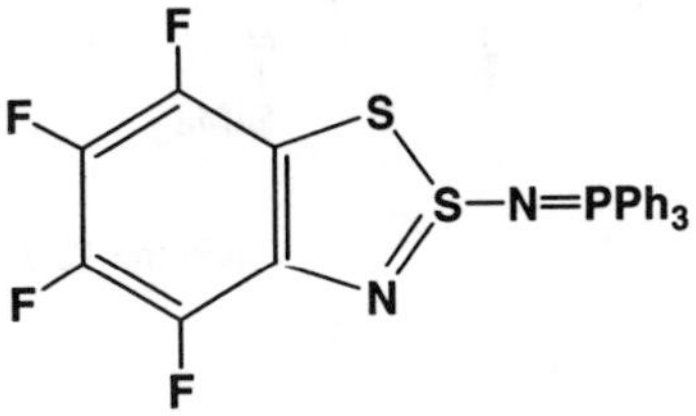

12.13

12.2.4 *Benzodithiadiazines, $C_6R_4S_2N_2$ (R = H, F)*

The antiaromatic twelve π-electron benzodithiadiazine **12.12** (R = H), an inorganic naphthalene analogue, is obtained as a volatile deep-blue solid by the reaction of $PhNSNSiMe_3$ with SCl_2, followed by an intramolecular ring closure with elimination of HCl (Scheme 12.2).[27]

Photoelectron spectra have confirmed the expected trends in the frontier orbitals.[28] The tetrafluoro derivative **12.12** (R = F) is prepared by treatment of $C_6F_5SNSNSiMe_3$ with CsF in acetonitrile (Scheme 12.2).[29] Several difluoro- and trifluoro-benzodithiadiazines have also been prepared by these methods.[30] In contrast to **12.12** (R = H), which has an essentially planar structure in the solid state,[27] the dithiadiazine ring in the tetrafluoro derivative is somewhat twisted.[29] In the gas phase, on the other hand, electron diffraction studies show that **12.12** (R = F) is planar whereas **12.12** (R = H) is non-planar.[31]

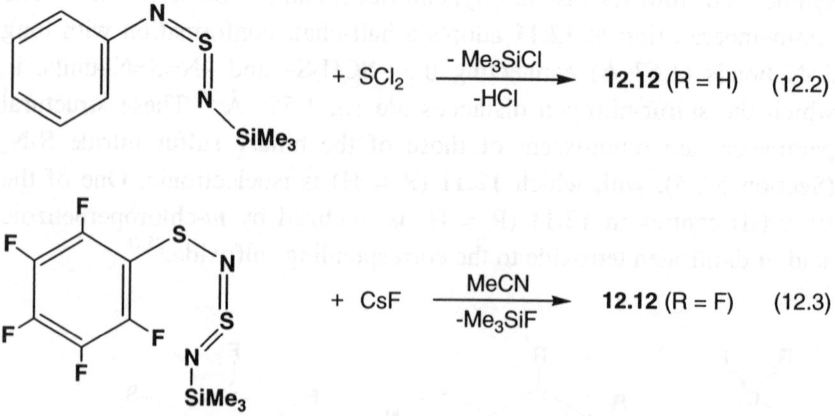

$$+ SCl_2 \xrightarrow[\text{-HCl}]{\text{- Me}_3\text{SiCl}} \textbf{12.12 (R = H)} \quad (12.2)$$

$$+ CsF \xrightarrow[\text{-Me}_3\text{SiF}]{\text{MeCN}} \textbf{12.12 (R = F)} \quad (12.3)$$

Scheme 12.2   Synthesis of benzodithiadiazines

Only a few reactions of benzodithiadiazines have been investigated. In common with dithiatriazines **12.8,** the anti-aromatic system **12.12** (R = H) undergoes a reversible *S,S'*-cycloaddition with norbornadiene.[27] The reaction of **12.12** (R = F) with triphenylphosphine results in a ring contraction to give the imino λ^5-phosphane **12.13**.[32]

12.3 Seven-membered Rings

12.3.1 *Trithiadiazepines, $(RC)_2N_2S_3$*

Trithiadiazepines are seven-membered, ten π-electron systems isoelectronic with $[S_4N_3]^+$ (Section 5.3.6). 1,3,5,2,4-Benzotrithiadiazepine (**12.14**, R = H) is obtained as bright-yellow crystals by the reaction of benzo-1,2-bis(sulfenyl chloride) with $Me_3SiNSNSiMe_3$ (Eq. 12.4).[27] The tetrafluoro derivative **12.14** (R = F) has been prepared by a similar procedure.[33] The isomeric 1,2,4,3,5-benzotrithiadiazepine (**12.15**) is formed in the reaction of $PhNSNSiMe_3$ and S_2Cl_2, followed by intramolecular cyclization (Eq. 12.5).[34]

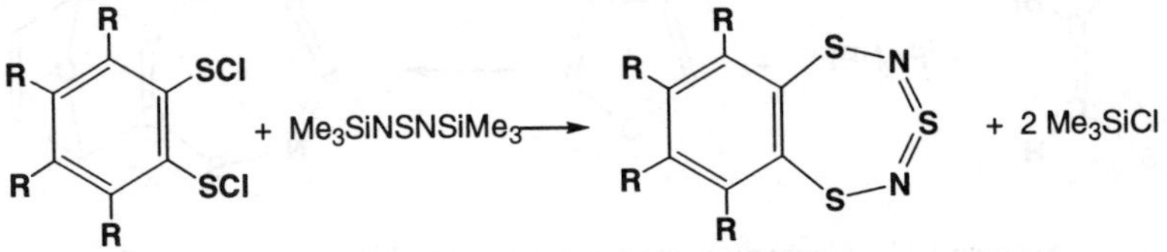

(R = H, F)   **12.14**

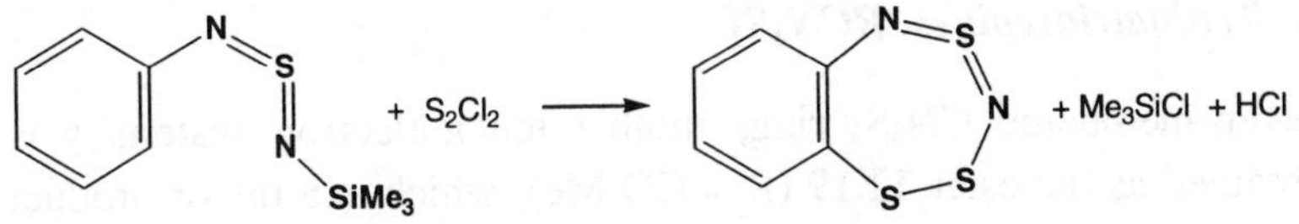

12.15

The parent trithiadiazepine $(HC)_2N_2S_3$ (**12.16**) is synthesized by chlorination of ethanedithiol to the trichloro derivative $Cl(SCl)CHCH_2SCl$, followed by cyclocondensation with $Me_3SiNSNSiMe_3$.[35] The seven-membered ring in **12.16** is planar and the bond lengths indicate complete delocalization.[36] It is inert to protic and Lewis acids and, as befits an aromatic system, it does not undergo cycloaddition reactions.[3] However, the benzo derivative **12.14** (R = H) undergoes reversible *S,S'*-cyclooaddition with norbornadiene.[27] The parent heterocycle **12.16** undergoes standard electrophilic aromatic substitution reactions at carbon. Thus, the disubstituted derivatives **12.17** (X = Br, NO_2, HgOAc) may be prepared by reactions of **12.16** with *N*-bromosuccinimide, $[NO_2][BF_4]$ or mercury(II) acetate, respectively.[3]

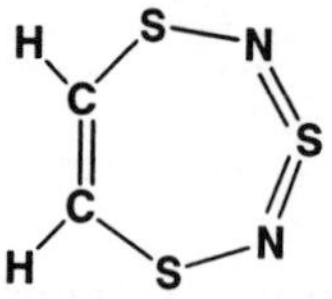

12.16	**12.17**	**12.18**	**12.19**
	(X = Br, NO$_2$, HgOAc)		

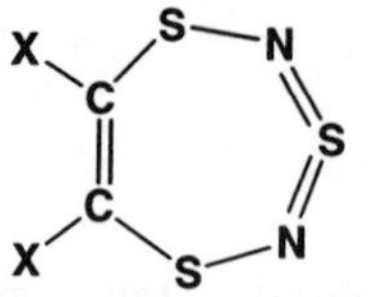

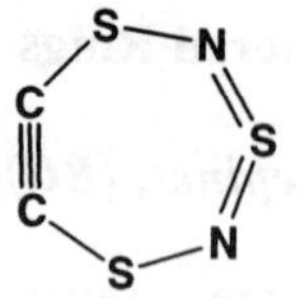

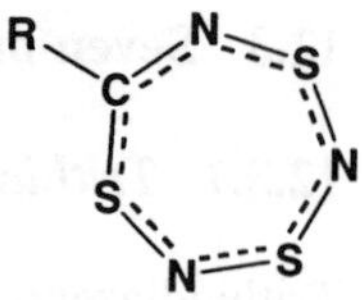

The heterocyclic aryne (trithiadiazepyne) **12.18**, generated by treatment of the monobromo derivative of **12.16** with a strong base, can be trapped as a Diels–Alder adduct with furan (Scheme 12.3).[37,38]

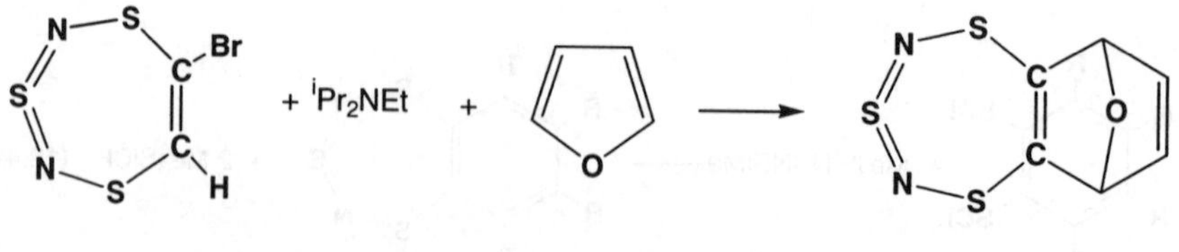

Scheme 12.3   Generation and trapping of a trithiadiazepyne

12.3.2   Trithiatriazepines, RCN$_3$S$_3$

The seven-membered CN$_3$S$_3$ ring, another ten π-electron system, was first obtained as the ester **12.19** (R = CO$_2$Me), which is a minor product of the reaction of S$_4$N$_4$ with dimethylacetylene dicarboxylate.[39a] It has a planar structure with bond lengths that indicate delocalization. The parent 1,3,5,2,4,6-trithiatriazepine **12.19**   (R = H) is obtained as a colourless solid by carefully heating the ester with aqueous HCl followed by decarboxylation.[39b]

12.4   Eight-membered Rings

12.4.1   Dithiatetrazocines, (RC)$_2$N$_4$S$_2$

1,5-Dithiatetrazocines, **12.20** (R = Ph) and **12.21**, were first prepared by the cyclocondensation of benzamidine or dimethylguanidine, respectively, with SCl$_2$ in the presence of a base.[40] The yields are low, however, and air oxidation of the corresponding dithiadiazolium salts [ArCN$_2$S$_2$]X in the presence of SbPh$_3$ is a better alternative for the synthesis of aryl derivatives of **12.20**.[41] By the use of two different dithiadiazolium salts, this method can be adapted to generate unsymmetrically substituted dithiatetrazocines that are of interest in the design of non-linear optical materials.[42] Cyclophanes containing one or two C$_2$N$_4$S$_2$ rings have also been synthesized.[43]

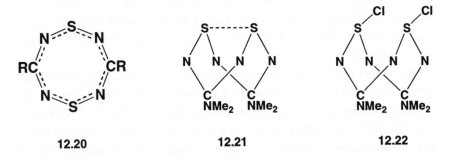

12.20   **12.21**   **12.22**

The structures of 1,5-dithiatetrazocines are remarkably dependent upon the nature of the group attached to carbon. With an aryl substituent on each carbon in **12.20**, the heterocyclic ring is planar and the bond lengths indicate a fully delocalized ten π-electron aromatic system (*cf.* [S$_4$N$_4$]$^{2+}$) (Section 5.3.7). Additional evidence of delocalization comes from the EPR spectrum of the electrochemically generated anion radical of **12.20** (R = Ph), which exhibits equal hyperfine coupling to all four nitrogen nuclei (Section 3.5).[44] By contrast, the bis(dimethylamino) derivative **12.21** adopts a folded structure with d(S•••S) = 2.43 Å.[40] The folding is explained by the destabilizing influence of the π-donor Me$_2$N substituents on the HOMO of the C$_2$N$_4$S$_2$ ring and subsequent second-order Jahn–Teller distortion (Section 4.6).[45-47] Dithiatetrazocines with exocyclic N(Me)Bu groups exhibit both *cis/trans* and ring inversion isomerism on the NMR time scale.[48]

Consistent with its aromatic character, the planar ring **12.20** (R = Ph) has high thermal stability and is chemically unreactive. For example, it does not react with *n*-butyllithium, *m*-chlorobenzoic acid or N$_2$O$_4$.[40,49] Furthermore, it exhibits no basic properties towards HClO$_4$.[40] By contrast, the folded ring **12.21** is readily oxidized and serves as an informative model for understanding the processes involved in the oxidation of sulfur-nitrogen heterocycles containing transannular S•••S bonds, *e.g.*, S_4N_4. Two types of stereochemistries are involved in these oxidation processes. Reaction with Cl_2 or Br_2 produces an *exo,endo* geometry, as in $(Me_2NC)_2N_4(SCl)_2$ (**12.22**), whereas fluorination or reaction with the radical $[(CF_3)_2NO]^•$ results in an *exo,exo* substitution pattern.[50] These different outcomes are the result of polar and radical oxidation mechanisms, respectively (Section 4.8.2). The folded ring system **12.21** acts as a *S,S'*-bidentate ligand towards platinum upon reaction with $Pt(PPh_3)_4$.[51] It also behaves as a weak Lewis base *N*-donor towards Pt(II).[52]

Eight-membered CNS heterocycles of the type $Ar_2C_2N_4S_2Ph_2$ (**12.23**) containing three-coordinate sulfur are generated in good yields by the cyclocondensation of trisilylated benzamidines $(4\text{-}XC_6H_4)CN_2(SiMe)_3$ (X = Br, CF_3) with PhSCl in a 1:3 molar ratio at very low temperature (*i.e.*, under kinetic control).[53] A small amount of the sixteen-membered ring $(4\text{-}BrC_6H_4)_4C_4N_8S_4Ph_4$ (**12.24**) has also been isolated from this reaction.

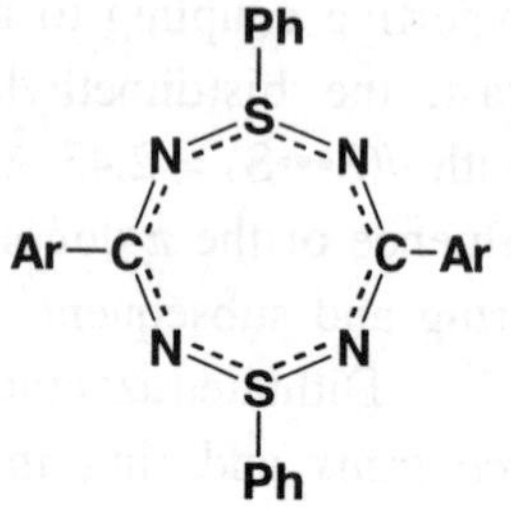

12.23
(Ar = 4-XC_6H_4; X = Br, CF_3)

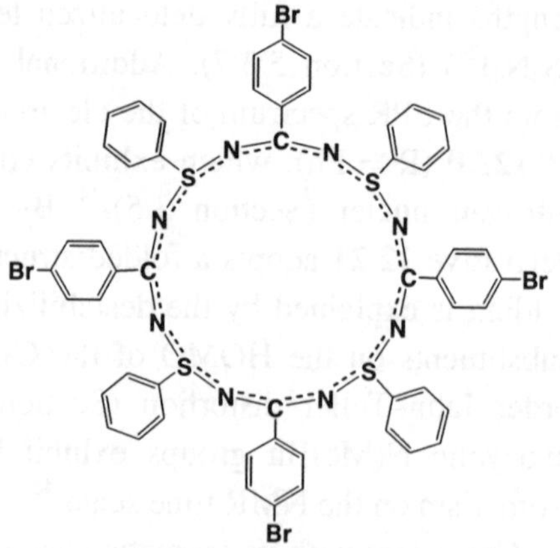

12.24

In the solid state the eight-membered rings **12.23** adopt boat conformations with the phenyl groups (attached to sulfur) in equatorial positions. Density functional theory calculations for the model system $H_2C_2N_4S_2H_2$ reveal that the observed C_{2v} geometry is a result of a second-order Jahn-Teller distortion of the planar (D_{2h}) structure.[53] The chair conformer is only *ca.* 2.3 kcal mol^{-1} higher in energy than the boat conformer. The sixteen-membered ring **12.24** has a cradle-like structure with S_4 symmetry. A fascinating property of the eight-membered rings is their *solid-state* photoisomerization to the purple diazenes PhSN=C(Ar)N=NC(Ar)=NSPh (Scheme 12.4).[54] For example, yellow crystals of **12.23** in a glass vial exposed to laboratory light become coated on the surface with purple crystals of the corresponding diazene within a few days. This transformation is also observed upon heating **12.23** to *ca.* 190°C. A combination of kinetic measurements and DFT calculations show that it is a first order process, which is thermally symmetry forbidden, but photochemically allowed.[54] The sixteen-membered ring **12.24** is unaffected by photolysis.

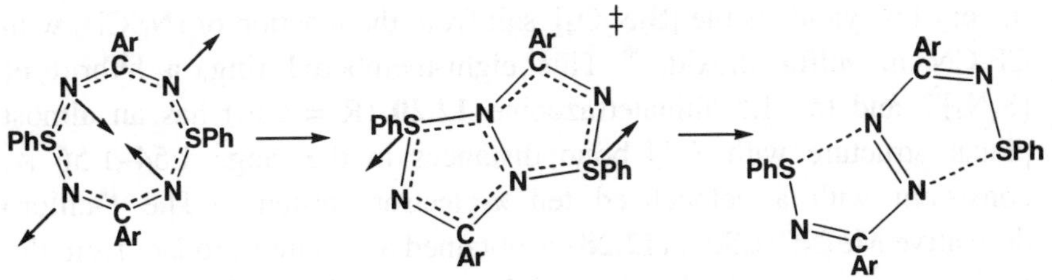

Scheme 12.4 Photochemical isomerization of a $C_2N_4S_2$ ring into a diazene

The oxidation of **12.23** or **12.24** with *m*-chloroperbenzoic acid occurs smoothly in dichloromethane to give the corresponding hybrid cyanuric-sulfanuric ring systems **12.25** and **12.26**, respectively.[55] In the eight-membered rings **12.25** the two oxygen atoms adopt *endo* positions leading to a twisted conformation for the $C_2N_4S_2$ ring. The sixteen-membered rings **12.26** retain the cradle conformation upon oxidation. The S–N bond distances are *ca.* 0.06 Å shorter in all the S(VI) heterocycles **12.25** and **12.26** compared to those in the corresponding sulfur(IV) systems, **12.23** and **12.24**, respectively. The thermolysis of

12.25 (Ar = 4-CF$_3$C$_6$H$_4$) at *ca.* 220°C occurs primarily by loss of a sulfanuric unit, NS(O)Ph, group to give the six-membered ring Ar$_2$C$_2$N$_3$S(O)Ph.[55]

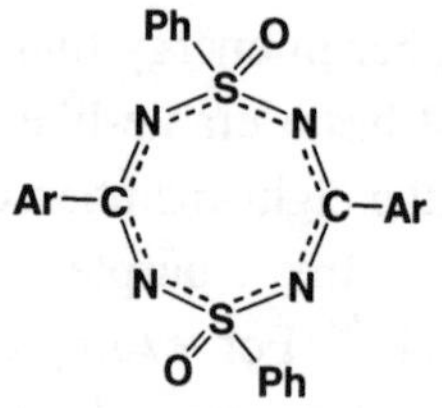

12.25 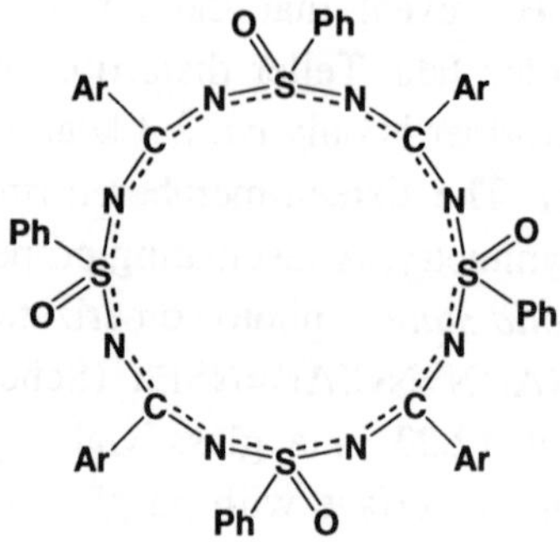   12.26

12.4.2 The trithiatetrazocine cation, [RCN$_4$S$_3$]$^+$

The trithiatetrazocine cation [CF$_3$CN$_4$S$_3$]$^+$ (**12.27**, R = CF$_3$) was isolated in very low yield, as the [S$_3$N$_3$O$_4$]$^-$ salt from the reaction of (NSCl)$_3$ with CF$_3$CN in sulfur dioxide.[56] This eight-membered ring, a hybrid of [S$_4$N$_4$]$^{2+}$ and the 1,5-dithiatetrazocine **12.20** (R = CF$_3$) has an almost planar structure with S–N bond distances in the range 1.54–1.56 Å, consistent with a delocalized ten π-electron system.[56] The *S*-chloro derivative Me$_2$NCN$_4$S$_3$Cl (**12.28**) is obtained as a minor product from the slow reaction of **12.9** (R = NMe$_2$) and Me$_3$SiNSO in carbon tetrachloride.[57] This heterocycle has a folded structure with d(S•••S) = 2.43 Å, identical to the value found for the dithiatetrazocine **12.21.** The chlorine atom occupies an *endo* position with d(S–Cl) = 2.23 Å, *cf.* 2.18 Å in **12.22.**

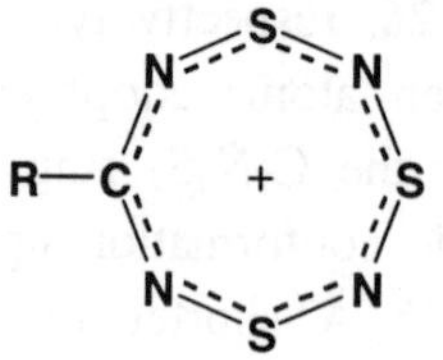

12.27 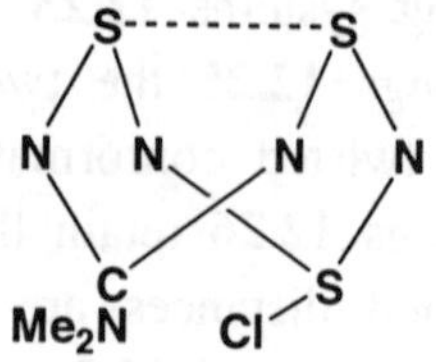   12.28

12.5   Bicyclic Ring Systems, RCN_5S_3

A wide range of derivatives of the bicylic ring systems RCN_5S_3 (**12.29**, R = alkyl, aryl, CF_3, NR'_2, Cl) have been prepared and structurally characterized.[58] Two synthetic routes are available. The first, which involves the reaction of S,S'-dichlorodithiatriazines with $Me_3SiNSNSiMe_3$, is best suited to derivatives in which R = Cl, CX_3 (X = F, Cl) or R'_2N (R' = alkyl) (Eq. 12.6).[57,58] Aryl derivatives, on the other hand, are obtained by the reactions of trisilylated benzamidines with $(NSCl)_3$ (Eq. 12.7).[13,14] A modification of the second route that employs lithium N,N'-bis(trimethylsilyl)amidines is necessary to prepare derivatives with electron-withdrawing aryl substituents such as C_6F_5 or $4\text{-}NCC_6H_4$.[58]

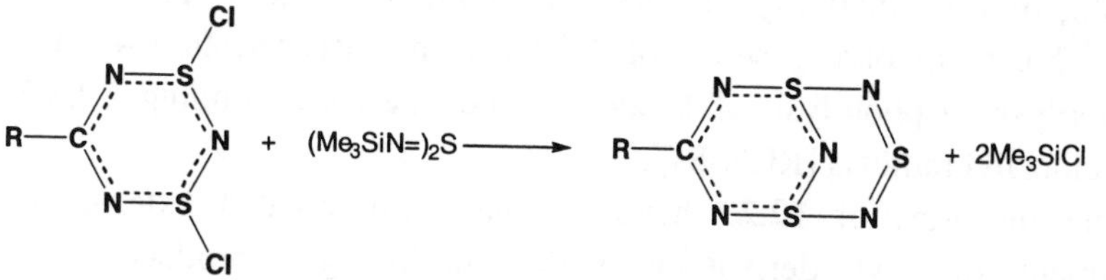

$$\text{(12.6)}$$

12.29

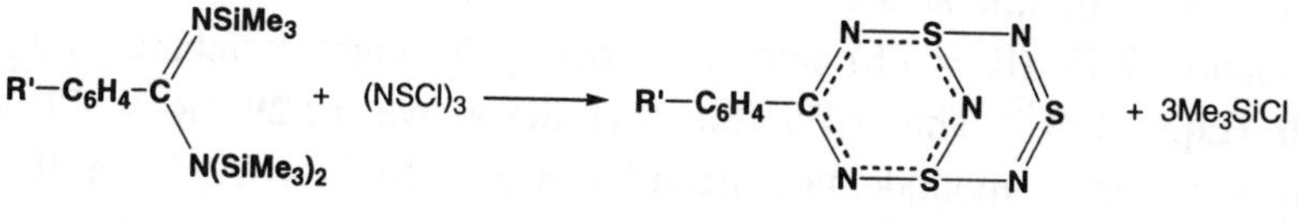

$$\text{(12.7)}$$

The molecular structure of the bicyclic framework in **12.29** is best described as a dithiatriazine bridged by an –N=S=N– group. The bond lengths in the latter unit are 1.54–1.55 Å, indicative of double bonds, while the "connecting" S–N bonds (1.72–1.75 Å) reflect single bond character and the S–N bonds in the remaining S_3N_3 unit fall within the intermediate range of 1.58–1.64 Å.[58,59] The exocyclic substituent R has only a minor effect on the structural parameters within the CN_5S_3 framework. However, the C–N bond lengths are slightly lengthened (and the NCN bond angle is reduced) by electron-donating substituents (R = NMe_2), whereas they are shortened (with a concomitant enlargement of the NCN bond angle) by electron-withdrawing substituents. The structural trends in **12.29** are quite well reproduced by a variety of theoretical calculations.[58] Interestingly, two fundamentally different packing arrangements are observed in the solid state: (a) stacking of RCN_5S_3 molecules and (b) face-to-face dimerisation of S_3N_3 subunits. Although these interactions are undoubtedly electrostatic in origin, there is no obvious correlation with the nature of the R substituent attached to carbon. The aryl-substituted derivatives **12.29** (R = aryl) form 2:1 inclusion complexes with aromatic fluorocarbons or hydrocarbons.[60] The parallel-displaced arrangement of the aromatic rings of the host and guest in these complexes is typical of the non-covalent π–stacking interactions of the arene-polyfluoroarene type.

In solution, the phenyl derivative **12.29** (R = Ph) is fluxional. The mechanism of the fluxional process has been shown by an ^{15}N NMR investigation of a partially ^{15}N-labelled sample, *i.e.*, $PhCN_2*N_3S_3$ (*N = 99% ^{15}N), to involve a series of 1,3-nitrogen shifts (Section 4.8.4).[61] Thermolysis or photolysis of **12.29** generates the corresponding 1,2,3,5-dithiadiazolyl radicals $[RCN_2S_2]^{\bullet}$.[58]

The reactions of **12.29** have not been investigated extensively. Treatment of aryl derivatives with chlorine gas produces *S,S'*-dichlorodithiatriazines, **12.9**,[13,14] whereas the CF_3 derivative is unaffected by Cl_2.[58] Nucleophilic attack by EPh_3 (E = P, As) transforms the bicyclic compound **12.29** (R = Ph) into the monocyclic eight-membered ring **12.30** (Eq. 12.8).[62] The trifluoromethyl derivative **12.29** (R = CF_3) behaves as an *N*-monodentate ligand towards Ni^{2+} or Ag^+ via the bridgehead nitrogen atom; an *N,N'*-bridging mode is also observed in the Ag^+ complex.[63] Consistently, calculations confirm that the largest negative charge resides on the bridgehead nitrogen atom. Protonation, on the other hand, occurs at a nitrogen atom linked to the carbon centre.[64] Interestingly, the Hg^{2+} cation converts **12.29** (R = CF_3, NMe_2, Ph) to the corresponding planar, eight-membered rings **12.27** (Section 12.4.2).[64]

$$Ph-C \cdots \quad + \; EPh_3 \longrightarrow \quad Ph-C \cdots S-NEPh_3 \qquad (12.8)$$

(E = P, As)

12.30

References

1. (a) J. L. Morris and C. W. Rees, *Chem. Soc. Rev.*, **15**, 1 (1986); (b) C. W. Rees, *J. Heterocycl. Chem.*, **29**, 639 (1992).

2. R. T. Oakley, *Prog. Inorg. Chem.*, **36**, 1 (1988).

3. T. Torroba, *J. Prakt. Chem.*, **341**, 99 (1999).

4. (a) P. J. Hayes, R. T. Oakley, A. W. Cordes and W. T. Pennington, *J. Am Chem. Soc.*, **107**, 1346 (1985); (b) R. T. Oakley, R. W. Reed, A. W. Cordes, S. L. Craig and J. B. Graham, *J. Am. Chem. Soc.*, **109**, 7745 (1987).

5. (a) R. T. Boeré and T. L. Roemmele, *Phosphorus, Sulfur and Silicon*, **179**, 875 (2004); (b) R. T. Boeré, R. T. Oakley, R. W. Reed and N. P. C. Westwood, *J. Am. Chem. Soc.*, **111**, 1180 (1989).

6. R. T. Boeré, A W. Cordes, P. J. Hayes, R. T. Oakley, R. W. Reed and W. T. Pennington, *Inorg. Chem.*, **25**, 2445 (1986).

7. (a) D. Fenske, C. Ergezinger and K. Dehnicke, *Z. Naturforsch.*, **44B**, 857 (1989); (b) A. W. Cordes, R. T. Oakley and R. W. Reed, *Acta Crystallogr.*, **C42**, 1889 (1986).

8. (a) J. Geevers, J. T. Hackman and W. P. Trompen, *J. Chem. Soc. C*, 875 (1970); (b) T. Chivers, D. Gates, X. Li, I. Manners and M. Parvez, *Inorg. Chem.*, **38**, 70 (1999); (c) T. Clark, T. Chivers, A. Lough and I. Manners, unpublished results (2003).

9. E. Lork, S-J. Chen, G. Knitter and R. Mews, *Phosphorus, Sulfur and Silicon*, **93-94**, 309 (1994).

10. T. V. V. Ramakrishna, A. J. Elias and A. Vij, *Inorg. Chem.*, **38**, 3022 (1999).

11. S-J. Chen, U. Behrens, E. Fischer, R. Mews, F. Pauer, G. M. Sheldrick, D. Stalke and W-D. Stohrer, *Chem. Ber.*, **126**, 2601 (1993).

12. E. Fischer, E. Jaudas-Prezel, R. Maggiulli, R. Mews, H. Oberhammer, R. Paape and W-D. Stohrer, *Chem. Ber.*, **124**, 1347 (1991).

13. (a) R. T. Boeré, C. L. French, R. T. Oakley, A. W. Cordes, J. A. J. Privett, S. L. Craig and J. B. Graham, *J. Am. Chem. Soc.*, **107**, 7710 (1985); (b) A. W. Cordes,

S. L. Craig, J. A. J. Privett, R. T. Oakley and R. T. Boeré, *Acta Crystallogr.*, **42C**, 508 (1986).

14. R. T. Boeré, J. Fait, K. Larsen and J. Yip, *Inorg. Chem.*, **31**, 1417 (1992).

15. A. J. Banister, M. I. Hansford, Z. V. Hauptmann, S. T. Wait and W. Clegg, *J. Chem. Soc., Dalton Trans.*, 1705 (1989).

16. H-U. Höfs, G. Hartmann, R. Mews and G. M. Sheldrick, *Z. Naturforsch.*, **39B**, 1389 (1984).

17. R. Maggiulli, R. Mews, M. Noltemeyer, W. Offerman, R. Paape and W-D. Stohrer, *Chem. Ber.*, **124**, 39 (1991).

18. T. Chivers, F. Edelmann, J. F. Richardson, N. R. M. Smith, O. True, Jr., and M. Trsic, *Inorg. Chem.*, **25**, 2119 (1986).

19. R. E. Hoffmeyer, W-T. Chan, J. D. Goddard and R. T. Oakley, *Can. J. Chem.*, **66**, 2279 (1988).

20. A. Apblett and T. Chivers, *Inorg. Chem.*, **28**, 4544 (1989).

21. J. B. Graham, A. W. Cordes, R. T. Oakley and R. T. Boeré, *Acta Crystallogr.*, **42C**, 992 (1986).

22. H. W. Roesky, P. Schäfer, M. Noltemeyer and G. M. Sheldrick, *Z. Naturforsch.*, **38B**, 347 (1983).

23. T. Chivers, J. F. Richardson and N. R. M. Smith, *Inorg. Chem.*, **25**, 47 (1986).

24. E. Jaudas-Prezel, R. Maggiulli, R. Mews, H. Oberhammer, T. Paust and W-D. Stohrer, *Chem. Ber.*, **123**, 2123 (1990).

25. R. M. Bannister, R. Jones, C. W. Rees and J. D. Williams, *J. Chem. Soc., Chem. Commun.*, 1546 (1987).

26. R. M. Bannister and C. W. Rees, *J. Chem. Soc., Perkin Trans.1*, 509 (1990).

27. A. W. Cordes, M. Hojo, H. Koenig, M. C. Noble, R. T. Oakley and W. T. Pennington, *Inorg. Chem.*, **25**, 1137 (1986).

28. A. P. Hitchcock, R. S. Dewitte, J. M. Van Esbroeck, C. L. French, R. T. Oakley and N. P. C. Westwood, *J. Electron Spectrosc. Related Phenom.*, **57**, 165 (1991).

29. A. V. Zibarev, Yu. V. Gatilov and A. O. Miller, *Polyhedron*, **11**, 1137 (1992).

30. A. Yu. Makarov, I. Yu. Bagryanskaya, F. Blockhuys, C. Van Alsenoy, Yu. V. Gatilov, V. Y. Knyazev, A. M. Maksimov, T. V. Mikhalina, V. E. Platonov, M. M. Shakirov and A. V. Zibarev, *Eur. J. Inorg. Chem.*, 77 (2003).

31. F. Blockhuys, S. L. Hinchley, A. Yu. Makarov, Yu. V. Gatilov, A. V. Zibarev, J. D. Woollins and D. W. H. Rankin, *Chem. Eur. J.*, **7**, 3592 (2001).

32. A. V. Zibarev, Y. V. Gatilov, I. Yu. Bagryanskaya, A. M. Maksimov and A. O. Miller, *J. Chem. Soc., Chem. Commun.*, 299 (1993).

33. I. Yu. Bagryanskaya, H. Bock, Y. V. Gatilov, A. Haas, M. M. Shakirov, B. Salouki and A. V. Zibarev, *Chem. Ber.*, **130**, 247 (1997).

34. A. Y. Makarov, M. M. Shakirov, K. V. Shuvaev, I. Yu. Bagryanskaya, Y. V. Gatilov and A. V. Zibarev, *Chem. Commun.*, 1774 (2001).

35. J. L.Morris and C. W. Rees, *J. Chem. Soc., Perkin Trans. 1*, 211 (1987).

36. J. L.Morris and C. W. Rees, *J. Chem. Soc., Perkin Trans. 1*, 217 (1987).

37. M. J. Plater and C. W. Rees, *J. Chem. Soc., Perkin Trans. 1* 301 (1999).

38. M. J. Plater and C. W. Rees, *Phosphorus, Sulfur and Silicon*, **43**, 261 (1989).

39. (a) S. T. A. K. Daley and C. W. Rees, *J. Chem. Soc., Perkin Trans. 1*, 203 (1987); (b) P. J. Dunn, J. L. Morris and C. W. Rees, *J. Chem. Soc., Perkin Trans. 1*, 1745 (1988).

40. I. Ernest, W. Holick, G. Rihs, D. Schomburg, G. Shoham, D. Wenkert and R. B. Woodward, *J. Am. Chem. Soc.*, **103**, 1540 (1981).

41. R. T. Boeré, K. H. Moock, S. Derrick, W. Hoogerdijk, K. Preuss, J. Yip and M. Parvez, *Can. J. Chem.*, **71**, 473 (1993).

42. A. D. Bond, D. A. Haynes and J. M. Rawson, *Can. J. Chem.*, **80**, 1507 (2002).

43. S. Dell, D. M. Ho and R. A. Pascal, Jr., *Inorg. Chem.*, **35**, 2866 (1996).

44. R. T. Boeré, personal communication, 2004.

45. R. T. Oakley, *Can. J. Chem.*, **62**, 2763 (1984).

46. R. Gleiter, R. Bartetzko and D. Cremer, *J. Am. Chem. Soc.*, **106**, 3437 (1984).

47. J. P. Boutique, J. Riga, J. J. Verbist, J. Delhalle, J. G. Fripiat, R. C. Haddon and M. L. Kaplan, *J. Am. Chem. Soc.*, **106**, 312 (1984).

48. R. A. Pascal, Jr. and R. P. L'Esperance, *J. Am. Chem. Soc.*, **116**, 5167 (1994).

49. M. Amin and C. W. Rees, *J. Chem. Soc. Perkin Trans. 1*, 2495 (1989).

50. R. T. Boeré, A. W. Cordes, S. L. Craig, R. T. Oakley and R. W. Reed, *J. Am Chem. Soc.*, **109**, 868 (1987).

51. T. Chivers, K. S. Dhathathreyan and T. Ziegler, *J. Chem. Soc., Chem. Commun.*, 86 (1989).

52. T. Chivers and R. W. Hilts, *Inorg. Chem.*, **31**, 5272 (1992).

53. T. Chivers, M. Parvez, I. Vargas-Baca, T. Ziegler and P. Zoricak, *Inorg. Chem.*, **36**, 1669 (1997).

54. T. Chivers, I. Vargas-Baca, T. Ziegler and P. Zoricak, *Chem. Commun.*, 949 (1996).

55. T. Chivers, M. Gibson, M. Parvez and I. Vargas-Baca, *Inorg. Chem.*, **39**, 1697 (2000).

56. H-U. Höfs, G. Hartmann, R. Mews and G. M. Sheldrick, *Angew. Chem., Int. Ed.*, **23**, 988 (1984).

57. T. Chivers, J. F. Richardson and N. R. M. Smith, *Inorg. Chem.*, **25**, 272 (1986).

58. C. Knapp, E. Lork, T. Borrman, W-D. Stohrer, and R. Mews, *Eur. J. Inorg. Chem.*, 3211 (2003).

59. R. Maggiulli, R. Mews, W-D. Stohrer, M. Noltemeyer and G. M. Sheldrick, *Chem. Ber.*, **121**, 1881 (1988).

60. C. Knapp, E. Lork, R. Mews and A. V. Zibarev, *Eur. J. Inorg. Chem.*, 2446 (2004).

61. K. T. Bestari, R. T. Boeré and R. T. Oakley, *J. Am. Chem. Soc.*, **111**, 1579 (1989).

62. R. T. Boeré, A. W. Cordes and R. T. Oakley, *J. Am. Chem. Soc.*, **109**, 7781 (1987).

63. R. Maggiulli, R. Mews, W-D. Stohrer and M. Noltemeyer, *Chem. Ber.*, **123**, 29 (1990).

64. C. Knapp, T. Borrman, E. Lork, P. G. Watson, W-D. Stohrer and R. Mews, *Phosphorus, Sulfur and Silicon*, **179**, 887 (2004).

Chapter 13

Heterocyclothia- And Selena-azenes

13.1 Introduction

Heterocyclothiazenes are S-N ring systems containing one (or) more other element(s). Cyclometallathiazenes, *i.e.*, cyclic systems containing a transition metal, are discussed in Section 7.3 and S-N heterocycles containing carbon are the subject of the previous two chapters. Mixed S/Se or S/Te rings are included in Section 8.10 in the examination of chalcogen-nitrogen halides. In this chapter heterocyclothiazenes containing other *p*-block elements will be considered. This class of ring systems is conveniently organized according to the coordination number of the chalcogen, beginning with two-coordinate chalcogen systems followed by those that involve three- or four-coordinate chalcogen centres. In each section the discussion is sub-divided according to ring size beginning with the smallest rings. The most extensively studied heterocycles of this type are hybrid P(V)N/SN ring systems. In addition to the fundamental interest in the structures and bonding in π-electron rich systems with two-coordinate sulfur or selenium, six-membered rings with three- or four-coordinate sulfur are important precursors to novel inorganic polymers via thermal ring-opening polymerization (Sections 14.4 and 14.5).

13.2   Ring Systems Containing Two-coordinate Chalcogen[1]

13.2.1   *Cyclic systems of phosphorus(V)*

As in the case of C-N-S heterocycles, the isolobal correspondence between S^+ and a $R_2P(V)$ group (Section 4.5) as a substituent in an S-N ring stimulated investigations of R_2PN/SN systems. These heterocyclothiazenes are hybrids of π-electron-rich and π-electron precise rings, *i.e.*, cyclothiazenes and cyclophosphazenes, $(NPR_2)_n$, respectively. The formal replacement of S^+ in the hypothetical $[S_3N_3]^+$ cation by PR_2 generates the six-membered ring $R_2PS_2N_3$ (**13.1**), an eight π-electron system. The first derivative **13.1** (R = Me_3SiNH) was obtained in 1976 from the reaction of S_4N_4 with $(Me_3Si)_2NP(NSiMe_3)_2$.[2] The reaction of the diphosphines R_2PPR_2 (R = Me, Ph) or Ph_2PH with S_4N_4 in boiling toluene also provides a source of **13.1** (R = Me, Ph)[3] together with the eight-membered rings 1,3-$R_4P_2N_4S_2$ (**13.2**, E = S) and 1,5-$R_4P_2N_4S_2$ (**13.3**, E = S).[4] The best route to **13.1** (R = Ph) is shown in equation 13.1.[5]

$$Ph_2P(NSiMe_3)(NHSiMe_3) + 1/2S_4N_4 \rightarrow Ph_2PN_3S_2 + (Me_3Si)_2NH \quad (13.1)$$

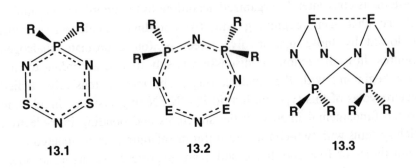

13.1   **13.2**   **13.3**

The heterocycles **13.1** are isolated as purple crystals (R = Me_3SiNH, Ph) or as a thermally unstable, purple oil (R = Me), which decomposes at room temperature to give **13.3** (R = Me, Ph).[6] The structure of **13.1** (R = Ph) consists of a six-membered ring in which the phosphorus atom lies 0.26 Å out of the S_2N_3 plane.[3] The S–N bond lengths fall within a narrow range (1.56-1.58 Å). By contrast, the electron-donating Me_3SiNH substituents on phosphorus give rise to localized π-bonding in the NSNSN fragment of the ring.[2b] *Ab initio* molecular orbital calculations confirm that the PN_3S_2 ring is an eight π-electron system. The purple colour ($\lambda_{max} \approx 550$ nm) is attributed to the $HOMO(\pi^*) \rightarrow LUMO(\pi^*)$ transition. The ^{15}N NMR spectra of $R_2P^{15}N_3S_2$ derivatives show the doublet-triplet pattern expected for the six-membered ring **13.1** with $^2J(^{15}N-^{15}N) \approx 3$ Hz.[3] Selenium analogues of **13.1** are unknown.

The cyclophosphadithiatriazines **13.1** form white crystalline adducts with norbornadiene, in which the alkene adds in a 1,3-fashion to give the *exo* adduct **13.4**.[3] This reversible adduct formation provides a convenient way of characterizing and storing the less stable derivatives of **13.1** (R = Me, F, OPh). The PN_3S_2 ring in **13.1** readily undergoes oxidative addition of Cl_2 (SO_2Cl_2 or $PhICl_2$) to give the S,S'-dichloro derivative (Section 13.3), which is converted into the bicyclic compound $R_2PN_5S_3$ (**13.5**) by reaction with $Me_3SiNSNSiMe_3$.[7] The difluoro derivative (**13.5**, R = F) is obtained from the reaction of PF_5 with $Me_3SiNSNSiMe_3$.[8] The $-N=S=N-$ unit bridges the PN_3S_2 ring in **13.5** (R = F) via relatively long S–N bonds (1.69 Å, *cf.* 1.55 Å in the $-N=S=N-$ bridge). Thus it is not surprising that mild heating (*ca.* 90°C) of **13.5** regenerates **13.1** (via loss of N_2S).

13.4   **13.5**   **13.6**

The PN_3S_2 ring systems exhibits Lewis base properties towards strong acids, *e.g.*, HBF_4,[9] or metal centres. In these interactions one of the nitrogens adjacent to phosphorus is the site of electrophilic attack. For example, the neutral 1:1 adduct **13.6** is obtained by displacement of one of the phosphine ligands in $PtCl_2(PEt_3)_2$ by **13.1** (R = Ph).[10]

Coordination results in a lengthening of the S–N bond involving the coordinated nitrogen atom by *ca.* 0.1 Å; this N atom is also displaced out of the PN_2S_2 plane by 0.63 Å. The Pt–N bond in **13.6** is weak and dissociation of the adduct occurs in solution.

The diphosphadithiatetrazocines **13.2** (E = S) and **13.3** (E = S) contain one more R_2PN unit in the ring than **13.1**. The structure of the orange-red 1,3-isomer (**13.2**, E = S, R = Ph) consists of an essentially planar N_3S_2 unit with the two phosphorus atoms located on opposite sides of the plane.[4] Like **13.1** this isomer forms a 1:1 adduct with norbornadiene via 1,3-addition across the sulfur atoms.

The 1,5-isomers **13.3** (E = S) are colourless, air-stable solids. They are prepared by the cyclocondensation reaction of $R_2PN_2(SiMe_3)_3$ with sulfur dichloride[5] or thionyl chloride.[11] A similar cyclocondensation process, using a mixture of $SeCl_4$ and Se_2Cl_2 as a source of selenium, produces a mixture of the isomers **13.2** and **13.3** (E = Se, R = Ph).[12] The structures of **13.3** (E = S, R = alkyl, aryl) are folded eight-membered rings with a cross-ring S•••S distance of *ca.* 2.50 Å.[3,4,11] This structural feature is the result of an intramolecular $\pi^*–\pi^*$ interaction (Section 4.6).[13] It gives rise to anomalous ^{31}P NMR chemical shifts that are *ca.* 80-100 ppm downfield compared to other P-N-S rings (or cyclophosphazenes).[14] Although the structure of the selenium analogue of **13.3** (R = Ph) is unknown, a Se•••Se bond distance of 2.59 Å has been determined in the bis-(η^1-*N*) adduct with $PtCl_2(PEt_3)$.[15] Interestingly, **13.3** (E = Se, R = Ph) is thermochromic in solution, being deep green at room temperature and yellow at –20°C. The EPR spectrum of this solution exhibits a 1:2:3:2:1 quintet, suggesting the formation of a radical with two equivalent nitrogens, *e.g.*, the four-membered ring $[Ph_2PN_2Se]^{•-}$.[12]

1,5-Diphosphadithiatetrazocines serve as excellent models for determining the identity of the initial products formed in the reactions of S-N compounds. The NPR_2 units act as an architectural brace as well as a source of structural information through ^{31}P NMR spectroscopy. Thus, in contrast to S_4N_4 (Section 5.2.6), the integrity of the eight-membered ring in **13.3** (E = S) is retained in reactions with either electrophiles or nucleophiles. For example, the oxidative addition of halogens occurs smoothly to give *exo,endo* dihalogenated derivatives $1,5-Ph_4P_2N_4S_2X_2$ (**13.7**, X = Cl, Br).[16] The dication $[Et_4P_2N_4S_2]^{2+}$ (**13.8**), prepared by the reaction of the S,S'-dichloro derivative $1,5\text{-Et}_4P_2N_4S_2Cl_2$ with $AlCl_3$, is planar with an average S–N bond length of 1.51 Å,[17] *cf.* 1.55 Å in $[S_4N_4]^{2+}$.

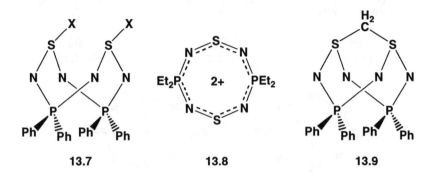

13.7   **13.8**   **13.9**

The 1,5-isomers **13.3** behave as weak Lewis bases in forming N-bonded adducts with protic or Lewis acids[18] and with platinum(II).[19] The cross-ring S•••S contact is retained in these adducts. The $P_2N_4S_2$ ring in **13.3** is also readily susceptible to nucleophilic attack at sulfur. For example, reactions with organolithium reagents proceed smoothly to give monolithium derivatives $Li[Ph_4P_2N_4S_2R]$, which have dimeric structures in the solid state.[20] Alkali-metal derivatives of the dianion $[Ph_4P_2N_4S_2]^{2-}$ may also be prepared by the reaction of **13.3** (E = S, R = Ph) with alkali-metal superhydrides $M[Et_3BH]$ (M = Li, Na], but these insoluble products have not been structurally characterized.[21] However, the reaction of $Li_2[Ph_4P_2N_4S_2]$ with diiodomethane produces **13.9** in which a methylene group bridges the two sulfur atoms.[22]

The combination of hard (N) and soft (S) coordination in the 1,5-$P_2N_4S_2$ ring system leads to a diversity of coordination modes in complexes with transition metals (Fig. 13.1).[23] In some cases these complexes may be prepared by the reaction of the dianion $[Ph_4P_2N_4S_2]^{2-}$ with a metal halide complex, but these reactions frequently result in redox to regenerate **13.3** (E = S, R = Ph). A more versatile approach is the oxidative addition of the neutral ligand **13.3** (E = S) to the metal centre.

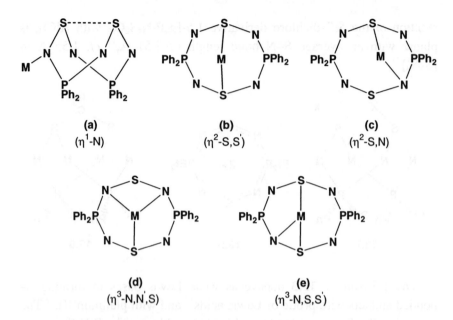

(a)
$(\eta^1\text{-N})$

(b)
$(\eta^2\text{-S,S}')$

(c)
$(\eta^2\text{-S,N})$

(d)
$(\eta^3\text{-N,N}',\text{S})$

(e)
$(\eta^3\text{-N,S,S}')$

Fig. 13.1   Coordination modes in metal complexes of $[Ph_4P_2N_4S_2]^{2-}$. The uncoordinated sulfur atom in (c) and (d) acts as a bridge to give dimeric structures

For example, the reactions of **13.3** (E = S, Se) with zerovalent platinum or palladium complexes, gives E,E'–bonded adducts **13.10** (Eq. 13.2).[24,25] The reaction of $Li_2[Ph_4P_2N_4S_2]$ with $cis\text{-PtCl}_2(PPh_3)_2$ may also be used to prepare **13.10** (E = S, M = Pt, R = Ph).

M(PPh$_3$)$_2$L$_2$ +   1,5 - Ph$_4$P$_2$N$_4$E$_2$

(M = Pt, L$_2$ = C$_2$H$_4$)   (E = S, Se)
(M = Pd, L = PPh$_3$)

$\longrightarrow$

(13.2)

13.10

Gentle heating of **13.10** causes reversible dissociation of one PPh₃ ligand to give the centrosymmetric dimer **13.11**, in which the monomer units exhibit N,E coordination (mode (c) in Fig. 13.1).[24] As expected for complexes of late transition metals, bonding to sulfur is stronger than the interaction with the nitrogen centres in complexes of the $[Ph_4P_2N_4S_2]^{2-}$ dianion. This structural feature is reflected in a facile 1,3-metallotropic shift between nitrogen centres in **13.11**, which is conveniently monitored by ³¹P NMR spectroscopy. The intermediate in this fluxional process likely involves *N,N',E*-coordination for both monomer units (mode (d) in Fig. 13.1). This bonding mode has been established in the monomeric units of the centrosymmetric ruthenium(II) dimer **13.12**.[26]

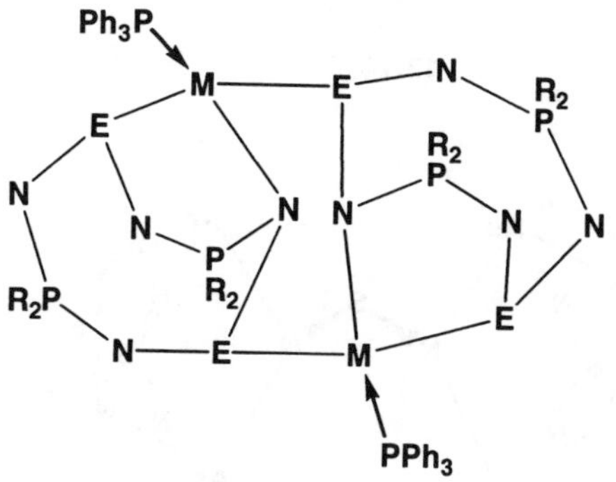

13.11

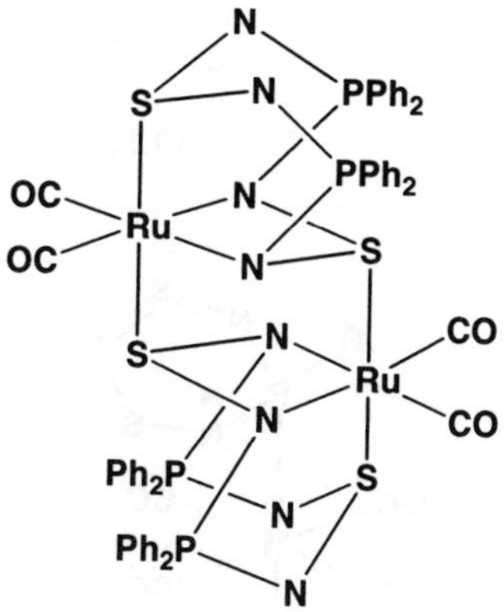

13.12

13.2.2 *Heterocyclothiazenes containing phosphorus(III) or other p-block elements*

A few examples of five-membered rings of the type EN_2S_2 (**13.13**, E = AsMe, SnR₂) are known. The arsenic derivative $MeAsN_2S_2$ is obtained as a red liquid from the reaction of $MeAsCl_2$ with $Me_3SiNSNSiMe_3$ in a 2:1 molar ratio.[27] The organotin derivative $Me_2SnN_2S_2$ is prepared by the reaction of S_4N_4 with $N(SnMe_3)_3$.[28] It is a useful reagent for the preparation of other heterocyclothiazenes. For example, the reaction with carbonyl fluoride produces OCN_2S_2 (Section 11.3.3).[29] Organotin derivatives, *e.g.*, $^nBu_2SnN_2S_2$, may also be synthesized by the reaction of diorganotin dichlorides with $[S_4N_3]Cl$ in liquid ammonia.[30] This derivative has been used to prepare cyclometallathiazenes MN_2S_2 (M = Rh, Ir), as illustrated in Eq. 13.3.

$$^nBu_2SnN_2S_2 + (\eta^5\text{-}C_5Me_5)IrCl_2(PPh_3) \rightarrow (\eta^5\text{-}C_5Me_5)IrN_2S_2 + {^nBu_2SnCl_2} + PPh_3 \text{ (13.3)}$$

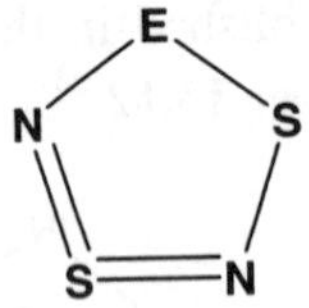

13.13

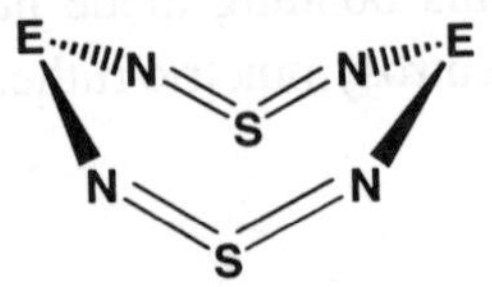

13.14

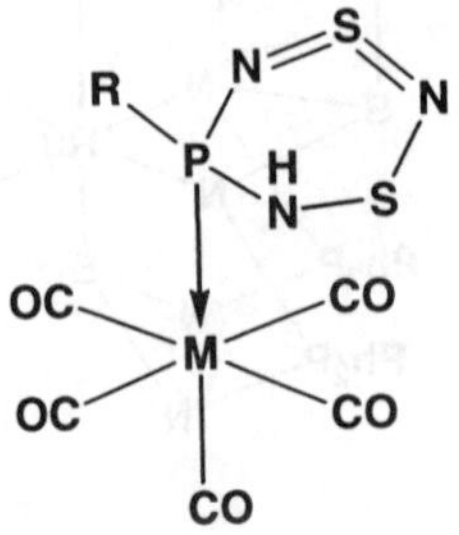

13.15

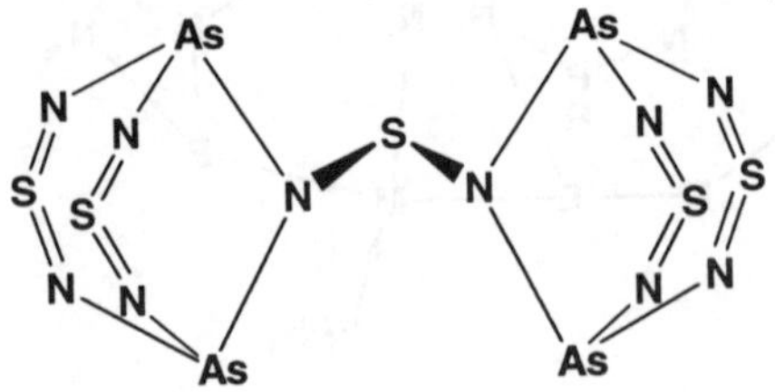

13.16

Eight-membered rings of the type $E(NSN)_2E$ (**13.14**) are known for E = BNR_2 (R = ^nPr, ^iPr, ^nBu),[31] SiR_2 (R = Me, ^tBu),[32] AsR (R = Me, ^tBu, Ph, Fc),[33] Sb (R = ^tBu).[34] These ring systems are generally obtained by the reaction of the appropriate organoelement halide [*e.g.*, R_2NBCl_2, Me_2SiCl_2, $RECl_2$ (E = As, Sb)] with $Me_3SiNSNSiMe_3$ or $K_2[NSN]$ (Section 5.4.1) in a 1:1 molar ratio (Eq.13.4). The reagent $KNSNSiMe_3$ has also been used to prepare the antimony system.[34] A related tricyclic compound MeSi(NSN)$_3$SiMe is prepared from MeSiCl$_3$ and Me$_3$SiNSNSiMe$_3$.[35]

$$Me_2SiCl_2 + 2Me_3SiNSNSiMe_3 \rightarrow$$

$$Me_2Si(NSN)_2SiMe_2 + 2Me_3SiCl \qquad (13.4)$$

The eight-membered rings **13.14** normally adopt boat conformations in the solid state with short S=N bond distances (1.51-1.52 Å) that are typical of sulfur diimides. There are no transannular S•••S contacts. The sole exception is the antimony derivative $^tBuSb(NSN)_2Sb^tBu$, which is a planar eight-membered ring.

Ring systems of the type **13.14** containing phosphorus(III) are not accessible by the route shown in Eq. 13.4. Furthermore, the reactions of RPCl$_2$ adducts of M(CO)$_5$ (M = Cr, W) with K$_2$[NSN] produce metal complexes of the protonated six-membered ring RPN$_3$S$_2$(H) (**13.15**) rather than the expected complexes of **13.14** (E = PR).[36] The latter may, however, be prepared by exploiting the base-promoted conversion of thionyl imides into sulfur diimides (Section 10.4) as exemplified in Eq. 13.5.[37] In contrast to the boat-shaped As$_2$N$_4$S$_2$ ring, which is unaffected by metal coordination, the P(III)$_2$N$_4$S$_2$ ring in the N^{i}Pr$_2$ derivative is essentially planar with the two Cr(CO)$_5$ substituents on opposite sides of the ring. The cage compound **13.16**, in which two As$_2$N$_4$S$_2$ rings are bridged by an =N–S–N= unit is formed in the reaction of K$_2$[NSN] with arsenic trihalides.[38]

$$2[Cr(CO)_5(RPCl_2)] + 4KNSO \rightarrow$$

$$[(CO)_5Cr[RP(NSN)_2PR]Cr(CO)_5] + 2SO_2 + 4KCl \quad (13.5)$$

$$(R = {}^tBu, N^iPr_2)$$

13.3 Ring Systems Containing Three-coordinate Chalcogen

Tellurium chemistry is often significantly different in comparison with selenium chemistry owing to the larger size and metallic character of tellurium (Section 1.1). As an illustration, the cyclocondensation of

Li[Ph$_2$P(NSiMe$_3$)$_2$] with ArTeCl$_3$ or TeCl$_4$ generates the four-membered ring **13.17** rather than the tellurium analogues of the eight-membered rings **13.3**.[39] The planar PN$_2$Te ring in **13.17** is the only known P-N-Te heterocycle. An analogous amidinate complex PhC(NSiMe$_3$)$_2$TeCl$_3$ has also been characterized (Section 11.2).[40]

13.17   **13.18**   **13.19**

Ring systems of the type (R$_2$PN)$_x$(NSX)$_y$ can be considered as hybrids of the well-known heterocycles (R$_2$PN)$_n$ (n = 3, 4 etc.) and (NSCl)$_3$. The *S,S'*-dichloro derivative **13.18** is prepared by oxidative addition of Cl$_2$ to **13.1**, using SO$_2$Cl$_2$ or PhICl$_2$.[7] Three approaches to *S*-monochloro derivatives **13.19** have been reported: the reactions of (a) S$_3$N$_2$O$_2$ with PCl$_5$,[41] (b) PCl$_5$ and Me$_3$SiNSNSiMe$_3$,[42] or (c) Ph$_2$PCl with S$_4$N$_4$ in a 3:1 molar ratio in boiling acetonitrile.[43] The first two methods produce **13.19** (R = Cl), whereas method (c) is potentially applicable to a wide range of R substituents. The sulfur atom in **13.19** (R = Ph) lies *ca.* 0.3 Å out of the NPNPN plane.[43] The P–N bond lengths linking the PNP unit to the NSN unit are *ca.* 0.08 Å longer than those within the PNP unit, suggesting a tendency towards localization of the π-bonding at opposite ends of the molecule. The S–Cl bond distance in **13.19** (R = Ph) is 0.21 Å longer than that in (NSCl)$_3$ indicating partial ionic character. The corresponding iodide (Ph$_2$PN)$_2$(NSI) has a long S–I bond (2.72 Å).[45] Halide ion acceptors, *e.g.*, SbCl$_5$, react readily with **13.19** (R = Cl, Ph) to give the corresponding cations. The tetrachlorinated derivative [Cl$_2$PN)$_2$(NS)]$^+$ is a planar six-membered ring.[42]

The treatment of **13.19** (R = Ph) with secondary amines produces the corresponding dialkylamino derivatives **13.20**.[46] The electron-donating

R$_2$N substituent promotes a ring expansion that converts **13.20** into the twelve-membered ring **13.21** either upon heating or on standing in acetonitrile solution.[47] Further heating produces the spirocyclic sulfur(VI) system **13.22**, which has high thermal stability. A related spirocyclic system, which consists of two eight-membered P$_2$N$_4$S$_2$ rings fused at a common sulfur atom, is formed when **13.19** (R = Ph) is treated with Me$_3$SiNSNSiMe$_3$.[43] These two rings are linked by a weak S•••S bond (2.37 Å) to give the tricyclic structure **13.23**. Treatment of **13.19** (R = Ph) with SbPh$_3$ in acetonitrile produces the twelve-membered ring **13.24,** which has a relatively short transannular S•••S bond [2.39 Å, *cf.* 2.50 Å in the eight-membered rings **13.3** (E = S)]. [48]

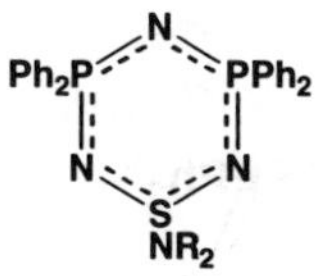

13.20 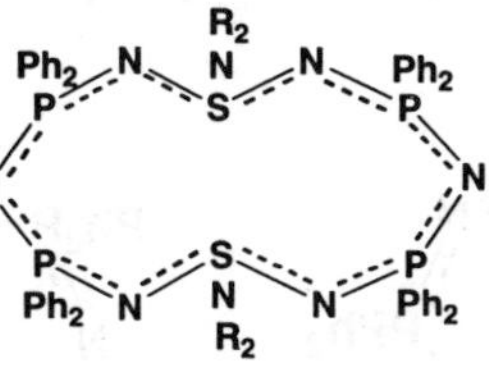 **13.21** 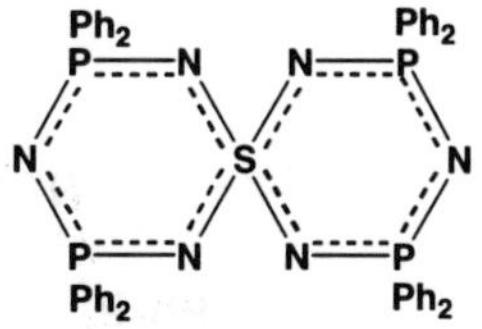 **13.22**

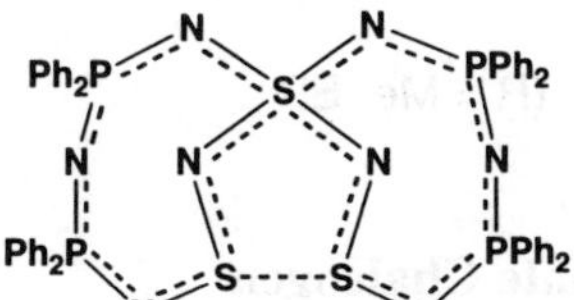

13.23 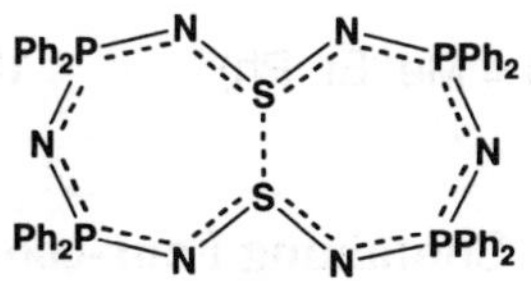 **13.24** 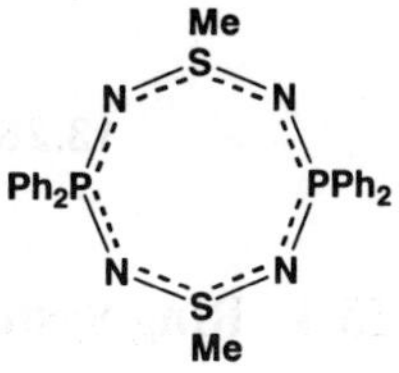 **13.25**

The ring system **13.18** (R = Cl) is a source of hybrid PN/SN polymers containing three-coordinate sulfur via a ring-opening polymerization process. This polymerization occurs upon mild thermolysis at 90°C (Section 14.4).[49]

Eight-membered rings of the type (R$_2$PN)$_2$(NSR)$_2$ (R = alkyl, aryl) can be synthesized in two ways. The *S,S,'*-dimethyl derivative **13.25** is obtained by the reaction of Li[Ph$_4$P$_2$N$_4$S$_2$Me] with methyl iodide.[50] The corresponding *S,S,'*-diphenyl derivative is produced by the treatment of R$_2$PN$_2$(SiMe$_3$)$_3$ with three equivalents of PhSCl.[51]

Selenium analogues of **13.25** may be prepared by the cyclocondensation reaction of $Ph_2PN_2(SiMe_3)_3$ with organoselenium trichlorides $RSeCl_3$ in acetonitile.[52] The formation of the *Se,Se'*-dialkyl derivatives **13.26** (R = Me, Et) is accompanied by small amounts of **13.3** (E = Se, R = Ph). The eight-membered $P_2N_4Se_2$ ring in **13.26** (R= Me) adopts a chair conformation similar to that found for the sulfur analogues. Although there are no close intermolecular Se•••N contacts, the *Se,Se'*-dialkyl derivatives **13.26** (R = Me, Et) undergo a remarkable *solid-state* transformation at room temperature to give the 1,3-isomers **13.27** (R = Me, Et). The *Se,Se'*-diphenyl derivative **13.26** (R = Ph) does not undergo this isomerization, but prolonged thermolysis at 140°C produces the diselenide PhSeSePh and cyclophosphazenes.

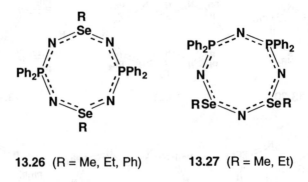

13.26 (R = Me, Et, Ph)   **13.27** (R = Me, Et)

13.4   Ring Systems Containing Four-coordinate Chalcogen

Ring systems of the type $[NS(O)Cl]_x(NPCl_2)_y$ have been known since 1963, and their chemistry has been investigated very thoroughly.[54] They are formally hybrids of the sulfanuric ring $[NS(O)Cl]_3$ and cyclophosphazenes $(NPCl_2)_n$. The best known of these heterocycles are the six-membered rings **13.28** and **13.29**. Higher oligomers are formed as by-products in the thermal ring-opening polymerization of **13.29** (Section 14.5). Small amounts of larger macrocycles in this homologous series have been isolated and structurally characterized by X-ray crystallography.[55] The twelve-membered ring exists as *cis* (**13.30a**) and *trans* (**13.30b**) isomers, while the twenty-four membered ring **13.31** is a rare example of this ring size for inorganic heterocycles.

13.28

13.29

13.30a

13.30b

13.31

Several preparative routes to the heterocycle **13.29** have been reported. These include the vacuum thermolysis of $Cl_3P=N-PCl_2=N-SO_2Cl$.[56] An alternative, low-yield synthesis is the [3 + 3] cyclocondensation of $[Cl_3P=N=PCl_3][PCl_6]$ and sulfamide $SO_2(NH_2)_2$.[57] The best method involves the reaction of sulfamide with PCl_5 followed by a [5 + 1] cyclocondensation reaction between the bis(phosphazo)sulfone $Cl_3P=N-SO_2-N=PCl_3$ and hexamethyldisilazane (Scheme 13.1).[58]

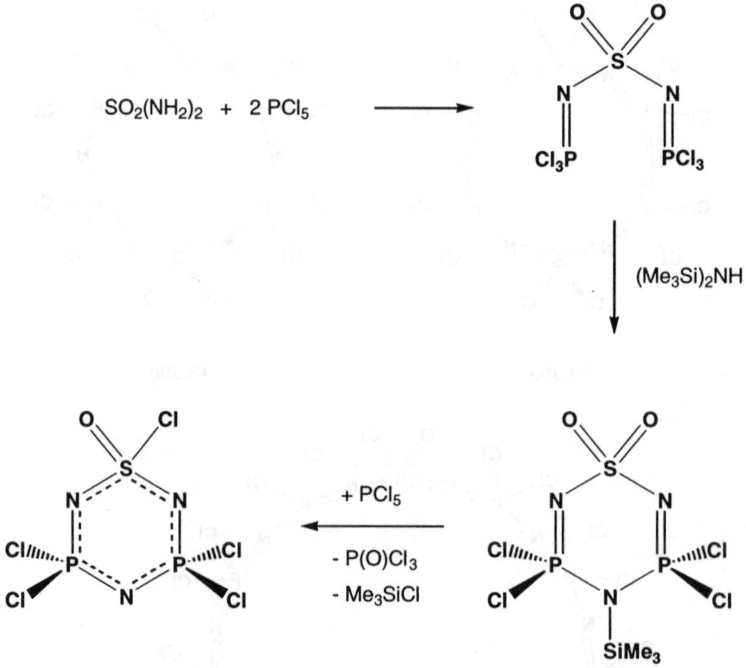

Scheme 13.1   Synthesis of $(NPCl_2)_2[NS(O)Cl]$

The hybrid ring systems **13.28** and **13.29** are colourless solids stable to moist air. The structures of a number of derivatives have been determined. The geometry around sulfur is influenced by the ligands attached to it. For $S(O)X$ centres (X = Cl, F) the halogen is in an axial position with respect to the mean plane of the ring, forcing the oxygen into an equatorial position. However, when Cl is replaced by NMe_2 or Ph the oxygen atom adopts the axial position. The S–N bonds are generally shorter than the P–N bonds; the variation in these bond lengths has been attributed to the greater electronegativity of sulfur compared to that of the phosphorus centers.

The reactions of these hybrid ring systems with nucleophilic reagents has been studied in considerable detail, with emphasis on the regiochemistry.[54] Broadly speaking, phenylation and fluorination take place preferentially at sulfur, whereas the phosphorus is attacked first in aminolysis and alcoholysis reactions. Thus, fluorination of **13.28** with an excess of AgF_2 in boiling CCl_4, or with SbF_3 at 85°C, causes replacement of the two chlorines on sulfur, but not those on phosphorus. The product is obtained as a mixture of *cis* and *trans* isomers in a 4:1 molar ratio. The PCl_2 site can be fluorinated by using KF in the presence of 18-crown-6. Friedel-Crafts phenylation of **13.28** produces *cis-* and *trans-* $(NPCl_2)(NSOPh)_2$ in excellent yields, whereas Ph_2Hg gives the monophenylated derivative.

References

1. T. Chivers, D.D. Doxsee, M. Edwards and R. W. Hilts, Diphosphadithia- and Diphosphadiselena-tetrazocines and Their Derivatives in R. Steudel (ed.) *The Chemistry of Inorganic Ring Systems*, Elsevier, pp. 271-294 (1991).

2. (a) R. Appel and M. Halstenberg, *Angew. Chem., Int. Ed. Engl.*, **15**, 695 (1976); (b) J. Weiss, *Acta Crystallogr.*, **B33**, 2271 (1977).

3. N. Burford, T. Chivers, A. W. Cordes, W. G. Laidlaw, M. C. Noble, R. T. Oakley and P. N. Swepston, *J. Am. Chem. Soc.*, **104**, 1282 (1982).

4. N. Burford, T. Chivers and J. F. Richardson, *Inorg. Chem.*, **22**, 1482 (1983).

5. T. Chivers, K. S. Dhathathreyan, S. W. Liblong and T. Parks, *Inorg. Chem.*, **27**, 1305 (1988).

6. N. Burford, T. Chivers, P. W. Codding and R. T. Oakley, *Inorg. Chem.*, **21**, 982 (1982).

7. N. Burford, T. Chivers, R. T. Oakley and T. Oswald, *Can. J. Chem.*, **62**, 712 (1984).

8. (a) H. W. Roesky and O. Petersen, *Angew. Chem., Int. Ed. Engl.*, **12**, 415 (1973); (b) R. Appel, I. Ruppert, R. Milker and V. Bastian, *Chem. Ber.*, **107**, 380 (1974).

9. T. Chivers, S. W. Liblong, J. F. Richardson and T. Ziegler, *Inorg. Chem.*, **27**, 860 (1988).

10. T. Chivers, R. W. Hilts, I. Krouse, A. W. Cordes, R. Hallford and S. R. Scott, *Can. J. Chem.*, **70**, 2602 (1992).

11. T. Chivers, M. Edwards and M. Parvez, *Inorg. Chem.*, **31**, 1861 (1992).

12. T. Chivers, D. D. Doxsee and M. Parvez, *Inorg. Chem.*, **32**, 2238 (1993).

13. (a) G. Chung and D. Lee, *Bull. Korean Chem. Soc.*, **21**, 300 (2000); (b) H. Jacobsen, T. Ziegler, T. Chivers and R. Vollmerhaus, *Can. J. Chem.*, **72**, 1582 (1994).

14. T. Chivers, M. Edwards, C. A. Fyfe and L. H. Randall, *Mag. Reson. Chem.*, **30**, 1220 (1992).

15. T. Chivers, D. D. Doxsee, R. W. Hilts, A. Meetsma, M. Parvez and J. C. van de Grampel, *J. Chem. Soc., Chem. Commun.*, 1330 (1992).

16. N. Burford, T. Chivers, M. N. S. Rao and J. F. Richardson, *Inorg. Chem.*, **23**, 1946 (1984).

17. M. Brock, T. Chivers. M. Parvez and R. Vollmerhaus, *Inorg. Chem.*, **36**, 485 (1997).

18. T. Chivers, G. Y. Dénès, S. W. Liblong and J. F. Richardson, *Inorg. Chem.*, **28**, 3683 (1989).

19. T. Chivers and R. W. Hilts, *Inorg. Chem.*, **31**, 5271 (1992).

20. T. Chivers, M. Edwards, R. W. Hilts, M. Parvez and R. Vollmerhaus, *Inorg. Chem.*, **33**, 1440 (1994).

21. T. Chivers, M. Edwards, X. Gao, R. W. Hilts, M. Parvez and R. Vollmerhaus, *Inorg. Chem.*, **34**, 5037 (1995).

22. T. Chivers, M.Cowie, M. Edwards and R. W. Hilts, *Inorg. Chem.*, **31**, 3349 (1992).

23. T. Chivers and R. W. Hilts, *Coord. Chem. Rev.*, **137**, 201 (1994).

24. T. Chicers, M. Edwards, A. Meetsma, J. C. van de Grampel and A. van der Lee, *Inorg. Chem.*, **31**, 2156 (1992).

25. T. Chivers, D. D. Doxsee and R. W. Hilts, *Inorg. Chem.*, **32**, 3244 (1993).

26. T. Chivers, R. W. Hilts, M. Parvez, D. Ristic-Petrovic and K. Hoffman, *J. Organomet. Chem.*, **480**, C4 (1994).

27. O. J. Scherer and R. Wies, *Angew. Chem., Int. Ed. Engl.*, **11**, 529 (1972).

28. C. P. Warrens and J. D. Woollins, *Inorg. Synth.*, **25**, 46 (1989).

29. H. W. Roesky and H. Wiezer, *Angew. Chem., Int. Ed. Engl.*, **12**, 674 (1973).

30. S. M. Aucott, A. M. Z. Slawin and J. D. Woollins, *Can. J. Chem.*, **80**, 1481 (2002).

31. G. Schmid, H. Gehrke, H-U. Kolorz and R. Boese, *Chem. Ber.*, **126**, 1781 (1993).

32. (a) G. Ertl and J. Weiss, *Z. Naturforsch.*, **29B**, 803 (1974); (b) M. Herberhold, S. Gerstmann, W. Milius and B. Wrackmeyer, *Z. Naturforsch.*, **48B**, 1041 (1993).

33. (a) O. J. Scherer and R. Wies, Angew. *Chem., Int. Ed. Engl.*, **10**, 812 (1971); (b) A. Gieren, H. Betz, T. Hubner, V. Lamm, M. Herberhold and K. Guldner, *Z. Anorg. Allg. Chem.*, **513**, 160 (1984); (c) N. W. Alcock, E. M. Holt, J. Kuyper, J. J. Mayerle and G. B. Street, *Inorg. Chem.*, **18**, 2235 (1979); (d) C. Spang, F. Edelmann, T. Frank, M. Noltemeyer amd H. W. Roesky, *Chem. Ber.*, **122**, 1247 (1989).

34. M. Herberhold and K. Schamel, *Z. Naturforsch.*, **43B**, 1274 (1988).

35. H. W. Roesky, M. Witt, B. Krebs, G. Henkel and H-J. Korte, *Chem. Ber.*, **114**, 201 (1981).

36. A. Gieren, C. Ruiz-Pérez, T. Hübner, M. Herberhold, K. Schamel and K. Guldner, *J. Organomet. Chem.*, **366**, 105 (1989).

37. T. Chivers, K. S. Dhathathreyan, C. Lensink, A. Meetsma, J. C. van de Grampel and J. L. de Boer, *Inorg. Chem.*, **28**, 4150 (1989).

38. M. Herberhold, K. Guldner, A. Gieren, C. Ruiz-Pérez and T. Hübner, *Angew. Chem., Int. Ed. Engl.*, **26**, 82 (1987).

39. T. Chivers, D. D. Doxsee, X. Gao and M. Parvez, *Inorg. Chem.*, **33**, 5678 (1994).

40. E. Hey, C. Ergezinger and K. Dehnicke, *Z. Naturforsch.*, **44B**, 205 (1989).

41. H. W. Roesky, *Angew. Chem., Int. Ed. Engl.*, **11**, 642 (1972).

42. S. Pohl, O. Petersen and H. W. Roesky, *Chem. Ber.*, **112**, 1545 (1979).

43. T. Chivers, M. N. S. Rao and J. F. Richardson, *J. Chem. Soc., Chem. Commun.*, 982 (1982).

44. N. Burford, T. Chivers, M. Hojo, W. G. Laidlaw, J. F. Richardson and M. Trsic, *Inorg. Chem.*, **24**, 709 (1985).

45. T. Chivers, M. N. S. Rao and J. F. Richardson, *J. Chem. Soc. Chem. Commun.*, 700 (1983).

46. T. Chivers and M. N. S. Rao, *Inorg. Chem.*, **23**, 3605 (1984).

47. T. Chivers, M. N. S. Rao and J. F. Richardson, *J. Chem. Soc., Chem. Commun.*, 702 (1983).

48. T. Chivers, M. N. S. Rao and J. F. Richardson, *J. Chem. Soc., Chem. Commun.*, 186 (1983).

49. J. A. Dodge, I. Manners, H. R. Allcock, G. Renner and O. Nuyken, *J. Am. Chem. Soc.*, **112**, 1268 (1990).

50. T. Chivers, X. Gao, R. W. Hilts, M. Parvez and R. Vollmerhaus, *Inorg. Chem.*, **34**, 1180 (1995).

51. T. Chivers. S. S. Kumaravel, A. Meetsma, J. C. van de Grampel and A. van der Lee, *Inorg. Chem.*, **29**, 4591 (1990).

52. T. Chivers, D. D. Doxsee, and J. Fait, *J. Chem. Soc., Chem. Commun.*, 1703 (1989).

53. T. Chivers, D. D. Doxsee, J. Fait and M. Parvez, *Inorg. Chem.*, **32**, 2243 (1993).

54. (a) J. C. van de Grampel, *Rev. Inorg. Chem.*, **3**, 1 (1981); (b) J. C. van de Grampel, *Coord. Chem. Rev.*, **112**, 247 (1992).

55. Y. Ni, A. J. Lough, A. L. Rheingold and I. Manners, *Angew. Chem., Int. Ed. Engl.*, **34**, 998 (1995).

56. H. H. Baalmann, H. P. Velvis and J. C. van de Grampel, *Rec. Trav. Chim. Pays-Bas*, **91**, 935 (1972).

57. U. Klingebiel and O. Glemser, *Z. Naturforsch.*, **27B**, 467 (1972).

58. D. Suzuki, H. Akagi and K. Matsumura, *Synth. Commun.*, 369 (1983).

Chapter 14

Chalcogen–Nitrogen Chains And Polymers

14.1 Introduction

The discovery of the metal-like behaviour and superconducting properties of the non-metallic polymer $(SN)_x$ in 1973 sparked interest in the area of sulfur–nitrogen chemistry.[1] One aspect of that endeavour invokes the notion that molecular chains that incorporate thiazyl units between organic substituents could serve as molecular wires in the development of nanoscale technology.[2] The combination of electron-withdrawing and electron-accepting groups at opposite ends of the chain may generate materials with large molecular dipoles. Subsequent crystallization of these polar molecules in acentric space groups may give rise to materials that exhibit physical properties ranging from non-linear optical behaviour to piezoelectric and ferroelectric effects. This chapter will begin with a description of the synthesis, structure and properties of $(SN)_x$ in order to provide the background for a discussion of sulfur–nitrogen chains involving two-coordinate sulfur.

Polymers involving sulfur in the +4 and, especially, +6 oxidation states have also attracted considerable attention. High molecular weight polymers containing repeating [NS(O)R] (R = Me, Ph) units have been characterized. Hybrid polymers involving combinations of PN and S(IV)N or S(VI)N units in the backbone have also been studied extensively.[3] As might be expected, the properties of these polymers resemble those of poly(phosphazenes) rather than $(SN)_x$. For example, poly(thionylphosphazenes) have been found to be useful as matrices for phosphorescent sensing devices with potential applications in the aerospace industry. The synthesis, structures and properties of these novel materials are discussed in the second half of this chapter.

14.2 Poly(sulfur nitride), (SN)ₓ, and Related Polymers[1]

Polymeric sulfur nitride (or polythiazyl) was first reported in 1910. It is prepared by the topochemical, solid-state polymerization of S_2N_2 at 0°C over several days.[4] Although the cracking of S_4N_4 provides an obvious source, the explosive nature of this starting material prompted a search for other precursors to S_2N_2 (Section 5.2.4). On the basis of EPR evidence the polymerization process is thought to involve the intermediate formation of a diradical generated by cleavage on one S–N bond (Figure 14.1). A time-resolved X-ray diffraction study shows that

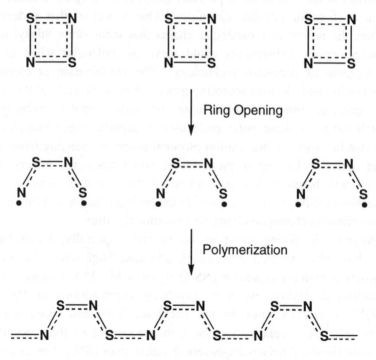

Fig. 14.1   Synthesis and structure of (SN)ₓ

this topochemical process is nondiffusive in nature and produces monoclinic β-(SN)$_x$ (90%) and orthorhombic α-(SN)$_x$ (10%).[5a] A recent proposal for the polymerization mechanism involves excitation of the square planar singlet S_2N_2 molecule to the triplet surface, followed by puckering of the triplet species and polymerization in a direction approximately perpendicular to the S_2N_2 plane.[5b]

In order to prepare thin films of (SN)$_x$ on plastic or metal surfaces, several processing techniques have been investigated, *e.g.*, the electroreduction of [S$_5$N$_5$]$^+$ salts.[6] Powdered (SN)$_x$ is prepared by the reaction of (NSCl)$_3$ with trimethylsilyl azide in acetonitrile.[7] The sublimation of (SN)$_x$ at 135°C and at pressure of 3×10^{-6} Torr. produces a gas-phase species, probably the cyclic [S$_3$N$_3$]$^{\bullet}$ radical, that reforms the polymer as epitaxial fibres upon condensation.[8]

Poly(sulfur nitride) is a shiny metallic solid consisting of highly oriented parallel fibres. The crystal structure reveals an almost planar *cis, trans* polymer with approximately equal adjacent S–N bond lengths of 1.63 and 1.59 Å.[4a] These values are intermediate between single and double bonds consistent with the delocalized electronic structure of (SN)$_x$. In the conventional bonding description (Section 4.3), each nitrogen and sulfur provide one and two electrons, respectively, in a *p* orbital for π-bonding. Consequently, the band structure in the polymer is 75% occupied. According to the spin-coupled valence bond method the π–system is composed of a singly occupied π orbital on each S atom and a lone pair centred around each N atom. As a good approximation, the polymer may be regarded as a one-dimensional chain of sulfur atoms. The single electron at each sulfur site gives rise to a half-filled band and, hence, metallic character.[9] The bond angles at N and S are about 120° and 106°, respectively. Poly(sulfur nitride) is a conducting material at room temperature. It becomes superconducting below liquid helium temperature. The conductivity at room temperature along the fibres is in the range $1- 4 \times 10^3$ Ω^{-1} cm^{-1} depending on the quality of the crystals, and this value increases by about two orders of magnitude at 4 K. The conductivity across the fibres is much smaller as a result of weak S•••S interactions (3.48 Å) between polymer chains. Typical values of the anisotropy ratio are 50 at room temperature and 1000 at 40 K.[1]

In addition to its conducting properties, the high electronegativity of $(SN)_x$, even greater than that of gold, leads to improvements in the efficiency of certain devices. For example, $(SN)_x$ can act as an efficient barrier electrode in ZnS junctions, increasing the quantum efficiency of the blue emission by a factor of 100 over gold.[10] It can also be used to increase the efficiency of GaAs solar cells by up to 35%. Metal ions interact more strongly with a poly(sulfur nitride) surface than with other metal electrodes. This property has stimulated investigations of possible applications of $(SN)_x$ as an electrode material.[12a] For example, $(SN)_x$ electrodes that have been modified by immersion in sodium molybdate have been shown to reduce acetylene to ethylene.[12b]

Although $(SN)_x$ does not react with water or acidic solutions, it slowly decomposes in alkaline solutions and it is readily oxidized. Partial bromination of $(SN)_x$ or powdered S_4N_4 with bromine vapour yields the blue-black polymer $(SNBr_{0.4})_x$ which has a room temperature conductivity of $2 \times 10^4 \ \Omega^{-1} \ cm^{-1}$.[13] The sulfur-nitrogen chain in this polymer is partially oxidized and the bromine is present as Br_3^- ions and intercalated Br_2 molecules.[14] The polymer $(SNBr_{0.4})_x$ is also produced by the reaction of $(NSCl)_3$ with Me_3SiBr in CH_2Cl_2 at $-60°C$.[15] Similar highly conducting, non-stoichiometric polymers can be obtained by treating S_4N_4 with ICl or IBr.[13]

The selenium analogue of poly(sulfur nitride) is unknown, but its properties would be of considerable interest. The potential precursor Se_4N_4 is highly explosive and thermolysis under vacuum gives only dinitrogen and elemental selenium.[16] Other possible precursors for $(SeN)_x$, such as $[NSe]^+$ or $[Se_5N_5]^+$, have not been isolated. The mixed chalcogen nitride $1,5-S_2Se_2N_4$ (Section 5.2.6) has been prepared and structurally characterized, but attempts to convert this cage to the hypothetical four-membered ring $SSeN_2$ and, hence, the polymer $(SNSeN)_x$, have been unsuccessful.[17]

Another possible modification of poly(sulfur nitride) that is expected to produce conducting polymers is the replacement of alternating sulfur in the thiazyl chain by an RC unit, *i.e.*, $[(R)CNSN]_x$. This type of polymer would have five π-electrons per four atoms in the repeating unit and, consequently, would have a partially occupied conducting band.[18] The prospect of tuning the electronic properties of this polymer by changing the substituent R attached to carbon is intriguing. However, despite the extensive studies of carbon-containing S–N heterocycles described in Chapters 11 and 12, polymers of this type have not been prepared.

Conducting polymers with *p*-phenylene groups in the backbone can be generated by the metathetical reaction shown in Eq. 14.1.[19] Doping of these polymers with acceptors such as I_2, Br_2, or AsF_5 increases the conductivity to *ca.* $10^{-4} \, \Omega^{-1} \, cm^{-1}$.

$$Me_3SiNSNC_6H_4NSNSiMe_3 + SCl_2 \rightarrow$$

$$1/x[-C_6H_4NSNSNSN-]_x + 2Me_3SiCl \tag{14.1}$$

The use of 1,4-$ClSC_6H_4SCl$ instead of SCl_2 in reaction 14.1 produces the polymer $[C_6H_4NSNSC_6H_4SNSN]_x$. Polymers with $C_6H_4SNSNSC_6H_4$ or $C_6H_4SNSNSC_6H_4SNSNSC_6H_4$ segments separated by flexible spacer groups have also been synthesized.[20] These insulators are converted to semiconductors when doped with Br_2. Similar polymers have been prepared from 1,3-$ClSC_6H_4SCl$.[21]

14.3 Sulfur–Nitrogen Chains

The simplest chain compounds of the type RS_xN_yR are the sulfur diimides RNSNR (Section 10.4) The other known thiazyl chains are summarized in Table 14.1. They can be conveniently classified as sulfur-rich, nitrogen-rich or even-chain species.

Table 14.1   Sulfur–Nitrogen Chains as a Function of Chain Length

Chain length	S-rich	N-rich	Even-chain
3	$[RSNSR]^+$	RNSNR	
4			RS_2N_2R
5	RS_3N_2R		
6			
7	$[RS_4N_3R]^+$	RN_4S_3R	
8			RS_4N_4R
9			RS_5N_4R

The electron-counting rules that are used to determine the number of π-electrons in S-N rings (Section 4.4) can also be applied to thiazyl chains. Each S atom donates two electrons to the π–system and each N is a one-electron donor. The acyclic chain structures favour an even number of π-electrons. Consequently, most of the known S–N chains contain an even number of N atoms and are neutral. The exceptions are the cations $[RSNSR]^+$ (R = Cl, Br) and $[RS_4N_3R]^+$. The cationic systems are stabilized by (a) removal of an antibonding electron from the π–system, (b) lowering of the energies of the filled molecular orbitals induced by the positive charge, and (c) the ionic contribution to the lattice energy in the solid state. Conversely, anionic systems are destabilized by addition of an electron to an antibonding level in the already electron-rich π–system. Thus, it is not surprising that no stable anionic chains are known. The only selenium analogues of these S-N chain species are the selenium diimides RNSeNR (Section 10.4). The mixed chalcogen species $Me_3SiNSNSeNSNSiMe_3$, with a single selenium atom in the middle of a seven-atom chain, has been prepared and structurally characterized.[22]

Early examples of the synthesis of S–N chains frequently involved the reaction of S_4N_4 with nucleophilic reagents. For example, trithiadiazenes ArS_3N_2Ar are obtained by treatment of S_4N_4 with Grignard reagents, *e.g.* $4\text{-}ClC_6H_4MgBr$,[23] and the trithiatetrazene $Ph_2CNSNSNSNCPh_2$ is prepared by the reaction of diphenyldiazomethane with S_4N_4.[24] In recent years rational syntheses have been developed for chain lengths up to ArS_5N_4Ar via condensation reactions involving the elimination of Me_3SiCl or $Me_3SiOSiMe_3$ (Scheme 14.1).[25-27] The reagent $(NSCl)_3$, in the presence of $AgAsF_6$, has also been used to extend chain length, *e.g.*, in the conversion of ArS_3N_2Ar into $[ArS_4N_3Ar][AsF_6]$ (Ar = Ph, $4\text{-}O_2NC_6H_4$).[28]

The S–N chains ArS_2N_2Ar' (**14.1**, Ar = $4\text{-}O_2NC_6H_4$, Ar' = $4\text{-}MeOC_6H_4$), PhN_4S_3Ph (**14.2**) and $ArS_4N_4SiMe_3$ (**14.3**, Ar = $4\text{-}O_2NC_6H_4$ adopt alternating *cis,trans* conformations similar to $(SN)_x$.[27,29] In contrast to $(SN)_x$, however, there is a distinct alternation of short and long sulfur–nitrogen bonds in the oligomeric chains, consistent with a more localized [–S–N=S=N–] structure. Other chain conformations are also observed, as illustrated by the structures of ArS_3N_2Ar (**14.4**, Ar = $4\text{-}ClC_6H_4$),[23]

[ArS$_4$N$_3$Ar]$^+$ (**14.5,** Ar = 4-MeC$_6$H$_4$),[26] ArS$_5$N$_4$Ar (**14.6,** Ar = 4-MeC$_6$H$_4$),[30] and ^{t}BuN$_4$S$_3$^{t}Bu (**14.7**).[31] The small energy difference between various isomers is indicated by the subtle conformational change that occurs upon the replacement of the Ph groups in **14.2** by ^{t}Bu substituents to give **14.7**. The selenium-containing chain (Me$_3$SiNSN)$_2$Se adopts the same conformation as **14.7**.[22] *Ab initio* calculations indicate that weak intramolecular S•••S interactions occur in **14.4**.[32,33]

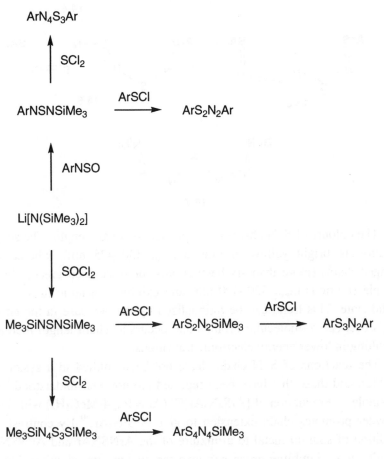

Scheme 14.1 Syntheses of sulfur–nitrogen chains

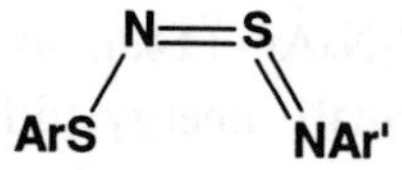

14.1

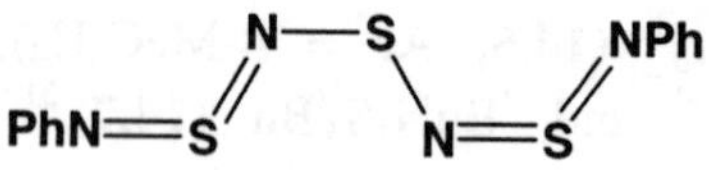

14.2

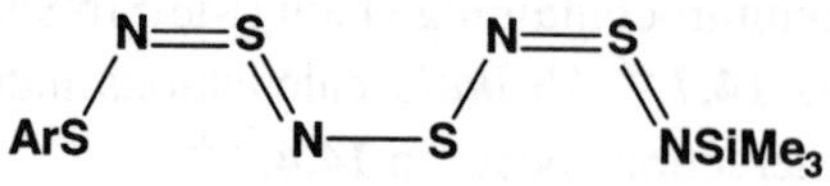

14.3

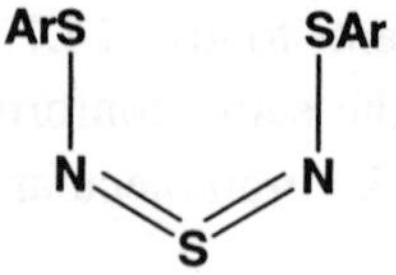

14.4

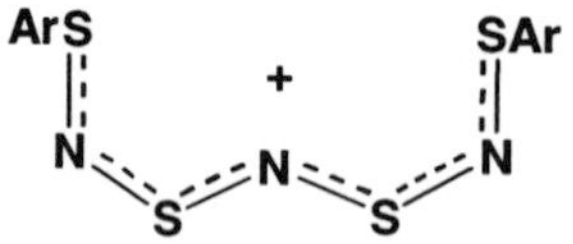

14.5

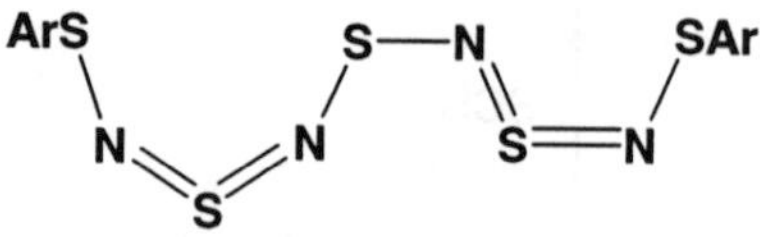

14.6

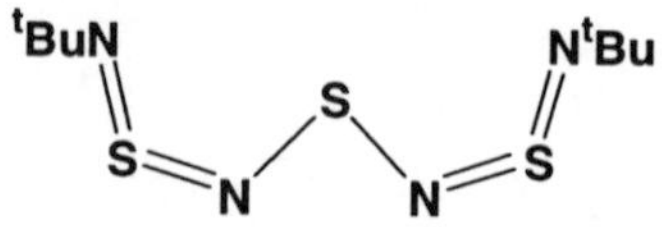

14.7

The colours of S–N chains are dependent on chain length. The shorter chains are bright yellow or orange (λ_{max} 330–475 nm), whereas the longer chains (more than six heteroatoms) produce deep green, blue or purple solutions (λ_{max} 520–590 nm) and exhibit a metallic lustre in the solid state. This trend can be rationalized by the decrease in the energy gaps between π–molecular orbitals that occurs as chain length increases resulting in lower energy electronic transitions.

The reactions of S–N chains have not been studied in a systematic fashion and those that have been reported are not well understood.[2] For example, the reduction of $[ArS_4N_3Ar]Cl$ (Ar = Ph, 4-MeC$_6$H$_4$) with silver powder promotes chain extension to give ArS_5N_4Ar.[30] By contrast, the addition of sodium metal to a mixture of the ArNSNAr and Ar'NSNAr' results in a scrambling process to give the unsymmetrical sulfur diimide ArNSNAr' (Section 10.4.4).[29] Another example of this type of transformation is the generation of unsymmetrical dithiadiazenes

ArNSNSAr' from a mixture of ArNSNAr and Ar'SNSNSAr'.[30] The limited data available indicate the occurrence of transformations analogous to the ring expansion or ring contraction processes observed for S–N ring systems.[34]

The incorporation of S–N chains between metal centres by the use of heteroaryl substituents in complexes of the type **14.7** has been proposed as a way to generate new materials that may function as molecular wires.[2] However, the synthesis of thiazyl chains bearing metal-binding sites has yet to be achieved.

14.4 Sulfur–Nitrogen Polymers Containing Three-Coordinate Sulfur(IV)

Thiazyl halides form cyclic oligomers, *e.g.*, $(NSCl)_3$ or $(NSF)_4$ (Sections 8.7 and 8.8), rather than polymers. However the thiazyl chloride unit NSCl has been incorporated into the hybrid polymer **14.8**, which is generated by a thermal ring-opening polymerization (ROP) process (Scheme 14.2).[3b,35] This process is conveniently monitored by ^{31}P NMR spectroscopy because, as a general guideline, polymerization is accompanied by an upfield shift of *ca.* 30 ppm in the ^{31}P NMR resonance, *i.e.*, from δ +24.5 for the six-membered ring to δ –4.6 for the polymer **14.8**.

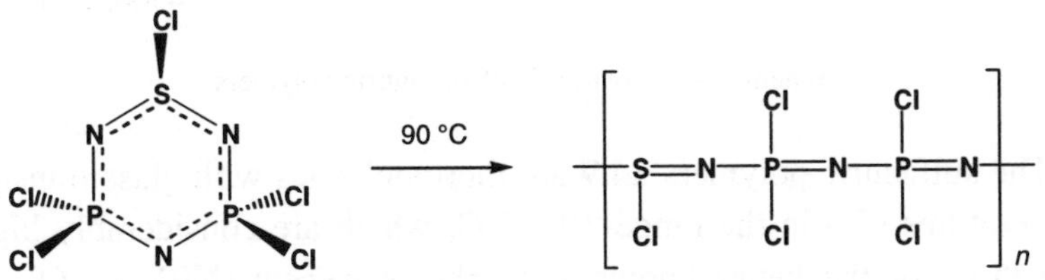

14.8

Scheme 14.2 Synthesis of poly(thiophosphazene)

Polymers containing three-coordinate sulfur(IV) are generally hydrolytically sensitive even when the chloro substituents are replaced by phenoxy groups.[36] Consequently, much more attention has been accorded to analogous polymer systems that contain four-coordinate sulfur(VI). Interestingly, nucleophilic substitution in **14.8** occurs preferentially at the sulfur sites.[36]

14.5 Sulfur–Nitrogen Polymers Containing Four-Coordinate Sulfur(VI)

Sulfanuric halides $[NS(O)X]_x$ (X = Cl, F) are normally isolated as cyclic trimers (x = 3) (Section 8.9). These ring systems do not undergo ring-opening polymerization; they decompose exothermically above 250°C.[37] Consequently, an alternative route to sulfanuric polymers of the type **14.9** had to be developed. High molecular weight polymers containing repeating [–N=S(O)R–] (R = Me. Ph) units are generated from an acyclic precursor by a condensation process in which the formation of a very strong Si–O bond provides a driving force for the elimination (Scheme 14.3).

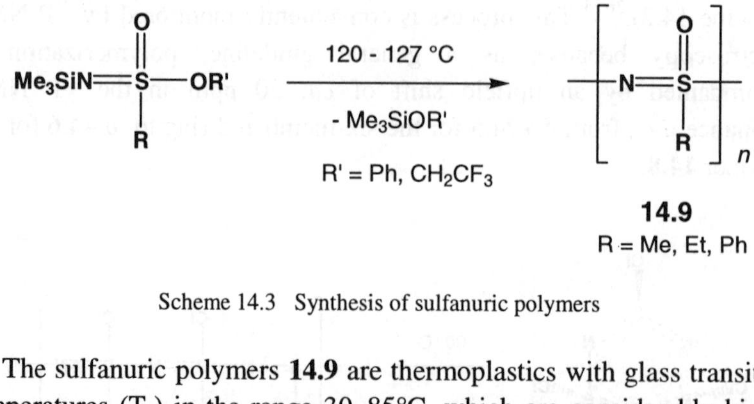

Scheme 14.3   Synthesis of sulfanuric polymers

The sulfanuric polymers **14.9** are thermoplastics with glass transition temperatures (T_g) in the range 30–85°C, which are considerably higher than those of the better known poly(phosphazenes) $(NPX_2)_n$ (T_g = – 96°C, X = F; –84°C, X = OEt; –66°C, X = Cl; –6°C, X = OPh). The higher T_gs of the sulfanuric systems are attributed to (a) increased intermolecular interactions as a result of the polar S=O groups and (b) greater side group–main chain interactions as a result of the smaller <NSN bond angle (calculated value 103°) compared to *ca.* 120° for <NPN in poly(phosphazenes).[37b]

Hybrid polymers that contain both [NS(O)R] and [NPR$_2$] units (R = Cl, F OAr, NHR) have been the subject of extensive investigations as a result of their promising properties.[3] The first example of this type of sulfanuric–phosphazene polymer **14.10a** was reported on 1991.[38] It was prepared by the thermal ring-opening polymerization of a cyclic precursor at 165°C (Scheme 14.4). The conversion of the six-membered ring into a polymer is accompanied by an upfield shift of ^{31}P NMR resonance of *ca.* 35 ppm. Small amounts of macrocycles {[NS(O)Cl](NPCl$_2$)$_2$}$_n$ (n = 2-6) are also formed during this process and the twelve (n = 2) and twenty-four-membered (n = 4) rings have been isolated and structurally characterized (Section 13.4).[39] The polymer is produced as a pale yellow elastomer by adding hexane to a cold, rapidly stirred CH$_2$Cl$_2$ solution of the product. The fluorinated derivative **14.10b** may be obtained in a similar manner by using a slightly higher temperature (180°C). Subsequently, it was found that this polymerization process occurs at room temperature in CH$_2$Cl$_2$ in the presence of GaCl$_3$ (10% of the stoichiometric amount) as a catalyst.[40] The yield of the polymer **14.10a** obtained by the latter method is essentially quantitative.

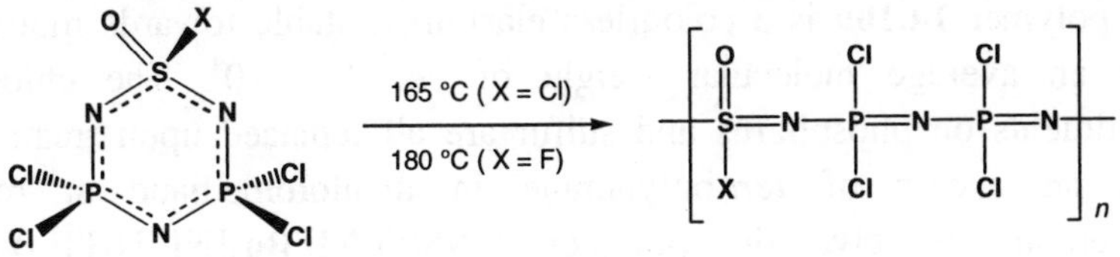

14.10a, X =Cl
14.10b, X =F

Scheme 14.4   Synthesis of poly(thionylphosphazenes)

The cation [NSO(NPCl$_2$)$_2$]$^+$ (**14.11**) is the proposed intermediate in this ring-opening polymerization process. This cation is extremely reactive, as illustrated by the isolation of the solvent-derived product **14.12** when it is generated by halide abstraction from the cyclic precursor with AlCl$_3$ in 1,2-dichloroethane.[40]

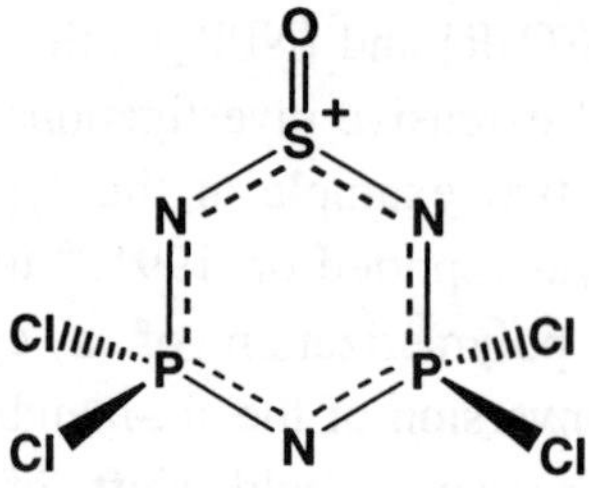

14.11 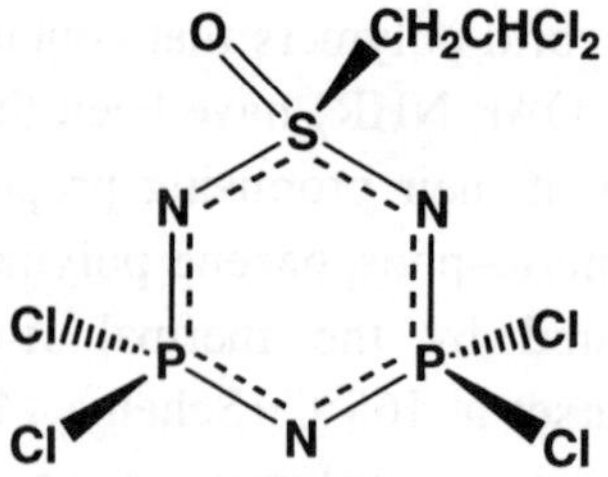   **14.12**

In order to obtain hydrolytically stable materials, the chlorine substituents in the polymer **14.10a** may be replaced by phenoxy or *tert*-butylamino groups.[41,42] The reaction of **14.10a** with sodium phenoxide in dioxane at room temperature is regiospecific. It produces the polymer $\{[NS(O)Cl][NP(OPh)_2]_2\}_n$ (**14.13**), *i.e.*, the chlorines attached to phosphorus are replaced by OPh groups, but the S–Cl bond remains intact. This observation is different from the behaviour of the analogous sulfur(IV) polymer **14.8** towards nucleophilic reagents. In that case substitution takes place preferentially at the sulfur centre (Section 14.4). The polymer **14.10a** is a colourless elastomer, stable towards moisture with an average molecular weight of *ca*. 3×10^4. The chlorine substituents on phosphorus and sulfur are all replaced upon treatment with an excess of *tert*-butylamine in dichloromethane at room temperature to give the polymer $\{[NS(O)NH^tBu][NP(NH^tBu)_2]_2\}_n$ (**14.14**), which has an average molecular weight of *ca*. 2.5×10^4.

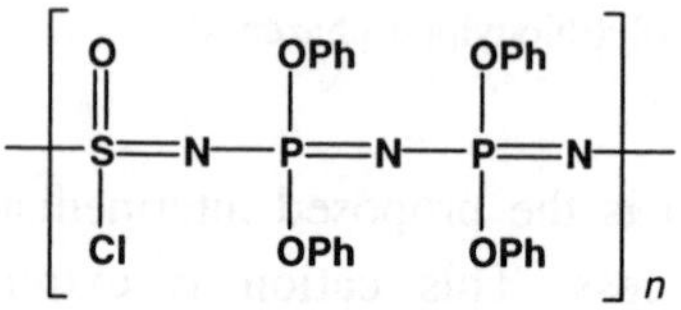

14.13 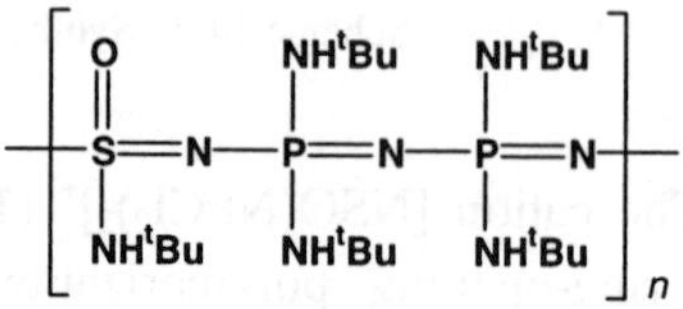   **14.14**

As expected, the T_gs of the hybrid sulfanuric–phosphazene polymers are much closer to the values reported for poly(phosphazenes) than those of sulfanuric polymers (*vide supra*). The values for the polymers **14.10a** and **14.10b** are –46°C and –56°C, respectively. *Ab initio* molecular orbital calculations, including geometry optimizations, on short-chain model compounds of **14.10a** and **14.10b** show that they adopt a non-planar *trans, cis* structure.[43] The calculations also reveal a decrease in the torsional barriers for rotation of the SNP and PNP bond angles when changing from Cl on sulfur in **14.10a** to F attached to sulfur in **14.10b**. This calculated increase in flexibility is thought to account for the lower T_gs of fluorine-containing poly(phosphazenes) as well as the hybrid polymer **14.10b**. The T_g of –16°C for the butylamino-substituted hybrid polymer **14.14** is significantly lower than the value of +8°C reported for the corresponding poly(phosphazene) $[NP(NH^tBu)_2]_n$ as a result of the presence of the small S=O group and only five ^tBuNH substituents in **14.14** compared to six ^tBuNH groups in the phosphazene polymer.

Polythionylphosphazenes have considerable potential as components of the matrices for phosphorescent oxygen sensors.[44] These sensors contain transition-metal-based dyes with oxygen-quenchable excited states dispersed in a polymer matrix. They can be used, for example, to determine the air pressure distribution on the wings of an aircraft in wind tunnel experiments. The high solubility and high diffusion coefficient for oxygen of the alkylamino derivatives of these hybrid polymers are crucial properties in this application.[45] Although **14.14** forms sticky films, the block co-polymer polythionylphosphazene-*b*-polytetrahydrofuran, prepared from the reaction of the cation of **14.10a** with THF, forms free-standing films.[46]

Hybrid polymers containing an equal number of alternating [NS(O)R] and [NPR₂] units in the backbone can formally be derived by ROP of the appropriate eight-membered ring. In practice, however, this approach is not successful because the ring strain in eight-membered rings is considerably smaller than that in six-membered inorganic ring systems. This type of copolymer has, however, been synthesized by the ingenious exploitation of condensation processes as exemplified in Scheme 14.5.[47] The formation of the acyclic precursor **14.15** is highly regiospecific with respect to the elimination of the Me₃Si group attached to the PN unit and the CF₃CH₂O group bonded to sulfur in the form of Me₃SiOCH₂CF₃. Molecular weight determinations indicate that the hybrid polymer **14.16** produced in this way consists of *ca.* 30 repeat units.

The ^{31}P NMR spectrum of the polymer **14.16** shows two singlets of approximately equal intensities that differ in chemical shift by only 0.17 ppm. This observation is attributed to the formation of equal amounts of isotactic (**14.16a**, both S=O groups on the same side of the polymer chain) and syndiotactic (**14.16b**, S=O groups on opposite sides of the polymer chain) forms of the polymer. The atacticity is confirmed by the ^{1}H and ^{13}C NMR spectra. Two resonances are observed for the Me$_2$P groups of **14.16a**, whereas **14.16b** gives rise to only one resonance.

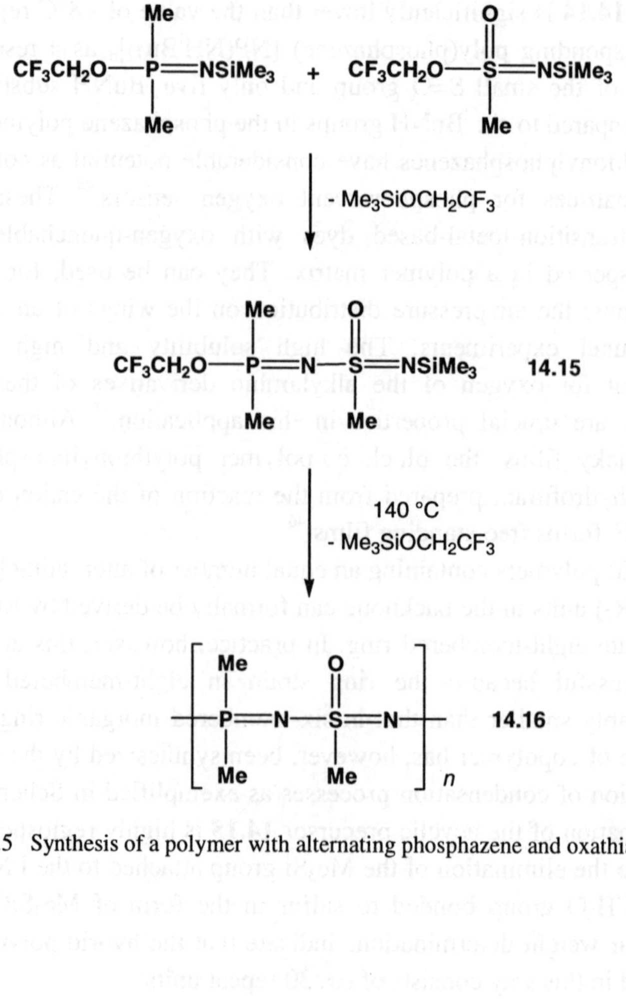

$$\underset{\underset{\textbf{Me}}{|}}{\overset{\overset{\textbf{Me}}{|}}{CF_3CH_2O-P}}=NSiMe_3 \quad + \quad \underset{\underset{\textbf{Me}}{|}}{\overset{\overset{O}{\|}}{CF_3CH_2O-S}}=NSiMe_3$$

$\Big\downarrow$ - Me$_3$SiOCH$_2$CF$_3$

$$\underset{\underset{\textbf{Me}}{|}}{\overset{\overset{\textbf{Me}}{|}}{CF_3CH_2O-P}}=N-\underset{\underset{\textbf{Me}}{|}}{\overset{\overset{O}{\|}}{S}}=NSiMe_3 \qquad \textbf{14.15}$$

$\Big\downarrow$ 140 °C
- Me$_3$SiOCH$_2$CF$_3$

$$\left[\underset{\underset{\textbf{Me}}{|}}{\overset{\overset{\textbf{Me}}{|}}{-P}}=N-\underset{\underset{\textbf{Me}}{|}}{\overset{\overset{O}{\|}}{S}}=N- \right]_n \qquad \textbf{14.16}$$

Scheme 14.5 Synthesis of a polymer with alternating phosphazene and oxathiazene units

14.16a   14.16b

References

1. (a) M. M. Labes, P. Love and L. F. Nichols, *Chem. Rev.*, **79**. 1 (1979); (b) A. J. Banister and I. B. Gorrell, *Adv. Mater.*, **10**, 1415 (1998).

2. J. M. Rawson and J. J. Longridge, *Chem. Soc. Rev.*, 53 (1997).

3. (a) I. Manners, *Coord. Chem. Rev.*, **137**, 109 (1994); (b) D. P. Gates and I. Manners, *J. Chem. Soc., Dalton Trans.*, 2525 (1997); (c) A. R. McWilliams, H. Dorn and I. Manners, *Topics. Curr. Chem.*, **220**, 141 (2002).

4. (a) C. M. Mikulski, P. J. Russo, M. S. Saran, A. G. MacDiarmid, A. F. Garito and A. J. Heeger, *J. Am. Chem. Soc.*, **97**, 6358 (1975); (b) M. J. Cohen, A. F. Garito, A. J. Heeger, A. G. MacDiarmid, C. M. Mikulski, M. S. Saran and J. Kleppinger, *J. Am. Chem. Soc.*, **98**, 3844 (1976).

5. (a) H. Müller, S. O. Svensson, J. Birch and Å. Kvick, *Inorg. Chem.*, **36**, 1488 (1997); (b) R. C. Mawhinney and J. D. Goddard, *Inorg. Chem.*, **42**, 6323 (2003).

6. A. J. Banister, Z. V. Hauptman, J. M. Rawson and J. T. Wait, *J. Mater. Chem.*, **6**, 1161 (1997).

7. A. J. Banister, Z. V. Hauptman, J. Passmore, C-M. Wong and P. S. White, *J. Chem. Soc., Dalton Trans.*, 2371 (1986).

8. W. M. Lau, N. P. C. Westwood and M. H. Palmer, *J. Am. Chem. Soc.*, **108**, 3229 (1986).

9. J. Gerratt, S. J. McNicholas, P. B. Karadakov, M. Sironi. M. Raimondi and D. L. Cooper, *J. Am. Chem. Soc.*, **118**, 6272 (1996).

10. A. E. Thomas, J. Woods and Z. V. Hauptman, *J. Phys. D*, **16**, 123 (1983).

11. M. J. Cohen and J. S. Harris, *Appl. Phys. Lett.*, **33**, 812 (1978).

12. (a) H. B. Mark, Jr., A. Voulgaropoulos and C. A. Meyer, *J. Chem. Soc., Chem. Commun.*, 1021 (1981); J. F. Rubinson, T. D. Behymer and H. B. Mark, Jr., *J. Am. Chem. Soc.*, **104**, 1224 (1982).

13. M. Akhtar, C. K. Chiang, A. J. Heeger, J. Milliken and A. G. MacDiarnid, *Inorg. Chem.*, **17**, 1539 (1978).

14. J. W. Macklin, G. B. Street and W. D. Gill, *J. Chem. Phys.*, **70**, 2425 (1979).

15. U. Demant and K. Dehnicke, *Z. Naturforsch.*, **41B**, 929 (1986).

16. T. M. Klapötke, Binary Selenium-Nitrogen Species and Related Compounds, in R.Steudel (ed.) *The Chemistry of Inorganic Ring Systems*, Elsevier, pp. 409-427 (1992).

17. A. Maaninen, R. S. Laitinen, T. Chivers and T. A. Pakkanen, *Inorg. Chem.*, **38**, 3450 (1999).

18. M-H. Whangbo, R. Hoffman and R. B. Woodward, *Proc. Roy. Soc. London A*, **366**, 23 (1979).

19. O. J. Scherer, G. Wolmershäuser and R. Jotter, *Z. Naturforsch.*, **37B**, 432 (1982).

20. J. C. W. Chien and S. Ramakrishnan, *Macromolecules*, **21**, 2007 (1988).

21. J. J. Longridge and J. M. Rawson, *Polyhedron*, **17**, 1871 (1998).

22. J. Konu, A. Maaninen, K. Paananen, P. Ingman, R. S. Laitinen, T. Chivers and J. Valkonen, *Inorg. Chem.*, **41**, 1430 (2002).

23. F. P. Olsen and J. C. Barrick, *Inorg. Chem.*, **12**, 1353 (1973).

24. E. M. Holt, S. L. Holt and K. J. Watson, *J. Chem. Soc., Dalton Trans.*, 1357 (1974).

25. J. Kuyper and G. B. Street, *J. Am. Chem. Soc.*, **99**, 7848 (1977).

26. J. J. Mayerle, J. Kuyper and G. B. Street, *Inorg. Chem.*, **17**, 2610 (1978).

27. A. V. Zibarev, Y. G. Gatilav and I. Y. Bagryanskaya, *Polyhedron*, **11**, 2787 (1992).

28. J. A. K. Howard, I. Lavender, J. M. Rawson and E. A. Swain, *Main Group Chem.*, **1**, 317 (1996).

29. K. Bestari, R. T. Oakley and A. W. Cordes, *Can. J. Chem.*, **69**, 94 (1991).

30. G. Wolmershäuser and P. R. Mann, *Z. Naturforsch.*, **46B**, 315 (1991).

31. W. Isenberg, R. Mews and G. M. Sheldrick, *Z. Anorg. Allg. Chem.*, **525**, 54 (1985).

32. R. Gleiter and R. Bartetzko, *Z. Naturforsch.*, **36B**, 492 (1981).

33. R. M. Bannister and H. S. Rzepa, *J. Chem. Soc., Dalton Trans.*, 1609 (1989).

34. (a) T. Chivers, *Chem. Rev.*, **74**, 341 (1985); (b) R. T. Oakley, *Prog. Inorg. Chem.*, **36**, 229 (1988).

35. J. A. Dodge, I. Manners, H. R. Allcock, G. Renner and O. Nuyhen, *J. Am. Chem. Soc.*, **112**, 1268 (1990).

36. H. R. Allcock, J. A. Dodge and I. Manners, *Macromolecules*, **26**, 11 (1993).

37. (a) A. K. Roy, *J. Am. Chem. Soc.*, **114**, 1530 (1992); (b) A. K. Roy, G. T. Burns, G. C. Lie and S. Grigoras, *J. Am. Chem. Soc.*, **115**, 2604 (1993).

38. M. Liang and I. Manners, *J. Am. Chem. Soc.*, **113**, 4044 (1991).

39. Y. Ni, A. J. Lough, A. L. Rheingold, and I. Manners, *Angew. Chem., Int. Ed. Engl.*, **34**, 998 (1995).

40. A. R. McWilliams, D. P. Gates, M. Edwards, L. M. Liable-Sands, I. Guzei, A. L. Rheingold and I. Manners, *J. Am. Chem. Soc.*, **122**, 8848 (2000).

41. Y. Ni, A. Stammer, M. Liang, J. Massey, G. J. Vansco and I. Manners, *Macromolecules*, **25**, 7119 (1992).

42. Y. Ni, P. Park, M. Liang, J. Massey, C. Waddling and I. Manners, *Macromolecules*, **29**, 3401 (1996).

43. R. Jaeger, J. B. Lagowski, I. Manners and G. J. Vansco, *Macromolecules*, **28**, 539 (1995).

44. Z. Pang, X. Gu, A. Yekta, Z. Masoumi, J. B. Coll, M. A. Winnik and I. Manners, *Adv. Mater.*, **8**, 768 (1996).

45. Z. Masoumi, V. Stoeva, A. Yekta, Z. Pang, I. Manners and M. A. Winnik, *Chem. Phys. Lett.*, **261**, 551 (1996).

46. R. Ruffulo, C. E. B. Evans, X. Liu, Z. Pang, P. Park, A. R. McWilliams, X. Gu, A. Yekta, M. A. Winnik and I. Manners, *Anal. Chem.*, **72**, 1894 (2000).

47. V. Chunechom, T. E. Vidal, H. Adams and M. L. Turner, *Angew. Chem., Int. Ed. Engl.*, **37**, 1928 (1998).

Chapter 15

Weak Intramolecular Chalcogen-Nitrogen Interactions

15.1 Introduction

In previous chapters it has been pointed out that weak intra- or inter-molecular chalcogen-chalcogen interactions are an important feature of chalcogen-nitrogen chemistry. Intramolecular contacts of this type result in bicyclic or cage structures, as described in Section 4.6. The predilection of chalcogen-nitrogen radicals to engage in intermolecular association has been employed in the design of new materials with unusual conducting or magnetic properties (Section 11.3). As a general rule both of these types of weak interactions can be described in terms of $\pi^*-\pi^*$ bonding involving singly occupied π–type orbitals. In this chapter a different type of weak interaction, *viz.* one that involves closed shell chalcogen and nitrogen atoms, will be considered.[1] This type of secondary bonding interaction may have an important influence on the structures and properties of certain chalcogen-nitrogen compounds, as illustrated by the chosen examples. The discussion will begin with an account of the evidence for the occurrence of these short contacts in the solid state, as determined from X-ray crystal structures. This will be followed by descriptions of the solution behaviour of such species, as disclosed by NMR studies, and the electronic nature of these bonding interactions as revealed by *ab initio* molecular orbital calculations. The subsequent sections will provide some examples of the influence of intramolecular E•••N (E = S, Se, Te) interactions on (a) the reactivity of chalcogen-nitrogen compounds and (b) the stabilization of reactive functional groups.

15.2 Molecular Structures

Weak intramolecular interactions between sulfur or selenium and nitrogen are a recurrent phenomenon in large biomolecules.[2] They may occur in the same residue or between neighbours of a peptide chain. The formation of four- or five-membered rings of the types **15.1** and **15.2**, respectively, is most common. A feature that is unique to proteins is the participation of sulfur atoms in bifurcated N•••S•••N contacts.

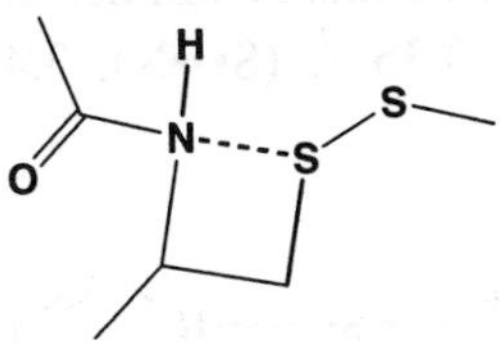

15.1	**15.2**	**15.3**

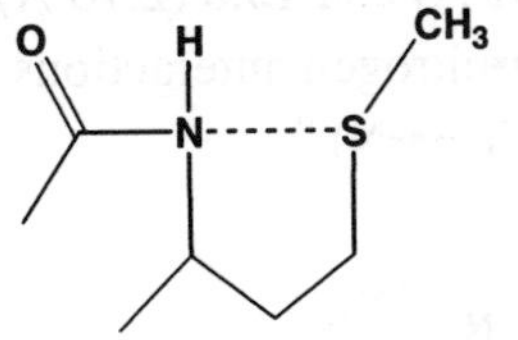

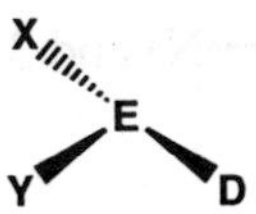

These interactions are most commonly observed for divalent chalcogen atoms and the nitrogen atom (the electron donor D) lies within the X–E–Y (E = S, Se, Te) plane, preferably along the extension of one of the covalent bonds as in **15.3**. This anisotropy is a clear indication that these short E•••N contacts have some bonding character, *i.e.*, they are subject to the geometric restrictions of orbital overlap. For example, in the diselenide **15.4** the nitrogen lone pairs are clearly oriented towards the Se–Se linkage.[3]

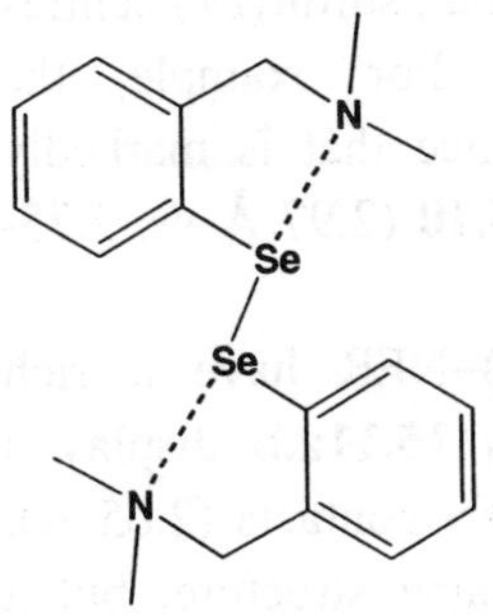

15.4	**15.5a**, X = Cl	**15.6**
	15.5b, X = Br	
	15.5c, X = I	

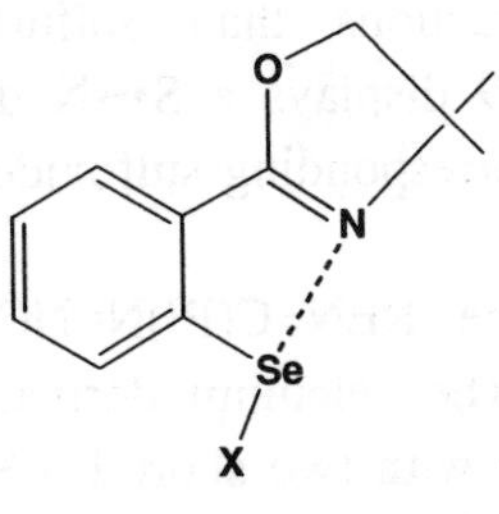

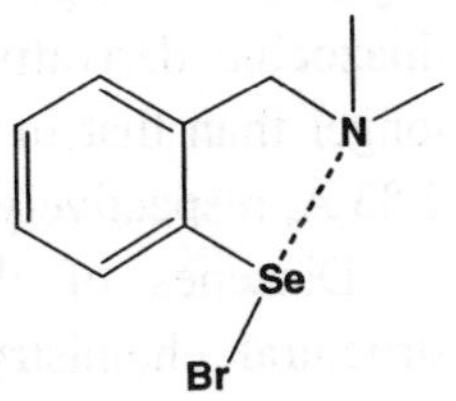

Several structural determinations indicate that the E•••N interaction is strongly influenced by the electronegativity of the opposite group X. In the series **15.5a-c** the Se•••N distances are 2.05 Å for X = Cl, 2.06 Å for X = Br and 2.13 Å for X = I.[4] In the diselenide **15.4**, the Se•••N distance of 2.86 Å is much longer than that in the corresponding bromide **15.6** (2.14 Å).[3] The strength of the chalcogen•••nitrogen interaction can also be enhanced by introducing substituents with an electron-withdrawing effect on the chalcogen, as demonstrated by comparison of the S•••N distances in **15.7** (2.61 Å) and **15.8** (2.10 Å).[5] The sums of van der Waals radii for chalcogen•••nitrogen interactions are 3.35 Å (S•••N), 3.45 Å (Se•••N) and 3.61 Å (Te•••N).[6]

15.7   **15.8**

The heavier chalcogens are more prone towards secondary interactions than sulfur. In particular, the chemistry of tellurium has numerous examples of "intramolecular coordination" in derivatives such as diazenes, Schiff bases, pyridines, amines, and carbonylic compounds.[7] The oxidation state of the chalcogen is also influential; sulfur(IV) centres engender stronger interactions than sulfur(II). For example, the thiazocine derivative **15.9** displays a S•••N distance that is markedly longer than that in the corresponding sulfoxide **15.10** (2.97 Å *vs.* 2.75-2.83 Å, respectively).[8]

Diazenes of the type REN=C(R')N=NC(R')=NER have a rich structural chemistry.[9,10] The selenium derivatives **15.11a,b** display a *cis,trans,cis* conformation with two short 1,5-Se•••N contacts (2.65 Å). Several sulfur analogues, *e.g.*, **15.1c**, have the same structure, but a different *cis,trans,cis* conformer **15.12** with two 1,4-S•••N contacts (2.83 Å) has also been characterized.[10a] A third type of diazene is the *trans,trans,trans* isomer **15.13a,b** with no intramolecular short contacts.[10b]

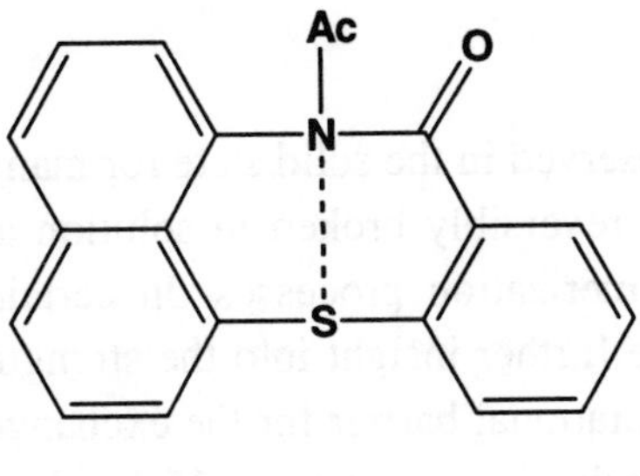

15.9

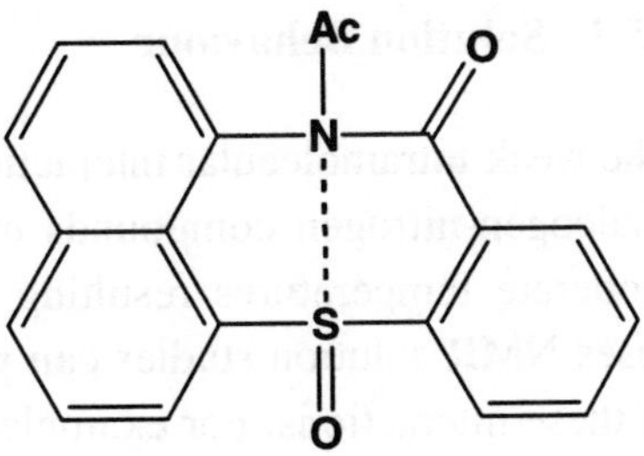

15.10

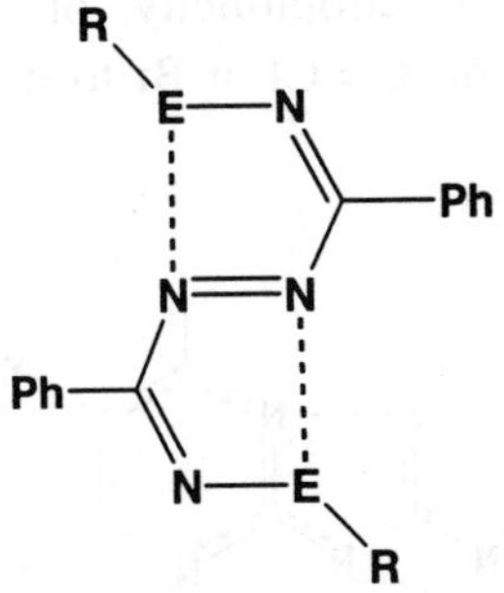

15.12

15.11a, E = Se, R = Me

15.11b, E = Se, R = Ph

15.11c, E = S, R = 4-CH$_3$C$_6$H$_4$

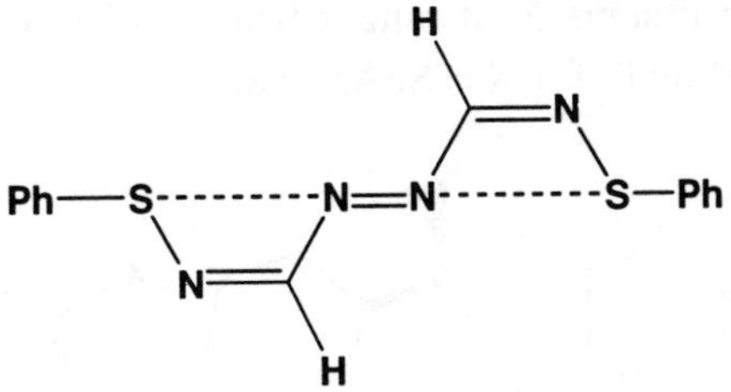

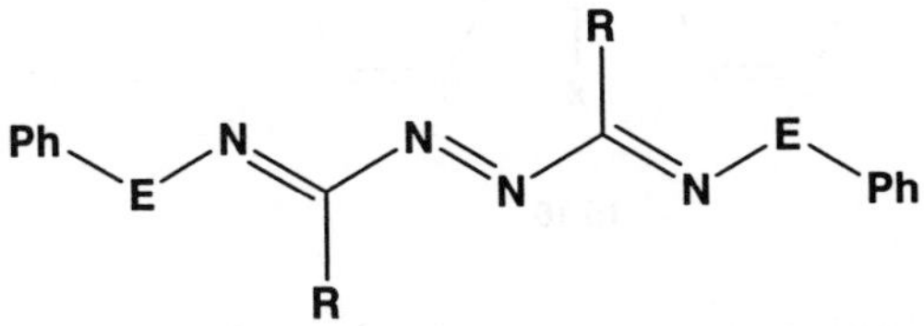

15.13a, E = S, R = 2-BrC$_6$H$_4$

15.13b, E = S, R = 2-CF$_3$C$_6$H$_4$

15.3   Solution Behaviour

The weak intramolecular interactions observed in the solid state for many chalcogen-nitrogen compounds may be reversibly broken in solution at moderate temperatures resulting in isomerization processes. In certain cases NMR solution studies can provide further insight into the strength of these interactions. For example, the rotational barrier for the exchange of the benzylic protons in 2-selenobenzylamine derivatives **15.14** (X = Br, Cl, OAc, CN, SPh, SeAr, Me) includes the dissociation of the Se•••N contact. The interaction energy estimated from variable temperature NMR studies can be correlated with the electrophilicity of the substituents X. It ranges from > 19 kcal mol^{-1} for X = Cl or Br to < 7.6 kcal mol^{-1} for X = SeAr or Me.[11]

15.14   **15.15**

15.16

These isomerization processes may be dependent on the nature of the solvent. For example, the rotational barrier of the tetrazathiapentalenes **15.15** (*ca.* 16 kcal mol^{-1}) is influenced by the donor or acceptor ability of the substituents X and Y through the S•••N short contacts.[12] Solvents with acidic protons increase the magnitude of the barrier, whereas solvents that are good Lewis bases decrease the size of the barrier, owing to solvation of the transition state.

The ^{15}N and ^{1}H (N–Me) NMR chemical shifts of the thiocine derivatives **15.16** (X = Cl, OMe, Me, O⁻), which display short S•••N distances (2.09 – 2.60 Å),[13] correlate with the electronegativity of X.[14]

The intramolecular Se•••N interactions in complexes of the type **15.5** result in a downfield shift in the ^{77}Se NMR resonances; the magnitude of this shift corresponds approximately to the strength of the interaction.[4]

15.4 Electronic Structures

A common interpretation of the interaction of chalcogens with nucleophiles considers donation of electron density from a lone pair on the donor atom into the σ^*(E–X) orbital (Figure 15.1). As the degree of covalency increases, a hypervalent three-centre four-electron bond is formed. Real systems fall somewhere between "secondary" ($n \rightarrow \sigma^*$) interactions and hypervalent (three centre – four electron) bonds. The two extremes can be distinguished by the correlation of X–E and E•••D distances.[15] In the hypervalent case both bond distances decrease simultaneously, whereas in the "secondary" bond the distances are anticorrelated. This concept has been applied in a study of selenoquinones **15.17** (R = Ph, Me) with short Se•••O contacts,[16] for

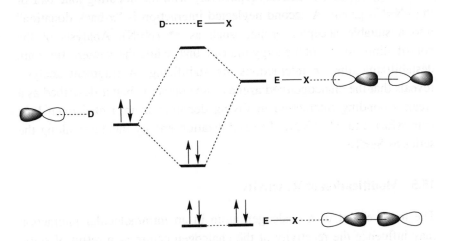

Fig. 15.1   Molecular orbital diagram of intramolecular donor (D) – chalcogen (E) interactions

which it explains and reproduces the trends observed for d(Se•••N) as a function of the substituents R.[11] The nature of the Se•••O interaction in 2-substituted bezeneselenyl derivatives of the type 2-YC$_6$H$_4$SeX (Y = CHO, CH$_2$OH, CH$_2$O^{i}Pr; X = Cl, Br, CN, SPh, SeAr, Me) has been investigated by ^{17}O NMR spectroscopy and by natural bond order (NBO) and atoms in molecules (AIM) theoretical methods.[17] The ^{17}O NMR chemical shifts reflect the strength of the Se•••O interaction. The NBO analysis shows that the stabilization energies due to an $n_O \rightarrow \sigma^*_{Se-X}$ orbital interaction correlate with Se•••O distances. The results also indicate that these interactions are predominantly covalent, rather than electrostatic, in character.[17]

Density functional theory calculations on model diazenes of the type **15.11**, **15.12**, **15.13** and other possible isomers show that the isomer **15.11**, for which the E•••N interactions generate five-membered rings, is the most stable.[10b] However, isomer **15.12**, which involves four-membered rings, is only 4.2 kcal mol^{-1} higher in energy. The other structurally characterized conformer **15.13** is the third most stable geometry. A symmetry-based analysis for **15.11** provides a detailed bonding description from which it is evident that the simplistic $n \rightarrow \sigma^*$ scheme neglects the lone pair of the chalcogen, which has an orientation and energy suitable to interact repulsively with the donating lone pair of the –N=N– group. A second neglected interaction is "π-back donation" into a suitable acceptor orbital, such as π^* (N=N). Analysis of the contributions to the total energy indicate that, while the σ interactions are destabilizing, the π interactions are stabilizing. A fragment analysis reveals that the chalcogen•••diazene close contact is better described as a weak π-bonding interaction involving donation from a chalcogen lone pair $\pi(E_{lp})$ into π^* (N=N). The stabilization energy increases along the series S<Se<Te.

15.5 Modification of Reactivity

The involvement of a chalcogen atom in an intramolecular interaction may influence the reactivity of the chalcogen centre as a result of steric or electronic effects. This feature may lead to unique reaction patterns. For example, the sulfur(II) centre involved in an intramolecular contact in **15.18** is remarkably resistant to oxidation. Whereas the other sulfur(II) atom is oxidized to a sulfone, an excess of 4-chloroperbenzoic acid reacts preferentially with the C=N double bond giving the epoxide **15.19**.[18] By contrast, the two sulfur atoms in the diazene **15.11c** are oxidized readily to sulfur(VI) with simultaneous reduction of the N=N double bond and formation of a thiatriazole ring to yield **15.20** (Ar = 4-CH$_3$C$_6$H$_4$).[19]

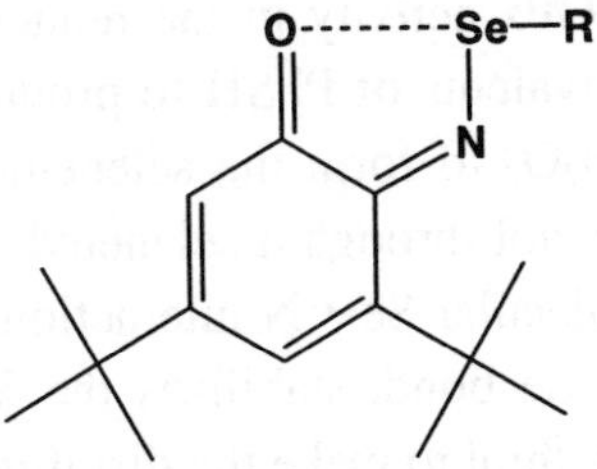

15.17 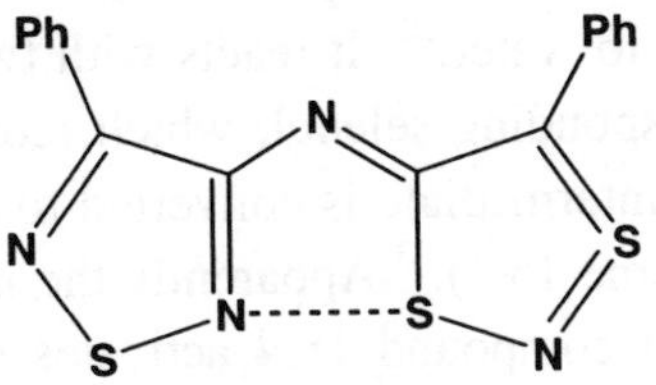   **15.18**

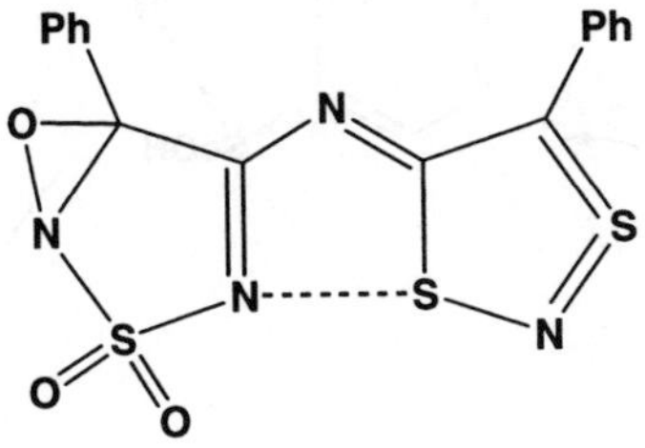

15.19

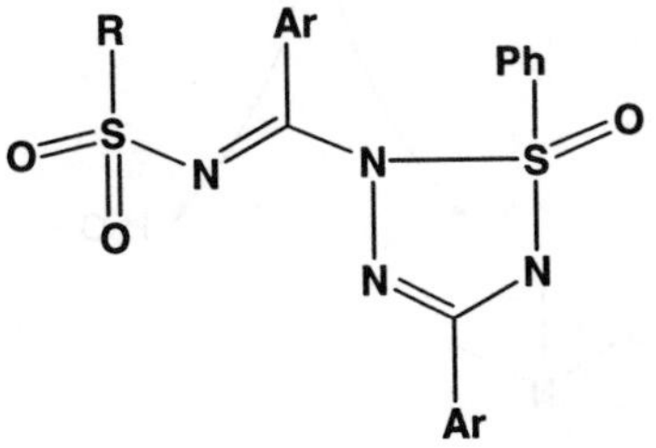

15.20 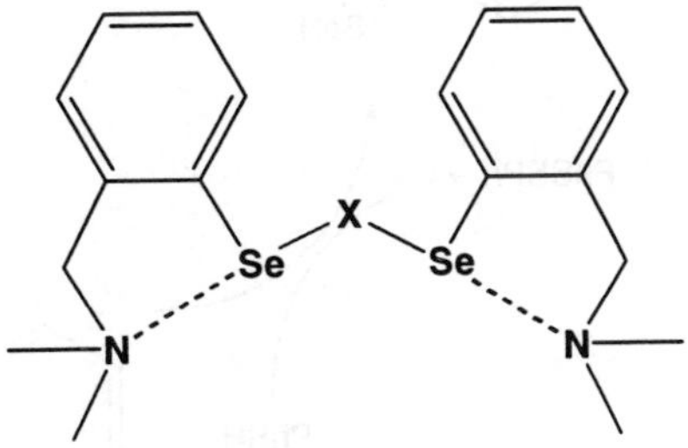   **15.21a**, X = S
                             15.21b, X = CH$_2$

The methodology of heteroatom-directed lithiation has been applied to the synthesis of a variety of organochalcogen compounds, including unstable, low-valent compounds (Section 15.6).[20] The Se•••N interactions in the diselenide **15.4** activate the Se–Se bond toward insertion of a sulfur atom or a methylene group to give the derivatives **15.21a,b**.[20]

Glutathione peroxidase (GPx) is a selenoenzyme that functions as an antioxidant by catalyzing the reduction of harmful peroxides by glutathione.[21] Compound **15.4** exhibits GPx activity in the reduction of H_2O_2 to water.[22] It reacts with two equivalents of PhSH to produce the corresponding selenol, which reduces H_2O_2 to form the selenenic acid. This intermediate is converted to the selenol through a selenenyl sulfide (Scheme 15.1).[23] Apparently the intramolecular Se•••N interaction in the model compound **15.4** activates the Se–Se bond, stabilizes the Se–OH group and facilitates the reaction with the thiol to make the disulfide.

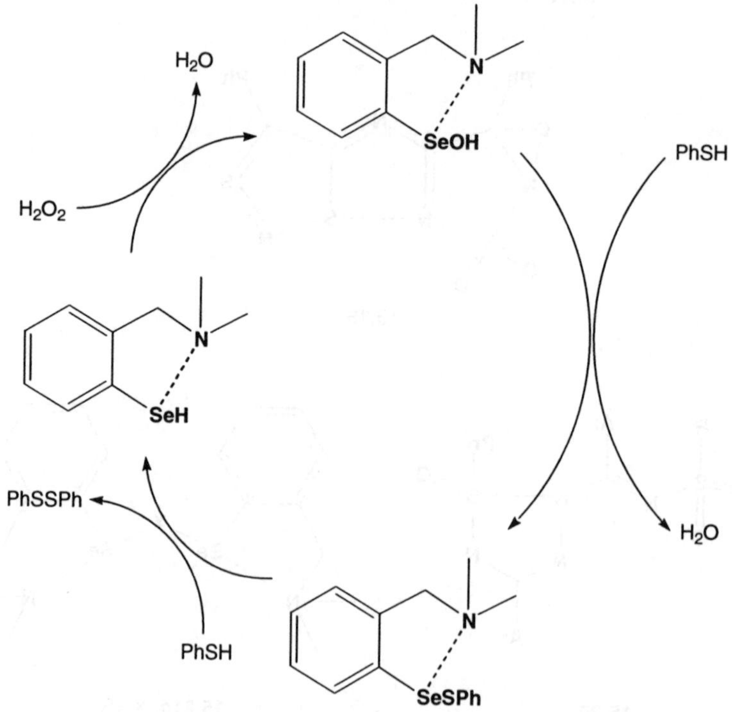

Scheme 15.1 Catalytic cycle for the conversion of a diselenide to a selenenic acid

The enzyme protein tyrosine phosphatase 1B (PTB1B) is a potential therapeutic agent for treating diabetes. X-Ray crystallographic studies reveal that the sulfur atom of the cysteine in the active site of this enzyme is covalently bonded to a nitrogen atom in the backbone of a neighbouring residue.[24] PTB1B uses its active-site cysteine to remove a phosphate group from a tyrosine on the insulin receptor. Its activity is turned off by H_2O_2-mediated oxidation of the cysteine to cysteine sulfenic acid, which rapidly converts to a sulfenyl amide species. It has been suggested that this sulfur•••nitrogen interaction protects cysteine from being further oxidized before it can return to the active thiol state.[24]

In some reactions intramolecular chalcogen•••nitrogen interactions may lead to stereochemical control. For example, selenenyl bromides react with C=C double bonds, providing a convenient method of introducing various functional groups. The reaction proceeds readily, but affords a racemic mixture. The modified reagent **15.22** contains a chiral amine in close interaction with the selenium atom. It reacts with olefins affording up to 97% *ee* of isomer **A** (Scheme 15.2).[25]

15.6   Stabilization of Reactive Functional Groups

Intramolecular chalcogen interactions may also stabilize reactive functional groups enabling the isolation of otherwise unstable species or their use as transient intermediates, especially in the case of selenium and tellurium. For example, tellurium(II) compounds of the type ArTeCl are unstable with respect to disproportionation in the absence of such interactions. The diazene derivative **15.23** is stabilized by a Te•••N interaction.[7] Presumably, intramolecular coordination hinders the disproportionation process. Other derivatives of the type RTeX that are stabilized by a Te•••N interaction include 8-(dimethylamino)-1-(naphthyl)tellurium bromide,[26] 2-(bromotelluro)-*N*-(p-tolyl)benzylamine,[27] and 2-[(dimethylamino)methyl]phenyltellurium iodide.[28] Intramolecular donation from a nitrogen donor can also be used to stabilize the Se–I functionality in related compounds.[4,29]

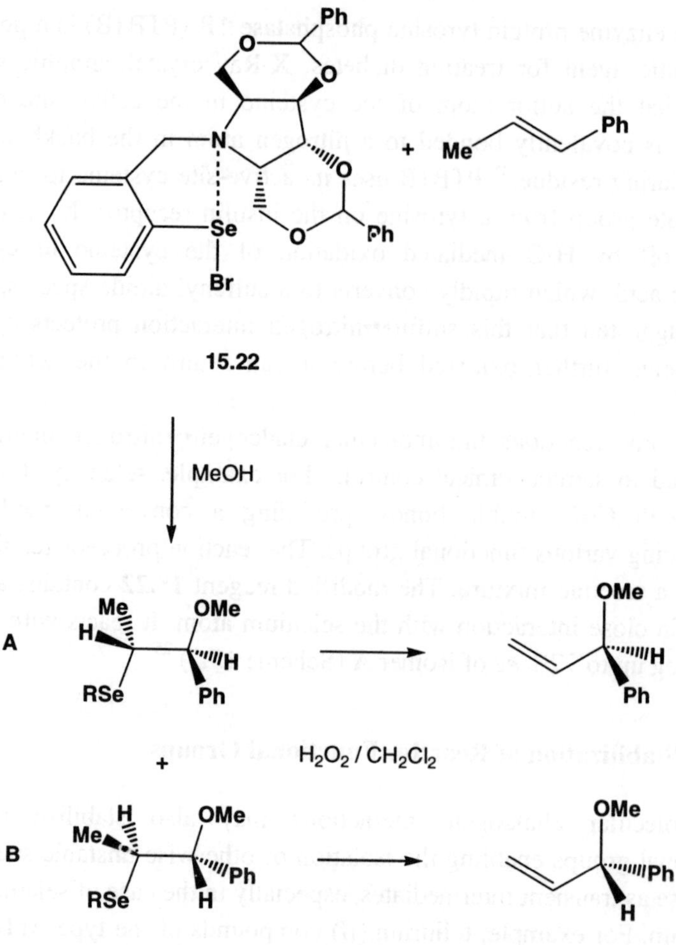

15.22

Scheme 15.2 Stereochemical control under the influence of an Se•••N interaction

Intramolecular heteroatom coordination may also influence the stabilities or structures of catenated tellurium compounds. For example, a rare example of a tritelluride, bis[2-(2-pyridyl)phenyl]tritelluride, is stabilized by a Te•••N contact of 2.55 Å.[30] The ditelluride (2-MeOC$_6$H$_4$COTe)$_2$ has an unusual planar structure. Although the C=O•••Te interaction is longer (3.11 Å) than the Me•••O contact (2.76 Å), *ab initio* molecular orbital calculations indicate that the planarity results predominantly from the former intramolecular connection.[31]

Transannular Te•••N interactions have also been employed to stabilize compounds of the type **15.24** with terminal Te=E (E = S, Se) bonds.[32] The Te=Se bond length in **15.24b** is 2.44 Å (*cf.* 2.54 Å for a Te–S single bond) and d(Te•••N) = 2.62 Å. Intramolecular coordination was also employed in the isolation of the first aryl-selenenium and – tellurenium cations **15.25a,b** as [PF$_6$]$^-$ salts.[33]

15.23   **15.24a**, E = S   **15.25a**, E = Se

                   15.24b, E = Se   **15.25b**, E = Te

The Se•••N interaction has been utilized in the stabilization of a transient selenenic acid ArSeOH.[34] Through such a reactive intermediate the diselenide **15.26** catalyzes the oxidation of alkenes to allylic esters or ethers in the presence of sodium persulfate.[35] Compound **15.26** also catalyzes the oxidation of thiols to disulfides by hydrogen peroxide serving as a model to study the role of the amino nitrogens located at the active centre of glutathione peroxidase.[11,36] Characterization of the intermediate steps by ^{77}Se NMR spectroscopy and kinetic studies indicate that the model behaves in the same way as the enzyme, although the latter possesses two nitrogens in proximity to the selenium of a selenocysteine. The proximal nitrogen is thought to play an additional role in activating the selenol into selenolate.

Although bulky aryl groups, *e.g.*, mesityl, are not effective in stabilizing arylselenium (II) azides, the use of intramolecular coordination in 2-Me$_2$NCH$_2$C$_6$H$_4$SeN$_3$ has enabled the first structural characterization of this reactive functionality.[37] The Se–N$_3$ (azide) bond length is 2.11 Å, while the intramolecular Se•••N distance is 2.20 Å, *cf.* 2.14 Å in the arylselenium bromide **15.5**, and 2.13 Å and 2.17 Å, respectively, in the corresponding chloride and iodide.[37] This arylselenium (II) azide is thermally unstable at 25°C; it decomposes with loss of N_2 to give the corresponding diselenide.

Metal selenolates of the type $M(SeAr)_2$ (M = Zn, Cd, Hg) are usually insoluble, polymeric compounds. Intramolecular Se•••N coordination has been employed to stabilize monomeric mercury selenolates, *e.g.*, **15.27**, but this approach was not successful for the zinc and cadmium derivatives.[38]

Chiral organoselenenyl halides may also be stabilized by intramolecular Se•••N interactions; ^{77}Se NMR chemical shifts indicate that these interactions are maintained in solution.[29b]

15.26   **15.27**

References

1. I. Vargas-Baca and T. Chivers, *Phosphorus, Sulfur and Silicon*, **164**, 207 (2000).

2. M. Iwaoka and S. Tomoda, *Bull. Chem. Soc. Jpn.*, **75**, 7611 (2002).

3. R. Kaur, H. B. Singh and R. P. Patel, *J. Chem. Soc., Dalton Trans.*, 2719 (1996).

4. G. Mugesh, A. Panda, H. B. Singh and R. J. Butcher, *Chem. Eur., J.* **5**, 1411 (1999).

5. G. L'abbe, L. van Meervelt, S. Emmers, W. Dehaen and S. Toppets, *J. Heterocycl. Chem.*, **29**, 1765 (1992).

6. N. W. Alcock, *Adv. Inorg. Chem. Radiochem.*, **15**, 1 (1972).

7. N. Sudha and H. B. Singh, *Coord. Chem. Rev.*, **135/136**, 469 (1994).

8. M. Kuti, J. Rábai, I. Kapovits, I. Jalsovski, G. Argay, A. Kálmán and L. Prákány, *J. Mol. Struct.*, **382**, 1 (1996).

9. (a) V. Chandrasekhar, T. Chivers, J. F. Fait and S. S. Kumaravel, *J. Am. Chem. Soc.*, **112**, 5371 (1990); (b) V. Chandrasekhar, T. Chivers, S. S. Kumaravel, M. Parvez and M. N. S. Rao, *Inorg. Chem.*, **30**, 4125 (1991).

10. (a) T. Chivers, B. McGarvey, M. Parvez, I. Vargas-Baca, T. Ziegler and P. Zoricak, *Inorg. Chem.*, **35**, 3839 (1996); (b) T. Chivers, I. H. Krouse, M. Parvez, I. Vargas-Baca, T. Ziegler and P. Zoricak, *Inorg. Chem.*, **35**, 5836 (1996).

11. M. Iwaoka and S. Tomoda, *J. Am. Chem. Soc.*, **118**, 8077 (1996).

12. K. Ohkata, M. Ohsugi, K. Yamamoto, M. Ohsawa and K. Akiba, *J. Am. Chem. Soc.*, **118**, 6355 (1996).

13. K. Akiba, K. Takee, Y. Shimizu and K. Ohkata, *J. Am. Chem. Soc.*, **108**, 6327 (1986).

14. K. Ohkata, M. Ohnishi, K. Yoshinaga, K. Akiba, J. C. Rongione and J. C. Martin, *J. Am. Chem. Soc.*, **113**, 9270 (1991).

15. G. A. Landrum and R. Hoffmann, *Angew. Chem., Int. Ed. Engl.*, **37**, 1887 (1998).

16. D. H. R. Barton, M. B. Hall, Z. Lin, S. Parekh and J. Reibenspies, *J. Am. Chem. Soc.*, **115**, 5056 (1993).

17. M. Iwaoka, H. Komatsu, T. Katsuda and S. Tomoda, *J. Am. Chem. Soc.*, **126**, 5309 (2004).

18. P. J. Dunn, C. W. Rees, A. M. Z. Slawin and D. J. Williams, *Chem. Commun.*, 1134 (1989).

19. V. Chandrasekhar, T. Chivers, L. Ellis, I. Krouse. M. Parvez and I. Vargas-Baca, *Can. J. Chem.*, **75**, 1188 (1997).

20. G. Mugesh and H. B. Singh, *Acc. Chem. Res.*, **35**, 226 (2002).

21. G. Mugesh and W-W. du Mont, *Chem. Eur. J.*, **7**, 1365 (2001).

22. G. Mugesh, A. Panda, H. B. Singh, N. S. Punekhar and R. J. Butcher, *Chem. Commun.*, 2227 (1998).

23. G. Mugesh, A. Panda, H. B. Singh, N. S. Punekhar and R. J. Butcher, *J. Am. Chem. Soc.*, **123**, 839 (2001).

24. (a) A. Salmee, J. N. Andersen, M. P. Myers, T-C. Meng, J. A. Hinks, N. K. Tonks and D. Barford, *Nature*, **423**, 769 (2003); (b) R. L. M. van Montfort, M. Congreve, D. Tisi, R. Carr and H. Jhoti, *Nature*, **423**, 773 (2003).

25. K. I. Fujita, K. Murata, M. Iwaoka and S. Tomoda, *Tetrahedron*, **53**, 2029 (1997).

26. S. C. Menon, H. B. Singh, J. M. Jasinski, J. P. Jasinski and R. J. Butcher, *Organometallics*, **15**, 1707 (1996).

27. A. G. Maslakov, W. R. McWhinnie, M. C. Parry, N. Shaikh, S. L. W. McWhinnie and T. A. Hamor, *J. Chem. Soc., Dalton Trans.*, 619 (1993).

28. R. Kaur, H. B. Singh and R. J. Butcher, *Organometallics*, **14**, 4755 (1995).

29. (a) W-W. du Mont, A. Martens-von Salzen, F. Ruthe, E. Seppälä, G. Mugesh, F. A. Devillanova, V. Lippolis and N. Kuhn, *J. Organomet. Chem.*, **623**, 14 (2001); (b) G. Mugesh, H. B. Singh and R. J. Butcher, *Tetrahedron: Asymmetry*, **10**, 237 (1999); (c) A. Panda, G. Mugesh, H. B. Singh and R. J. Butcher, *Organometallics*, **18**, 1986 (1999).

30. T. A. Hamor, N. Al-Salim, A. A. West and W. R. McWhinnie, *J. Organomet. Chem.*, **310**, C5 (1986).

31. O. Niyomura, S. Kato and S. Inagaki, *J. Am. Chem. Soc.*, **122**, 2132 (2000).

32. H. Fijihara, T. Uehara and N. Furukawa, *J. Am. Chem. Soc.*, **117**, 6288 (1995).

33. H. Fujihara, H. Mima and N. Furukawa, *J. Am. Chem. Soc.*, **117**, 10153 (1995).

34. M. Iwaoka and S. Tomoda, *Phosphorus, Sulfur and Silicon*, **67**, 125 (1992).

35. M. Iwaoka and S. Tomoda, *Chem. Commun.*, 1165 (1992).

36. M. Iwaoka and S. Tomoda, *J. Am. Chem. Soc.*, **116**, 2557 (1994).

37. T. M. Klapötke, B. Krumm and K. Polborn, *J. Am. Chem. Soc.*, **126**, 710 (2004).

38. R. Kaur, H. B. Singh, R. J. Patel and S. K. Kulshrestha, *J. Chem. Soc., Dalton Trans.*, 461 (1996).

Subject Index